MW00769068

JUST
ENOUGH
WIRELESS
COMPUTING

ISBN 0-13-099461-8

90000

9 780130 994615

Selected Titles from the

YOURDON PRESS SERIES

Ed Yourdon, *Advisor*

JUST ENOUGH SERIES

DUÉ Mentoring Object Technology Projects
HAYES Just Enough Wireless Computing
MOSLEY/POSEY Software Test Automation
RUSSELL/FELDMAN IT Leadership Alchemy
THOMSETT Radical Project Management
ULRICH Legacy Systems: Transformation Strategies
YOURDON Managing High-Intensity Internet Projects

YOURDON PRESS COMPUTING SERIES

ANDREWS AND STALICK Business Reengineering: The Survival Guide
BOULDIN Agents of Change: Managing the Introduction of Automated Tools
COAD AND MAYFIELD with Kern Java Design: Building Better Apps and Applets, Second Edition
COAD AND NICOLA Object-Oriented Programming
COAD AND YOURDON Object-Oriented Analysis, Second Edition
COAD AND YOURDON Object-Oriented Design
COAD WITH NORTH AND MAYFIELD Object Models, Strategies, Patterns, and Applications,
    Second Edition
CONNELL AND SHAFER Object-Oriented Rapid Prototyping
CONSTANTINE The Peopleware Papers: Notes on the Human Side of Software
CONSTANTINE AND YOURDON Structure Design
DEGRACE AND STAHL Wicked Problems, Righteous Solutions
DEMARCO Controlling Software Projects
DEMARCO Structured Analysis and System Specification
FOURNIER A Methodology for Client/Server and Web Application Development
GARMUS AND HERRON Measuring the Software Process: A Practical Guide to Functional
    Measurements
HAYES AND ULRICH The Year 2000 Software Crisis: The Continuing Challenge
JONES Assessment and Control of Software Risks
KING Project Management Made Simple
PAGE-JONES Practical Guide to Structured Systems Design, Second Edition
PUTNAM AND MEYERS Measures for Excellence: Reliable Software on Time within Budget
RUBLE Practical Analysis and Design for Client/Server and GUI Systems
SHLAER AND MELLOR Object Lifecycles: Modeling the World in States
SHLAER AND MELLOR Object-Oriented Systems Analysis: Modeling the World in Data
STARR How to Build Shlaer-Mellor Object Models
THOMSETT Third Wave Project Management
ULRICH AND HAYES The Year 2000 Software Crisis: Challenge of the Century
YOURDON Byte Wars: The Impact of September 11 on Information Technology
YOURDON Death March: The Complete Software Developer's Guide to Surviving "Mission
    Impossible" Projects
YOURDON Decline and Fall of the American Programmer
YOURDON Modern Structured Analysis
YOURDON Object-Oriented Systems Design
YOURDON Rise and Resurrection of the American Programmer
YOURDON AND ARGILA Case Studies in Object-Oriented Analysis and Design

JUST ENOUGH WIRELESS COMPUTING

Ian S. Hayes

PRENTICE HALL PTR
UPPER SADDLE RIVER, NJ 07458
WWW.PHPTR.COM

Library of Congress Cataloging-in-Publication Data

Hayes, Ian S.
   Just enough wireless computing / Ian Hayes.
      p. cm. -- (Just enough series)
   ISBN 0-13-099461-8
   1. Mobile computing. I. Title II. Series.

   QA76.59 .H39 2002
   004.165--dc21

                                                                            2002029311

Editorial/Production Supervision: Wil Mara
Composition: Aurelia Scharnhorst
Acquisitions Editor: Paul Petralia
Editorial Assistant: Rick Winkler
Marketing Manager: Debby van Dijk
Manufacturing Manager: Alexis Heydt-Long
Art Director: Gail Cocker-Bogusz
Cover Design Director: Jerry Votta
Cover Designer: Nina Scuderi

© 2003 Pearson Education, Inc.
Publishing as Prentice Hall PTR
Upper Saddle River, NJ 07458

All rights reserved. No part of this book may be reproduced, in any form or by any means, without permission in writing from the author and publisher.

The publisher offers discounts on this book when ordered in bulk quantities. For more information contact: Corporate Sales Department, Prentice Hall PTR, One Lake Street, Upper Saddle River, NJ 07458. Phone: 800-382-3419; FAX: 201-236-7141; E-mail: corpsales@prenhall.com.

Printed in the United States of America

10 9 8 7 6 5 4 3 2 1

ISBN 0-13-099461-8

Pearson Education LTD.
Pearson Education Australia PTY, Limited
Pearson Education Singapore, Pte. Ltd
Pearson Education North Asia Ltd
Pearson Education Canada, Ltd.
Pearson Educación de Mexico, S.A. de C.V.
Pearson Education—Japan
Pearson Education Malaysia, Pte. Ltd
Pearson Education, Upper Saddle River, New Jersey

To my wife, Anne.
Her love and support made this book possible.

ABOUT THE SERIES

In today's world of ever-improving technology—with computers that get faster, cheaper, smaller, and more powerful with each passing month—there is one commodity that we seem to have less and less of: time. IT professionals and managers are under constant pressure to deliver new systems more quickly than before; and one of the consequences of this pressure is that they're often thrown into situations for which they're not fully prepared. On Monday, they're given a new assignment in the area of testing, or risk management, or building a new application with the latest tools from IBM or Microsoft or Sun; and on Tuesday, they're expected to be productive and proficient. In many cases, they don't have time to attend a detailed training course; and they don't have time to read a thousand-page *War and Peace* tome that explains all the details of the technology.

Enter the Just Enough Series of books from Yourdon Press. Our mission is, quite literally, to provide just enough information for an experienced IT professional or manager to be able to assimilate the key aspects of a technology and begin putting it to productive use right away. Our objective is to provide pragmatic "how-to" information—supported, when possible, by checklists and guidelines and templates and wizards—that can be put to practical use right away. Of course, it's important to know that the refinements, exceptions, and extensions do exist; and the Just Enough books provide references, links to Web sites, and other resources for those who need it.

Over time, we intend to produce "just enough" books for every important aspect of IT systems development: from analysis and design, to coding and testing. Project management, risk management, process improvement, peopleware, and other issues are also covered while addressing several areas of new technology, from CRM to wireless technology, from enterprise application integration to Microsoft's .NET technology.

Perhaps one day life will slow down, and we'll be able to spend as much time as we want, learning everything there is to be learned about IT technologies. But until that day arrives, we only have time for "just enough" information. And the place to find that information is the Just Enough Series of computer books from Prentice Hall PTR/Yourdon Press.

ABOUT THE SERIES EDITOR

Edward Yourdon is an internationally recognized consultant, lecturer, and author/coauthor of more than 25 books, including *Managing High-Intensity Internet Projects, Death March, Time Bomb, The Rise and Resurrection of the American Programmer, Modern Structured Analysis*, and others. Widely known as the lead developer of the structured analysis/design methods in the 1970's and the popular Coad/Yourdon object-oriented methodology in the early 1990's, Edward Yourdon brings both his writing and technical skills as Series Editor while developing key authors and publications for the Just Enough Series/Yourdon Press.

Contents

CHAPTER 3
How Others Are Using Wireless   *39*

CHAPTER 4
Recognizing an Opportunity   *91*

CHAPTER   6
Justifying the Solution   *163*

CHAPTER 11
Wireless Networks *281*

CHAPTER 12

Wireless Applications   *319*

CHAPTER 13
Support *345*

APPENDIX A
Wireless Business Requirements Questionnaire *363*

Preface

Over the course of the past forty-plus years, we have seen computers shrink from the size of a room to the size of a wallet, and telephones move from walls and desks into our pockets. We have witnessed the rise and growing strength of a worldwide network for information interchange. And now we are seeing the convergence of these trends, forging a powerful platform from which we can reshape the way we do business and live our lives. This platform is wireless technology. The possibilities created by its arrival are the topic of this book.

Wireless technology first entered my life through a mobile phone, the same way it has probably entered the lives of millions of others. As a management consultant and frequent traveler, the advantages of being able to communicate easily and conveniently in almost any location are powerful and instantly addictive. No more telephone booths in airports. A means of notifying clients when hopelessly stuck in a traffic jam. Combine that telephone with my PDA, add the ability to surf the Internet and send and receive e-mails and we are starting to reach consultant nirvana! But as powerful as these benefits are, they barely scratch the surface of what is possible from wireless technology.

Much of my work revolves around making companies more effective, finding ways to better serve customers, enhance efficiency, and increase revenues. Process improvements are inevitably the basis for these benefits. The real power of wireless technology is its ability to support substantive process improvements. This power becomes clear when delving into the groundbreaking applications created by the wireless pioneers. Firms like Penskc Logistics, Honeywell, and UPS are using wireless technology in innovative ways to streamline operations and offer new services. The more I studied these firms, the more excited I got about the potential of the technology. Clearly, wireless technology could benefit almost any company in almost any industry. Yet surprisingly, companies are adopting wireless technology at a much slower than expected rate. Why?

The world of wireless is just too confusing. Walk into any bookstore. Pick up any trade publication. Listen to any solution provider pitch. It is a mind-boggling landscape of hype, acronyms, and endless technical details. And nothing is standing still. Networks, devices, and applications are evolving at a dizzying clip. Vendors seem to appear and disappear every few minutes. Standards are in flux and there are no runaway market leaders. Is wireless technology really ready for prime time? Who has time to make sense of this market? No wonder so many managers and executives remain on the sidelines. But the benefits of wireless technology are real. The companies that overcome its challenges are reaping these benefits and gaining a substantial competitive advantage.

My goal in writing this book is to help your company capture this advantage by providing a simple framework for navigating the wireless maze. From examples of real-life wireless solutions in use, to technology overviews and a step-by-step approach for devising your own solutions, this book will give you a foundation for launching a wireless initiative within your company. In today's busy world, few managers and executives have the time for in-depth research, yet they still need to make knowledgeable strategic decisions on the use of technology within their companies. In keeping with its title, this book seeks to educate you and give you just enough essential information to begin exploring mobile and wireless computing with intelligence and confidence. I hope it meets your expectations.

Audience for This Book

This book was written for anyone interested in understanding how and where to apply wireless technology within their company or organization. To serve the widest audience, the book minimizes the use of technical jargon and focuses on how to use wireless technology to gain real business benefit. As such, the book's contents should be especially useful to executives and managers who need a quick education on the practical aspects of wireless technology.

The book contains three sections to meet the varying needs of its audience. The first section focuses on the benefits and uses of wireless technology and contains extensive examples describing how companies are currently deploying wireless solutions. The second section offers a framework, complete with a questionnaire and checklists, to guide readers through the process of selecting and assembling a wireless solution. The third section provides technical overviews on various aspects of wireless technology including management issues, solution considerations, devices, networks, applications, and support requirements.

The book should appeal to the following categories of readers.

- *Business Executives and Managers* Executives and line managers will find the first section useful for learning about the potential of wireless technology and gaining specific ideas for solutions within their areas. Skimming the second section should offer an understanding of the scope and effort of launching a wireless initiative.

- *IT Professionals* IT professionals, such as CIOs, CTOs, project managers, information analysts, and software engineers, will find all three sections of the book useful. The third section should provide a foundation for further technical research and the second section, along with the questionnaire, checklists, and solution provider list in the appendices, will be especially useful for IT professionals charged with project implementation.

- *IT Solution Providers* Wireless software, hardware, and service providers will find the examples in the first section of the book useful as proof of the viability of wireless technology. The second section, and particularly the chapter on solution justification, will be valuable for assisting their customers in launching wireless initiatives.

- *Analysts and Media* The "just enough" concept employed by this book will help industry analysts, investment analysts, trade press editors, and writers gain quick insight into a variety of wireless topics. The book organizes many complex and seemingly unrelated components into a simple framework.

Summary of Contents

This book is organized into three topical sections to help readers focus on their areas of greatest interest.

The first section includes Chapters 2 and 3 and provides an overview of wireless technology and its benefits as well as extensive examples and case studies showing how companies are currently deploying wireless solutions. Readers new to wireless technology should begin with Chapter 2. More experienced readers may wish to move directly to the example applications in Chapter 3. The example solutions are a valuable source of application ideas, and the in-depth case studies describe how three companies implemented their chosen solutions.

The second section includes Chapters 4 through 7 and offers a framework, complete with a questionnaire and checklists, to guide readers through the process of selecting and assembling a wireless solution. This framework covers capturing business requirements (Chapter 4), solution definition (Chapter 5), solution justification (Chapter 6), and solution implementation (Chapter 7).

The third section provides more detailed information on various aspects of wireless technology. This section serves as a reference on specific topics and as a resource for supporting the wireless decision framework described in the second section. Its chapters cover management issues (Chapter 8), solution considerations (Chapter 9), devices (Chapter 10), networks (Chapter 11), applications (Chapter 12), and support (Chapter 13). Readers can use a chapter as a whole to gain a high-level view of a particular topic, or use the information from their "cheat sheets" to narrow their research to the subset of options that support their desired solution.

The book contains thirteen chapters and three appendices. These chapters cover the gamut of topics needed to research and deploy a wireless solution.

Chapter 1, "Introduction," introduces wireless technology and describes how to use this book. Its goal is to demystify a broad, highly complex and rapidly changing technology by avoiding unnecessary jargon and categorizing concepts in a way that enables readers to focus on areas relevant to their immediate needs. This approach provides novice readers with a starting point for exploring wireless technology while offering a more knowledgeable reader a means of quickly finding the topics and information of greatest interest.

Chapter 2, "How Wireless Can Help," explores how wireless technology can help readers achieve their business goals. It examines opportunities enabled by the technology and the types of benefits that can be obtained. It discusses how wireless technology provides these benefits and describes situations most amenable to wireless solutions. It shows how a process-based approach to applying wireless technology offers the greatest business benefits and simplifies technology selection.

Chapter 3, "How Others Are Using Wireless," uses examples and case studies to explore how companies are already using wireless applications. It has three goals: demonstrate that wireless technology is real and usable, illustrate the range of possible wireless applications, and offer templates for readers contemplating similar applications. Applications are organized by business objectives rather than technical design, thus enabling readers to hone in on the types of applications that most directly apply to their needs.

Chapter 4, "Recognizing an Opportunity," helps readers determine if a wireless solution is appropriate for their business objectives and, if so, translate those business objectives into a set of requirements for a wireless application. It advocates a simple top-down approach for identifying and capturing the business requirements for wireless solutions. It describes how to recognize business opportunities where wireless technology may be useful. It explains the process for moving from a business objective to an implementation strategy and introduces the "Five W's" approach to capture functional requirements in a form amenable to wireless solution design.

Chapter 5, "Defining a Solution," describes the process for turning business requirements into solution requirements. It uses the answers to the Why, Who, What, When, and Where questions from Chapter 4 to provide a framework for winnowing your wireless decisions into a manageable number. It explains how to develop specific requirements for devices, applications, data, and wireless networks. Comparing these requirements against the tables and other component-specific information in the second half of this book will enable readers to identify the wireless options that best apply to their needs.

Chapter 6, "Justifying the Solution," guides the reader through the process of estimating the cost of a wireless solution, and determining and quantifying potential benefits. It offers a four-step process that quantifies benefits, computes the short and long-term expenses, produces a ROI and cash flow analysis, and builds support for the proposed solution by demonstrating that its benefits are achievable.

Chapter 7, "Implementing the Solution," covers how to plan the implementation, manage the project, and redesign the underlying business processes. It offers implementation and deployment tips and techniques and describes how to get assistance from the right wireless service provider. Its goal is to provide readers with "just enough" information to understand relevant implementation issues and avoid major pitfalls.

Chapter 8, "Management Considerations," presents the topmost management issues that the reader must be prepared to deal with when pursuing a wireless project. It discusses business and legal issues affecting wireless solutions, from policies and standards to liability concerns, and approaches to take to deal with those issues. The goal of this chapter is to forewarn and forearm readers as they undertake a wireless implementation, so they can take steps to avoid potential future problems.

Chapter 9, "Solution Considerations," introduces the foremost issues affecting the design and implementation of a wireless solution. From development cycles to extensibility concerns to security issues, this chapter examines the top issues that readers must face as they begin to develop wireless solutions. The goal of the chapter is to review several major challenges in designing and developing a wireless solution, and present the reader with options for dealing with those issues.

Chapter 10, "Wireless Devices," provides an overview of the types of devices used in wireless solutions. It also discusses issues commonly encountered in using these devices, and considerations for determining which particular devices to use in a given solution. The goal of the chapter is to help readers choose the right device for their wireless solution by presenting them with a menu of options, cautioning them about strengths and constraints, and advising them of important factors affecting their ultimate selection of device.

Chapter 11, "Wireless Networks," summarizes the types of networks that may be involved in a wireless solution, presents issues associated with using each network type, and discusses considerations that help to determine the right network choice for a particular wireless solution. The goal of the chapter is to help readers understand their network options, and narrow their choices to those that best fit their needs.

Chapter 12, "Wireless Applications," presents the various kinds of wireless applications that may be involved in a wireless solution. It introduces universal wireless application development principles, provides an overview of the components comprising an end-to-end wireless application, discusses application design considerations, and reviews some common application development approaches. The goal of the chapter is to give readers sufficient information about wireless applications and design issues to help them hone in on their best options.

Chapter 13, "Support," examines the different support issues that surround a wireless solution. It explores the technical, business, and human aspects of operating and maintaining a wireless solution, noting issues, problems, and sources of assistance. The goal of the chapter is to alert readers to potential support issues and offer suggestions for overcoming those issues.

The appendices provide three types of supporting materials: a questionnaire and "cheat sheets" for capturing business and technical requirements, a list of solution providers who offer the types of software, hardware, and services mentioned in the book, and a glossary of terms.

A Final Note

This book is meant to be a primer on wireless technology, its uses and implementation issues. As a primer, its contents are necessarily high-level and omit details that are likely to be required when designing, constructing, and deploying a wireless initiative. Furthermore, the wireless industry is evolving at an ever-increasing pace, and while the author has invested considerable effort to ensure that the contents of this book are accurate and up-to-date, changes in the industry may obsolete portions of its content at any time. For these reasons, the author strongly advises consulting with topical experts and conducting further research before progressing on any wireless initiative.

No warranties can or will be made with respect to any data, explanations, or opinions expressed herein.

Ian S. Hayes
Clarity Consulting, Inc.
South Hamilton, MA

Acknowledgments

Though credited to a single author, this book is the result of much hard work and good ideas from many individuals and companies. Without these collective efforts and generous contributions of time and advice, I could not have written it. The final synthesis falls to me, and I bear sole responsibility for any errors or omissions.

First and foremost, I owe a tremendous debt of gratitude to Anne Hayes, my wife and business partner. She was closely involved with all aspects of this book, conducting research and interviews, assisting in the writing, editing my torturous prose, and putting up with many nights and weekends of hard work. Thank you Anne. I hope it was worth it!

Writing a book is a monumental task that begins with a good topic and a reason to launch the project. I am grateful to two individuals for inspiring me to get started. By involving me in the early stages of his company, John Keane, CEO of ArcStream Solutions, provided the topic. His enthusiasm about the potential of wireless technology is infectious and triggered my research in the area. Shortly thereafter, Ed Yourdon, famed IT author, consultant, and lecturer, provided the reason. He approached me with an intriguing idea—writing a book for his latest publication series based on the concept of "just enough" information for harried managers. With its huge potential limited by its complexity and market confusion, wireless technology is the perfect topic for the "just enough" concept. The match was made and the result is in your hands.

The ideas in this book were not developed in a vacuum. Rather, they have evolved over many years from writing projects, research, many collective discussions and experience from consulting engagements. I would like to acknowledge the contributions of the following individuals: Eleanor Carscaddon, Dr. John Halamaka, Kevin Hickey, Dave Kistler, Matt Theodores, and Andrew Weniger. It is impossible to acknowledge everyone by name, but I want to express my gratitude to my clients, conference attendees, fellow consultants, and solution providers that have also contributed to these ideas. Hopefully, this book will provide some measure of repayment for these favors.

The ability to interview individuals at companies who have deployed successful wireless projects has been extremely valuable, forming the basis for the case studies in Chapter 3 and contributing heavily to the rest of the book. I want to thank the following individuals for graciously agreeing to interviews and reviewing their case studies: Kerry Reedy of Atlantic Envelope Company, Shawna Todd and Santosh Patel of Honeywell, and Kevin Lamanna and Louise Moyer of Penske Logistics.

I am highly indebted to Sven Ingard of ArcStream Solutions and Jean-Luc Valente of Radioscape, for their review and advice on the book's content. Their insight and commentary has been invaluable and has immeasurably improved the book.

Finally, I'd like to thank my editors at Prentice Hall PTR, Paul Petralia and Wil Mara, for their considerable efforts in bringing out this book. Book projects are never easy, but it helps to have the assistance of pros.

Introduction

Are you getting ready to launch your company's first wireless application? Are you looking for a bit more information about a promising but confusing new technology? In either case, you probably have more questions than answers. Browse any business magazine, trade journal, or technology publication and you will find plenty of coverage on wireless technology and the powerful applications it enables. On the one hand, it is clear that companies are gaining advantage from their wireless investments. On the other hand, many articles treat wireless technology futuristically, describing how great life will be when the next generation of capabilities arrives. Anecdotally, you hear tales about inadequate bandwidths, poor coverage, and gaps in security—all imply that wireless technology is "not yet ready for prime time." Is wireless technology for real or not? Can you do anything with it right now? Is it possible to build and deploy a production quality wireless application that performs more than toy functions? How do you evaluate whether wireless technology can help your organization?

Undoubtedly, you've discovered that finding simple answers to your questions is difficult. Shelves are full of highly detailed technical books packed with an endless stream of options and acronyms, but no information on how to get started. As a manager or executive, you don't need an in-depth description of network encryption algorithms or the nuances of TDMA versus CDMA technologies underlying wireless wide area networks. You need to know whether, and how, wireless technology can solve one or more of your important business problems. And if wireless technology fits the bill, you want to know the fastest and easiest way to launch your project. In short, you don't want a wireless engineering course, you want just enough information to be dangerous. This book was written for you.

This chapter launches our explorations by introducing wireless technology and describing how to use this book. Its goal is to demystify a broad, highly complex, and rapidly changing technology by avoiding unnecessary jargon and categorizing concepts in a way that enables you to focus on areas relevant to your immediate needs. If you are a wireless novice, start from the beginning to build a foundation for later chapters. If you are ready to jump into defining your own wireless solution, skip to the second half of the chapter to understand the book's layout. The book is organized to help readers quickly find "just enough" information on the topics of their choice.

1.1   What Is Wireless Technology?

Wireless technology refers to the hardware and software that allows information to be transmitted between devices without the use of physical (wired) connections. Typical examples of wireless technology within our homes include radios, cellular telephones, and remote garage door openers. Wireless technology can be as simple as a TV remote control transmitting channel and volume changes to a TV set or as complex as a national network supporting millions of users through satellites, transmission towers, gateways, and sophisticated software applications.

Wireless technology aims to give its users access to needed information no matter where they are. It fosters information exchange and collaboration where physical co-location is not feasible. And wireless technology has long aided in tracking, locating, and managing valuable, movable assets such as cargo containers, laboratory equipment, and even taxicabs.

Wireless technology is composed of the following major components:

- *Devices* Devices are pieces of equipment that can send (transmitters), receive (receivers), or send and receive (transceivers) wireless information. Devices include mobile telephones, pagers, personal computers, remote sensors, handheld units or other types of equipment that capture, display, store, manipulate, print, or relay voice or electronic data.

- *Networks* A network consists of the software and hardware used to transmit information between devices. This communication can occur at short range using infrared technology, at a wider range using a high-speed wireless local area network (LAN) within a building, or at extra-terrestrial distances using satellites.

- *Applications* Applications refer to the software that operates on a wireless device, a host system, or a combination of the two to provide specific functionality to the device's user. Applications automate information capture, extract, and format data for display, perform calculations, and handle error checking and navigation between functions. Example applications include e-mail, package tracking, and credit card validation.

A **wireless solution** refers to a combination of device(s), application(s), network(s), and data source(s) that enable the mobile execution of one or more business functions. Examples of wireless solutions range from sales force automation tools that permit salespeople to respond more quickly to their customers to remote monitors that check the inventory and operation of ice vending machines.

At its simplest, a wireless solution is merely a means to eliminate wires. For instance, wireless LANs perform the same tasks as their wired counterparts, and, aside from the convenience of no wires, the average user is unlikely to notice a difference between the two. At the other extreme, a wireless solution may enable an entirely new service or permit a business function to be performed in a novel way at a remote location. For example, a wireless solution can keep business travelers instantly aware of changes to the status of their flights regardless of their location.

Although they are frequently confused and erroneously interchanged, the terms **mobile** and **wireless** are not the same. **Mobile** refers to people, items, or activities that move freely from place to place. By eliminating the need for physical connections, wireless technology enables mobility. A wireless solution, however, is not necessarily mobile. Office workers can use wireless LANs as part of their jobs, and may place and receive calls on a cell phone while sitting at a desk. Similarly, mobile workers, devices, and applications do not necessarily use wireless capabilities. A mobile user may use a personal digital assistant (PDA) primarily for its information organization capabilities, relying on periodic synching through a physical cradle, and a mobile laptop user may rely solely on dial-up connections to download and upload data.

1.2   Why Is Wireless Technology Interesting? _____

Ever since the invention of the radio, humans have been fascinated by wireless technology. There is something magical about signals containing voice, music, or data traveling through thin air to our radios, telephones, or PDAs. Most of us don't realize how fully wireless technology permeates our lives. When is the last time you walked over to your television to change channels? How many cordless telephones are in your house or apartment? Buoyed by increasing coverage, higher reliability, and greater sound quality, mobile telephones have become ubiquitous. To understand its impact on our lives, try to imagine cars without wireless technology. There would be no radio, no mobile phone, and, in newer vehicles, no OnStar or similar system providing directions and roadside assistance. If you travel to Europe or Japan, wireless technology is even more pervasive. Japanese consumers can use their mobile phones at vending machines to purchase sodas that are charged to their monthly telephone bill.

Why are we so attracted to wireless technology? Mystique aside, it provides two compelling advantages: freedom and convenience. Freedom from wired tethers that bind us to a specific location. Freedom to communicate wherever and whenever we want. Freedom to take our laptops outside to watch our children play as we check e-mail. Freedom to work effectively wherever our job takes us. The convenience of having access to information when we need it. The convenience of moving equipment without having to restring wires. The convenience of being able to perform tasks remotely, whether we are changing TV channels from our sofa or checking the condition of a shipment in a container far out at sea.

Wireless technology has the power to improve the way we live and work in many ways. Wireless technology brings the ability to close a sale on the spot at any customer location by having instant answers to the customer's queries. It eliminates the need to send people to dangerous locations to check equipment and gather data. It is the ability to provide communications to remote third-world areas without the overwhelming expense and effort of installing and maintaining thousands of miles of telephone wires.

Recent advances in technology have vastly expanded wireless capabilities and have opened the door for a new generation of ever more powerful wireless services and applications. It is now possible to process credit card transactions, trade stocks, surf the Internet, and track freight almost anywhere. Business activities that were previously tied to office desks can be moved closer to the customer, improving service and increasing revenues while reducing overhead. As the pace of technology evolution accelerates, so too will potential uses of wireless technology within your company.

1.3   Who Uses Wireless Technology?

Who uses wireless technology? We all do. Wireless technology is everywhere. We use wireless technology every time we use a remote control, cordless stereo speakers, or cellular telephone. Wireless applications appear in hospitals, manufacturing plants, hotels, trucks, and executive suites. Wireless tools are already in the hands of delivery people, equipment installers, insurance agents, stockbrokers, physicians, and pharmacists. Telemetry applications allow communications with containers in transit, sensors in oil rigs, and utility meters in residential neighborhoods.

Mobile phones, with their voice capabilities, currently constitute the largest wireless application by far. Giga Information Group reports annual worldwide sales of 400 million mobile telephones in both 2000 and 2001 and forecasts that the number of mobile phone users worldwide will reach 1.25 billion by the end of 2003.[1] Wireless data applications are growing at an even faster rate. In-Stat/MDR expects the number of business wireless data users to grow from 6.6 million at the end of 2001 to more than 39 million in 2006.[2] Wireless e-mail is the most common data application as reported by the World Information Technology and Services Alliance and Wireless IT Research Group.[3] Of the firms surveyed in their study, 36% currently used wireless e-mail, with an additional 49% planning on providing it in the future.

While the above figures address the most obvious uses of wireless technology, there is no shortage of more creative applications. Companies such as American Airlines, Ford Motor Company, Cadbury, Bank of America, and Anderson Trucking Services use wireless applications to track assets, inform customers, market their products, and offer new and innovative services. Wireless applications are found in industries from transportation to insurance and utilities. Consider some of the innovative applications already in the field. Sears Roebuck equips its field workers with wireless notebooks to create and close work orders, order parts, generate bills, and accept payments. Florida Power & Light uses smart meters to allow wireless readings of electric usage by passing vans. Transportation companies such as UPS, FedEx, and Penske Logistics are using sophisticated wireless tracking solutions to monitor and convey the shipment status of individual packages in real time. Financial services firms such as Fidelity Investments, Charles Schwab, and E*TRADE allow their investors to check account status, make trades, and receive wireless alerts on their PDAs. These applications are not limited to the private sector. Government

1. The Shape of Things to Come: The Evolving Device Market, Carl Zetie and Ken Smiley, Giga Information Group, October 5, 2001.
2. Wireless Data Adoption in the Enterprise, an In-Stat/MDR report, as reported at *www.instat.com*, March 5, 2002.
3. Study by World Information Technology and Services Alliance and Wireless IT Research Group, as reported on *www.allnetdevices.com*, March 1, 2002.

agencies across the U.S. are using wireless applications to automate inspection services, notify constituents of impending blackouts and traffic conditions, and support law enforcement activities. These samples only scratch the surface of the types of applications that are operating and providing real business value today. Many more examples and detailed case studies are found in Chapter 3.

The possibilities for voice- and data-based wireless applications are endless, and each advance in wireless capabilities enables a new generation of business functionality. As a result, new applications are appearing at an ever-increasing rate. Mobile workers, such as salespeople, field service technicians, and delivery people, are an obvious target for new wireless applications. In-Stat/MDR estimates that there are 78 million remote and mobile workers in the U.S. today, and that number is growing.[4] Wireless technology applications can arm these workers with tools and data access capabilities that were previously limited to desk-bound employees. These same capabilities will free office workers to become mobile. Wireless applications will allow employees from executives to purchasing agents to spend more time away from their desks and in front of customers, suppliers, and other constituents. Finally, installing a wireless LAN in key areas, such as conference rooms and training centers, can give employees access to corporate intranets and e-mail systems, allowing them to work conveniently and productively with their own equipment. Already over $1 billion annually, the wireless LAN market is predicted to exceed $5 billion by 2005.[5]

1.4 Why Now?

Why is wireless technology suddenly a hot topic? The underlying technology is hardly new, and devices such as laptops, cell phones, and pagers have been mobile worker staples for years. In the past, cost and network limitations have restricted the use of wireless technology to voice (telephony) and paging (short text message) applications. Recent convergence of a number of factors has produced a quantum leap in capabilities and possible applications, leading to a newfound interest in all things wireless. These factors include:

- The emergence of increasingly powerful handheld devices, such as PDAs
- Improvements in network data exchange quality, reliability, and bandwidth
- Better coverage by wide area wireless network providers
- Maturing device, network, and data exchange standards
- The preeminence of the Internet as a medium for corporate data exchange

---

4. Cahners-In-Stat/MDR, January 20, 2002 press release.
5. Wireless LANs—Poised for Untethered Growth, Jeff Abramowitz, Executive Director, Wireless LAN Association, 2001.

The result? Companies can now offer unwired workers access to the quality, quantity, and types of information formerly available only to their wired counterparts. These capabilities enable companies to unchain legions of workers from their desks, provide access to critical information where and when needed, and redesign a host of processes to improve efficiency, customer service, and enhance revenues. Future advances in wireless technology will further its appeal and unleash future generations of ever more powerful applications.

1.4.1 Wireless Today

Like so many technological advances in recent years, the wireless market is subject to an almost overbearing level of hype and confusion. As is true for any technology breakthrough, there is a world of difference between current and claimed capabilities. To date, the number one application of wireless technology has been voice transmission—making telephone calls using cell phones—an application that hasn't really had much affect on IT organizations per se. However, as described in the previous section, the number of data applications is growing rapidly. Wireless technology is no longer futuristic; successful wireless strategies and applications exist today. From financial services, to sales and field support, to healthcare services, wireless applications have delivered a host of tangible benefits. Burgeoning demand for data-based wireless applications promises to create a wireless mania within IT organizations over the next few years.

Despite recent advances, wireless technology is not yet mature. Wireless carriers, currently operating on the second generation of digital network capabilities, are in the midst of a performance upgrade on the way to implementing third generation capabilities over the next several years. Clashing standards and protocols, although improving, remain common. Developers are still learning the best ways to capture and present information on wireless devices. Vendor products don't always work as advertised and are rarely as "plug and play" as claimed. In short, the wireless market is clearly in flux, but the possibilities are great and the ability to derive benefits is real.

Even with its limitations, current wireless technology is sufficiently powerful and robust to immediately support many categories of high-payback business applications. Companies are already using wireless applications to improve the mobility of their employees, to offer instantaneous access to important data, to improve the safety of workers, to enhance the accuracy of field-collected data, to generate new revenues, and to administer more personal and effective customer service. The ideal wireless application candidates will be those that can provide immediate value and payback with today's capabilities and can be expanded or merged into more powerful applications as technology advances.

1.4.2   Wireless in the Future

Read any business journal article or analyst report to find glowing predictions on the future of wireless technology. Converging standards and next generation networks promise to provide quantum gains in coverage, bandwidth, and reliability. Broad-band, high-speed transmission will permit massive volumes of data to be trans-ported to and from mobile sites. As wireless carriers move toward a single network standard, coverage will improve nationally and globally. Internet connectivity will become ubiquitous, whether in a remote village in India or in the Rocky Mountains.

Certainly, the future will bring great advances to the wireless landscape. Wireless carriers are making tremendous investments in their networks and rising consumer and business interest is drawing many entrants to the wireless market. Current eco-nomic conditions notwithstanding, the open question is when, not if, the requisite advances will occur. Market pressures are forcing consolidations, with the welcome side effect of reducing the number of competing standards and incompatible proto-cols. One after another, thorny issues in areas such as wireless security and device management are being solved. The number and quality of tools available to develop, test, deploy, and manage wireless devices and applications increases almost daily.

Wireless devices are becoming more powerful and features are converging. In the not too distant future, the concept of having a separate telephone, PDA, and pager will seem hopelessly quaint for most business purposes. The availability of cheaper and more powerful wireless devices will spawn ever more sophisticated and robust applications. While relatively rare now, packaged wireless applications will become more common and available for more specialized business functions. Wider deploy-ments of wireless applications will mean shortages in skilled implementers for the foreseeable future, but an increase in the total number of trained individuals.

Does the future sound enticing? It follows the typical curve of any major technology as it matures. A look backward at the personal computer industry over the past twenty plus years should provide an indication of what is in store for the wireless technology industry. The key question is when to leap into the technology. Although their capabil-ities are laughable by today's standards, most companies got tremendous benefit from their PC investments even in the early 1980s. Imagine what your company would have lost if it waited until PCs were fully evolved and the PC market was stable. It might still be waiting today…

1.4.3   Why Consider Wireless Technology Now?

As a business executive, the challenge is knowing when to jump on a new technology, when to wait and watch, and when to ignore it altogether. Will your company gain tremendous strategic advantage by implementing a groundbreaking wireless application or will it spend needless time and money working through issues that others will shortly solve? When a new technology appears and promises to change our lives and reshape the way we do business, we have to question whether it is really ready for prime time. Wireless technology presents us with just this dilemma. Despite its high potential, companies have been deploying wireless technology at a slower than expected pace. Many are sitting on the sidelines unsure of how and when to deploy wireless technology to their benefit.

While this reticence is understandable, behind the hype and confusion, wireless technology is much more than a new set of toys for technologists and salespeople. Chapter 3 describes many companies that are already making productive use and gaining competitive benefit from their wireless investments. These forward-thinking companies will have a significant head start in deploying future versions of the technology. As is true for any rapidly evolving technology, being an earlier adopter means taking some risk and navigating a confusing and complex landscape of networks, devices, tools, services, and consulting firms. Hopefully, this book will help to simplify some of this complexity.

Exploring wireless technology now gives your company the ability to tap wireless capabilities while preparing to exploit new capabilities as they arise. Just as the Internet created incredible opportunities for the first companies that recognized its potential, wireless technology will bring first mover advantage to those companies with the foresight and imagination to create new ways of doing mobile business. Chances are, you'll come up with a variety of valuable applications. Moving now rather than waiting will allow your company to capture the full business benefits of those applications while gaining valuable experience in building, deploying, and maintaining wireless applications. This implementation experience provides a foundation for rapidly deploying future applications and is a competitive advantage in its own right.

1.5   How to Use This Book

Let's face it. The world is already a complicated and confusing place. We are inundated with articles, books, web sites, and information from all sources. Who has the time and energy to wade through the mountains of hype, mind-numbing details, and conflicting research that typify any new technology and are found abundantly in the wireless arena? All we want is just enough information to know how and where to start. The last thing we want to do is read through a dozen technical books to construct our own "big picture" view of the entire wireless field and extract the one or two items from each book that are immediately relevant. Five hundred possible options may exist, but right now, we care only about the five that apply to our current problem. We can research the details later. At this moment, we need just enough information to be dangerous. Get us started and we can take it from there.

This book was written to meet these objectives. Although it contains a wealth of material, it is organized to present just enough information on each topic to meet your needs. Whether you want to understand quickly if wireless technology can help your organization, cut your five hundred options down to five, or learn how to estimate the benefits of a wireless solution, this book can help you. Designed to meet the needs of business and IT executives, managers, and project leaders, this book is a quick and easy to follow, business-oriented primer on wireless technology. Given the breadth of the technology and its applications, the book necessarily simplifies each topic to focus on the key points that you need to start and guide your wireless explorations. Forewarned is forearmed; a little knowledge will go a long way in preparing your organization for the inevitable influx of wireless applications.

1.5.1 The Goal—Just Enough Information

The primary goal of this book is to help you relate and apply wireless technology to real life business issues. In keeping with the "just enough" concept, this book won't try to make you an expert on all aspects of wireless technology, but instead will offer high-level guidance in multiple areas. Its major objectives are to:

- Describe the benefits possible from wireless technology and help identify solutions that offer the specific benefits you seek
- Offer examples and case studies of real life, currently implemented wireless solutions across a variety of industries as proof of feasibility and as a source of ideas
- Identify and distill key wireless concepts to arm you with enough information about the technology and implementation options to ask intelligent questions of vendors, co-workers, and staff
- Help you identify and rapidly focus on the subset of technology that directly applies to your needs
- Provide a framework for launching a wireless project that covers the major steps from conception through implementation and deployment
- Direct you to sources for additional details as you need them

1.5.2 Scope

Wireless technology is such a broad topic that covering all of its aspects is impossible, even at a high level. For this reason, the book focuses primarily on applications involving wireless data exchange between people and particularly between mobile workers and their companies. Other areas such as telemetry (machine to machine communication), telematics (automotive wireless applications), and voice communications have their own unique characteristics and requirements. These areas are discussed where appropriate, but not emphasized. In addition, although wireless adoption is widespread throughout the world, to keep its size reasonable, this book limits its focus to developments in the U.S. market.

The rapid evolution of the wireless market causes technology and solution provider references to become quickly obsolete. For this reason, the book is intentionally business-oriented and delves into technological details only where necessary to support the objectives stated above. Wherever possible, the book strives to focus on concepts that will endure through technology changes. Some detailed references are unavoidable, and those details are subject to change. Always double-check with more recent sources before relying on details in any publication. The book's web site *www.justenoughwireless.com* will contain updates where appropriate. It will also contain the current version of the book's solution provider list.

Certain areas, such as wireless application development and information infrastructure design, are difficult to cover in a book such as this. Unlike network and device decisions, these areas are infinitely flexible, taking their shape from the creativity of their developers. Many books have been, and will be, written on wireless application development techniques and will give the topic more justice than is possible in this book. For these areas, the book concentrates on the major issues and options that managers should consider when evaluating solutions and leaves the details on development tools and methodologies for others.

1.5.3 Using the Book

This book is specifically designed to help readers quickly find "just enough" information on topics of interest. It supports these topics with extensive illustrations, examples, and case studies. The book is organized into three major sections.

The first section, which includes Chapters 2 and 3, provides an overview of wireless technology and its benefits, as well as extensive examples and case studies showing how companies are currently deploying wireless solutions. Readers new to wireless technology should begin with Chapter 2. More experienced readers may wish to move directly to the example applications in Chapter 3. The example solutions are a valuable source of application ideas, and the in-depth case studies describe how three companies implemented their chosen solutions.

The second section, which includes Chapters 4 through 7, offers a framework, complete with a questionnaire and checklists, to guide readers through the process of selecting and assembling a wireless solution. This framework covers capturing business requirements (Chapter 4), solution definition (Chapter 5), solution justification

(Chapter 6), and solution implementation (Chapter 7). Readers seeking to understand the issues involved in designing and implementing a wireless solution should skim the chapters in this section. Readers in the process of implementing a wireless application may want to follow the full process as shown in Figure 1.1. Guided by Chapter 4, this process uses the questionnaire contained in Appendix A to understand functional needs, the characteristics of potential users, and likely usage patterns to create the business requirements for a wireless solution. These business requirements are turned into technical requirements using the "cheat sheet" checklists and steps described in Chapter 5. These checklists are designed to work with the appropriate chapters in the third section of the book.

The third section, which includes Chapters 8 through 13, provides more detailed information on various aspects of wireless technology. These chapters cover management issues (Chapter 8), solution considerations (Chapter 9), devices (Chapter 10), networks (Chapter 11), applications (Chapter 12), and support (Chapter 13). This section serves as a reference on specific topics and as a resource for supporting the wireless decision framework described below. Readers can use a chapter as a whole

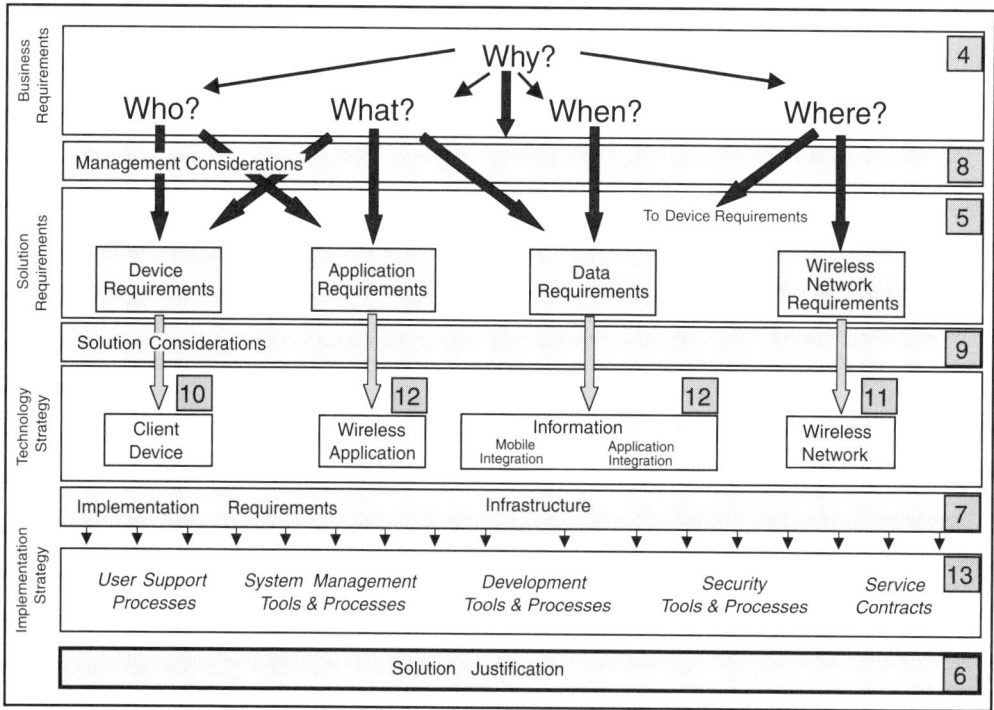

FIGURE 1.1
The Wireless Decision Process with Chapter References

to gain a high level view of a particular topic, or use the information from their "cheat sheets" to narrow their research to the subset of options that support their desired solution.

The book's appendices contain a questionnaire, cheat sheet checklists, a glossary of terminology, and an extensive list of wireless solution providers.

The book's associated web site, *www.justenoughwireless.com*, will contain updates, corrections, and breaking news. Depending on the level of reader interest, this web site will be updated periodically and offer current information on a variety of wireless technical topics.

How Wireless
Can Help

2

Let's cut to the bottom line. Every manager and executive contemplating wireless technology asks the same questions. What can wireless technology do for me? Given all of the business challenges and technology options before me, why should I explore wireless technology? Will it really change the way I do business or is it just another technical detail that can be handled by my network people? What kind of benefits can I expect from my wireless technology investments? The answers to these important questions depend on our creativity and how 7we choose to view wireless technology.

If we view wireless technology as simply a way of transferring information between devices without the need for a physical connection, it is strictly a cost/convenience option that can be left to technical specialists. Our work processes basically remain the same and our wireless investments are mostly justified by reductions in infrastructure costs. This viewpoint has merit, particularly in environments such as older buildings, rural areas, or third-world countries, where the installation of wired infrastructures is difficult and costly. As wireless adoption increases, wider availability and dropping costs will enhance and extend its cost/convenience advantages.

The technical benefits of wireless technology pale, however, in comparison to the benefits available from process improvements enabled through mobility and on-demand information access. If we view wireless technology as a tool to support greater process transformations, a seemingly trivial technical solution can provide significant business benefits. By providing mobility, wireless technology enables companies to simplify, improve, or extend business processes to save time or money, enhance the quality of service, or offer new products and services.

Mobility enables information to be delivered where needed and where it can bring the greatest value. It allows tasks that were previously tied to a desk to be brought to the customer. It enables better business decisions through immediate access to critical information. It can turn downtime to productive time. In the world of process improvements, business benefits accrue from the reshaped processes rather than directly from the technology. In fact, wireless technology may be a relatively insignificant component of a much larger solution, but it is the catalyst that allows the larger solution to occur.

The benefits of wireless technology range from subtle to far-reaching, depending on how and where the technology is applied. For example, a simple application such as wireless access to e-mail may be a "nice to have" feature for certain company employees, but an invaluable business tool for a traveling executive. If that e-mail application enables field service workers to receive and relay orders and service status to their office, it may change job responsibilities and reshape the way the company provides customer support. Creativity and the ability to move beyond the obvious will identify breakthrough opportunities to enhance our businesses. While for some business activities, benefits will be minimal beyond potential cost savings and increased convenience, other activities will be significantly transformed. These opportunities will be large enough to launch a host of new companies and to reshape practices in many industries. As wireless technology becomes increasingly ubiquitous, most of us will take its availability and benefits for granted, however, astute companies will capitalize on the windows of opportunity created along the way.

This chapter explores how wireless technology can help you achieve your business goals. It examines opportunities enabled by the technology and the types of benefits that can be obtained. It discusses how wireless technology provides these benefits and describes situations most amenable to wireless solutions. It shows how a process-based approach to applying wireless technology offers the greatest business benefits and simplifies technology selection. This chapter presents these topics at a general level to be applicable to the widest range of business needs and to encourage creative thinking about new possibilities. The next chapter of the book will explore specific examples of wireless applications by category and industry.

2.1   Capabilities and Opportunities _____

The best means of assessing what wireless technology can offer you and your company is to forget about the technology! Consider its core capabilities from a benefits standpoint. Although wireless implementations vary widely in scope and technical specifications, they share the ability to provide a baseline set of benefits. Rather than questioning whether or not wireless technology is appropriate for your company or for a specific business activity, consider whether any of the capabilities shown in Figure 2.1 and discussed below would be appropriate for your needs. For example, would increased mobility make your insurance adjusters more productive, or would better tracking capabilities significantly reduce shipping losses? It is likely that these capabilities would benefit many activities, processes, and functions across your company. They are the building blocks for gaining broader business benefits. For instance, providing tools to allow your sales force to spend more time in front of customers (mobility), combined with instant access to customer and product information (data exchange) could redefine the way you serve your customers (workflow) resulting in higher revenues, lower sales costs, and greater customer satisfaction. If the capabilities below are attractive, wireless technology is worth exploring. The next section will explore the types of business benefits that can be obtained from these capabilities.

2.1.1   Gain Mobility

Mobility is the greatest and most obvious benefit provided by wireless technology. It allows activities that were formerly tied to physical locations to be performed almost anywhere. Greater mobility is an extremely powerful capability for businesses.

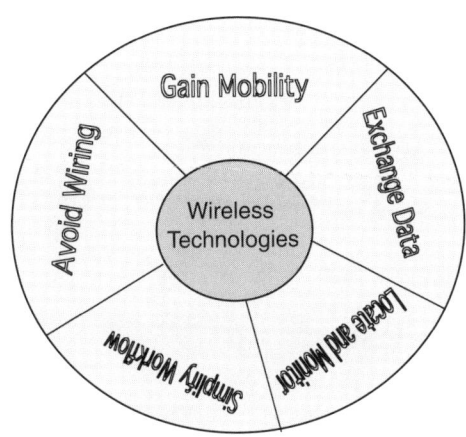

FIGURE 2.1
Wireless Opportunities

Information and computing power can be brought to the task location instead of forcing the task to be located at the site of the computing power. The implications of this shift range from subtle to enormous. Simply consider how mobile telephones have changed the way we conduct our personal and professional lives. Mobile phones enable executives and sales professionals to conduct business wherever they are, eliminating downtime during travel and increasing responsiveness to customers and issues. Mobility means more face-to-face contact with customers and partners. Can you imagine your sales force without access to their mobile phones? Can you also imagine how much more effective your sales force could be if it had mobile access to your company's key systems and information resources? Wireless technology's ability to provide mobility brings your company the following capabilities.

- *Mobilize tools and information* Industry analysts estimate that the population of mobile workers in the U.S. alone is approaching 40 million and growing. Mobile workers cross many industries and include salespeople, real estate agents, insurance adjusters, medical technicians, and equipment installers. While we typically think of mobile workers as people who travel some distance to perform their jobs, consider internal jobs, such as custodians, floor salespeople, production supervisors, and emergency room doctors, whose assignments keep them moving throughout the day. Currently, these individuals either do without access to the systems and resources available to their non-mobile counterparts or they have to return periodically to their office or other fixed location for access. Wireless technology can bring that access to them, where and when needed.

- *Free desk bound workers* Mobility isn't only for mobile workers. Mobility allows a new range of functions to be performed onsite and in real time. Physical presence enhances the effectiveness of many functions. Many tasks and job assignments are desk bound primarily because of their need for access to corporate systems and information. Using wireless technology to mobilize access to corporate systems can free workers to perform their assignments in the most optimal location. For example, purchasing agents spend much of their day tied to their desks reviewing and approving purchase orders. If they could perform these reviews using a mobile device, they could spend more time out of the office meeting with suppliers, building relationships, and negotiating better deals.

- *Enhance resource flexibility* Mobility gives your company the power to place the right resources in the right place at the right time. It allows you to redistribute operations to gain efficiencies or react to changing conditions. At the technical level, equipment can be brought to where it provides the greatest benefit. For example, mobilizing store checkout stands allows additional checkout stands to be set up to support holiday rushes and store sales events. At a process level, functions and jobs can be repartitioned. For instance, providing field service technicians with the ability to create and deliver invoices on site at the completion of an assignment shifts the billing function from the office to the field. At the people level, an individual can work anywhere that is served by their wireless network.

- *Move activities closer to the customer* In a business environment where customer service and personalization provide crucial competitive advantages, mobility allows you to bring many functions closer to the customer. The result is greater responsiveness and stronger relationships. For example, access to a customer's insurance records, databases of auto repair costs, and check printing equipment can allow an insurance adjuster to settle a claim immediately at the site of a minor accident. An equipment salesperson can research and respond to an inquiry on a back order instantly during a meeting in the customer's office.

2.1.2 Exchange Data When and Where Needed

While mobility allows activities to be performed where needed, flexible and immediate data exchange provides the value. Depending on the device and network options selected, wireless technology can support a wide variety of data exchange options, ranging from collecting usage data when driving past a water meter to full Internet access on a laptop. While other technical solutions, such as a handheld device with a synching cradle, can offer mobility, wireless solutions bring immediacy and greater accuracy to data exchanges. A mobile worker can instantly access the most current data, submit a status report or credit card scan, or be notified about a new assignment. If up-to-the-second data currency is required, for activities such as stock transactions and credit card authorizations, wireless technology is the only mobile option. Depending on an application's needs, wireless data can be "pulled," by linking to the source and requesting the desired information, or "pushed" by sending an alert to a user's device or automatically refreshing data. Wireless data exchange is already powerful and its uses will expand rapidly over the coming years as network coverage and throughput increases. Wireless data exchange provides companies with the following capabilities.

- *Improve communications* Whether through voice, e-mail, or forms-based data exchange, the quality and frequency of information exchange increases. An emergency room doctor can be notified of the results of lab tests immediately upon completion. A service worker can be rerouted, along with the appropriate information, to a higher priority assignment. The quality of sales reports increases when the salesperson is able to submit updates at the completion of a sales call.

- *Increase the quality of customer interactions*   Having instant access to information such as profiles, account history, and current order status, has the ability to significantly improve the quality of interactions with customers, suppliers, and other third parties. Access to this information enables richer and more varied communications. For example, a salesperson can accomplish multiple activities such as checking inventories, generating quotes, taking orders, and resolving issues during a customer meeting.

- *Enable better decision-making*   Decision-making is always improved by access to accurate and current information. Executives in the midst of negotiations can easily call up needed information or contact their offices through instant messaging or e-mail. Field workers diagnosing problems can access a full range of on-line manuals and diagnostic tools to help them perform their tasks.

- *Open new channels to reach customers*   One of the more controversial, but commercially valuable, features of wireless communications is the ability to push advertising and other marketing messages to mobile devices. For example, a store could broadcast its specials to mobile users within a defined geographic region. One consumer-products company in Europe launched a promotion where, in exchange for providing their mobile phone numbers, randomly selected customers could win a shopping spree at a nearby store. The company transmitted the offer, prize notification, and gift certificate directly to the consumer's mobile device.

- *Reduce downtime*   Wireless access to e-mail, voice mail, the Internet, and corporate applications allows workers and executives to make use of what would otherwise be downtime. For example, executives are increasingly turning to e-mail appliances as a means for processing the staggering loads of electronic communications common in the business world.

2.1.3 Locate and Monitor

While the previous capabilities focus on enhancing the effectiveness of people, wireless technology is equally useful for locating and monitoring a wide range of assets. These capabilities reduce losses from theft and damage, collect information from remote or difficult-to-reach locations, enhance safety, and enable a new wave of personalized services. Wireless tags permit the tracking of assets from cattle to container shipments. Wireless transmitters send storm data from buoys far at sea. Telematics systems in cars provide driving directions and enable rescue personnel to locate the vehicle in case of an accident. Advances in device design, location technology, and network coverage are fueling a rapid increase in new location and monitoring capabilities, including the following applications.

- *Track people, devices, and shipments*   Through a combination of tracking devices and wireless applications, companies have the ability to locate and even communicate with tangible assets. At the simplest level, these applications allow companies to trace shipments from point of origin to final destination. More advanced applications relay the specific location of the shipment, monitoring its condition (for example, ensuring that refrigeration equipment is operating) and providing notification of tampering or attempted theft. Tracking the location of its workers enables one utility company to quickly direct assistance in case of an emergency.

- *Improve inventory controls*   Companies often invest significant capital in parts and equipment inventories for their field service organizations. The cost of maintaining this inventory is high and few service organizations have systems capable of managing parts logistics in real time, resulting in widespread inefficiencies. Using wireless technology to integrate logistics, inventory management, and service parts planning systems provides better control over inventory and can significantly lower total inventory costs. Used in combination with bar coding, scanners, and other tracking techniques, these applications give companies an instant and accurate perspective on inventory levels and asset locations.

- *Collect data from remote sources*   Wireless applications can collect data from locations that are too dangerous, difficult, or costly to access by other means. Applications can collect billing data, operating conditions, scientific measures, or even relay requests for service. For example, oil companies use wireless technology to monitor oil rig equipment in the North Sea. Many local water departments collect water meter readings using wireless transmitters attached to residential water meters. This system is activated as a meter reader drives by and saves the time and expense of trying to schedule personal meter reader visits.

- *Tailor information to location*   Bolstered by the E911 mandate, which requires network operators to support emergency services by locating users of mobile devices to within 125 meters, wireless applications can learn the location of their users. This knowledge allows applications to tailor their information to the needs of the user, such as providing a list of restaurants within a mile of their location, or offering local traffic reports and driving directions. Current location-based applications are just scratching the surface of what will ultimately be possible as device location capabilities become ubiquitous. The opportunities for developing new services and tailored marketing campaigns appear endless.

2.1.4 Simplify Workflow

Wireless capabilities can significantly improve the performance of existing tasks. Moreover, a whole new level of benefits can be obtained if those capabilities are used as a basis for process improvement. Many company workflows and job responsibilities are based on the constraints imposed by paper or wired processes. Wireless technology offers the opportunity to redesign and simplify those processes to be faster, cheaper, and more responsive. Section 2.3 discusses process improvement in greater detail and provides an example of process improvement in action. Wireless technology provides the capabilities to accomplish the following work flow improvements.

- *Eliminate redundant activities*   Wireless technology provides a convenient way to eliminate activities that unnecessarily duplicate work effort. For example, mobile workers from inspectors to doctors on their rounds capture data on paper forms. Clerical workers are required to enter the information from these forms into computer systems. This reentry of data is costly, time consuming, and error-prone. Using a wireless device for the original data capture eliminates the need to reenter the data, increases data accuracy, provides immediate access to results, and frees clerical resources for other tasks.

- *Reduce cycle times*   Immediate data exchange can reduce cycle time by allowing multiple workflow steps to be accomplished in a single session. For instance, if a service worker can generate an invoice at the completion of a service call, the length of the standard billing cycle is shortened and several internal steps and data hand-offs are eliminated. While cycle times can be reduced on a situational basis, such as a salesperson having access to order status during a sales call, greater benefits can be achieved by redesigning the process to shift responsibilities on a permanent basis.

- *Integrate activities and services*   Combining information from multiple application sources often provides much greater value than the sum of the individual parts. Unfortunately, valuable data is often scattered across "stovepipe" systems including front-end (order entry, customer care, billing) and back-end (scheduling, inventory management, supply-chain management, accounting) applications. Much as web applications are inspiring companies to integrate their data in new and creative ways, wireless applications offer a similar opportunity to generate breakthrough improvements. The earlier example of an auto insurance adjuster illustrates the value of combining access to multiple customer, claims, and repair applications to offer an innovative service to customers involved in an accident. Similarly, doctors are starting to use drug prescription applications that offer advice about the appropriate drugs, check for possible interactions, and submit the order directly to a pharmacy.

- *Redistribute tasks*   By unlocking a given task from a physical location, wireless technology allows job responsibilities to be redistributed for greater efficiency. For example, rental car staff can use mobile devices to check-in and present a receipt to drivers as they drop off their cars. Standing next to the car, the staff can quickly and easily check mileage, fuel levels, and damage. Previously, check-in personnel would relay this information to the desk agent for processing, resulting in extra steps, and delays for the customer who had to wait not only for the information, but had to stand in line along with customers seeking to rent a car. By redistributing the work tasks, drivers receive faster service and desk agents can focus on selling rentals, insurance, and upgrades.

2.1.5   Alternative to Wired Connections

Wireless networks provide an attractive alternative for conventional wired networks in circumstances where physical constraints or convenience factors make wired solutions costly or impractical. Whether setting up a network for a small office or exchanging data with remote locations, there are wireless options. These options will become increasingly attractive as the technology advances, improving transmission throughput and quality, promoting mass adoption, and lowering costs.

- *Avoid wiring/rewiring costs and impact*   Wiring standard networks can be costly and impractical in certain situations. For example, many office buildings already have a maze of wires in their floors and walls representing many generations of network technologies. Tracing existing wires or adding new lines becomes increasingly cumbersome and difficult. In other cases, building design or aesthetic considerations make wired networks unattractive. In manufacturing facilities or production lines with moving equipment or complex set-ups, wireless connections are simpler to implement and safer for workers. Under these circumstances, the higher "per unit" cost of a wireless solution is more than offset by the time and cost of installing physical lines.

- *Convenience*   IT organizations are increasingly drawn to wireless local area networks (WLANs) for convenience. They allow IT staffers to relocate equipment at will, without the time and effort of rewiring. WLANs can be set up and moved quickly and easily, making them attractive for situations such as trade shows, temporary offices, and seasonal cash register stations. WLANs are especially attractive for conference rooms, where they enable attendees to bring laptop computers and have instant access to their network files. Starbucks, a gourmet coffee shop, is offering WLAN access to the Internet in some of its locations to attract traveling workers.

- *Access to new locations*   Wireless technology allows data exchange with locations that cannot otherwise connect to physical networks. Ships at sea, passengers in airliners, and travelers in remote locations can tap into wireless networks for voice and data communications. Wireless technology is especially attractive in third-world countries as a means to bypass the effort and expense of installing and maintaining telephone lines across inhospitable terrain.

2.2   Benefits

The previous section answers the questions about what wireless technology can do for you and your company. This section describes the types of benefits that you can expect to receive from your wireless investments. Wireless technology benefits are as diverse as its possible uses. The benefits shown in Figure 2.2 and listed below are necessarily general; specific benefits will vary by the nature of the function being addressed, the quality of the wireless solution, and the effectiveness of its use. Some applications, such as providing busy executives with wireless e-mail access, provide obvious benefits, but are hard to quantify tangibly. Conversely, other applications, such as those addressing field service workers, can provide easily measured benefits in many of the categories described below. Chapter 6 covers techniques for estimating the benefits that you may receive from a specific application.

2.2.1   Improve Productivity

Wireless technology is an effective tool for increasing the productivity of employees at every level of the organization. On-demand availability of beneficial information, better communications capabilities, automation of tedious tasks, and access to the

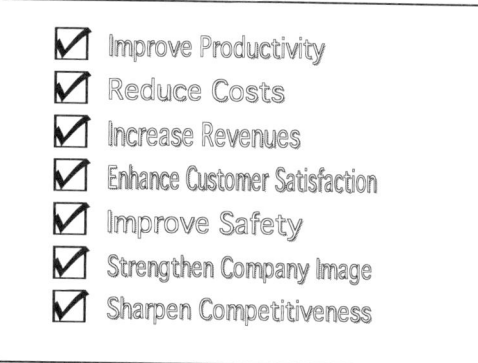

FIGURE 2.2
Wireless Benefits

right tools power a broad spectrum of productivity enhancing applications that cross industries and work functions. Better communications translates into better coordination of scarce resources along with more efficient scheduling and routing. Wireless forms improve clerical productivity by eliminating the drudgery and inaccuracy of paperwork and manual data collection. Being able to solve a problem on site, without waiting and without multiple visits, allows field service workers to serve more customers per day. Eliminating non-productive data entry and logging tasks allows these workers to spend less time at their desks and more time in the field. Allowing warehouse pickers to match items to orders with a combination of bar code scanners and handheld devices rather than paper forms allows fulfillment centers to process more orders with greater accuracy, and eliminates the need for customer returns. Automating paperwork allows salespeople to spend more time on sales calls, and access to the right information during those calls helps them close more business. Handheld devices help doctors reduce paperwork and other administrative burdens and avoid time wasting tasks such as hunting down test results and researching drug interactions, allowing doctors to handle more patients. Access to e-mail, appointment calendars, and other mobile productivity tools enable executives to make better use of their time and allow them to take advantage of otherwise unproductive breaks during the day.

2.2.2 Reduce Costs

In the right situations, wireless technology can offer cost advantages over the expense and effort of wired installations, but its greatest cost benefits arise from its business applications. Through its process improvement, efficiency, and asset monitoring capabilities, wireless technology can help your company reduce its costs, redirect personnel, and make better use of its assets. The examples below illustrate just a few of the ways that wireless technology can reduce costs.

- *More effective use of people* Wireless-enabled productivity improvements reduce the cost to perform an individual task, eliminate unnecessary and redundant tasks, and decrease overhead, thereby allowing mobile employees to spend more time on value-adding activities. By automating data capture, wireless applications eliminate the immediate cost of manual data entry and the secondary cost of data entry errors. Making more effective use of technicians can produce considerable cost savings. Field service organizations can improve overall technician utilization through better scheduling, and by enabling dispatchers to make real-time schedule adjustments and convey them to technicians in the field. Wireless monitoring devices save the cost of sending a worker to collect data from remote or difficult-to-reach locations.

- *More effective use of assets* By closely managing parts and equipment inventory, wireless applications help field service organizations do a better job of managing these assets, saving money spent replacing lost and mislaid parts and significantly decreasing the amount of capital invested in these inventories. Wireless tracking of shipments and valuable capital equipment reduces losses from theft, tampering, and misplacement. Monitoring devices installed on remote equipment can notify the owner when service is needed, thereby avoiding costly repairs.

2.2.3 Increase Revenue

Wireless technology can help companies generate additional revenues in a variety of ways. It can improve sales productivity, enabling salespeople to generate more revenue per capita. It can create new sources of revenue through new products and services. It can gain more revenue from existing sources either by expanding market share or by more efficiently capturing service revenues. And it can gain financial benefits from reducing billing and payment cycles.

- *Offering new products and services*   The current generation of wireless applications and services has only begun to tap the potential of the technology. Advances in technology, greater network coverage and capacity, and higher rates of business and consumer adoption will enable new generations of ever more powerful products and services. Offering new products and services made possible by wireless applications creates new sources of revenue and directly contributes to the bottom line. For example, General Motor's (GM's) OnStar telematics service gains immediate revenue when sold as an option on a new car and also collects a continuing stream of subscription fees. A wireless service that lets golfers on the course order items from the snack bar boosts food service revenues.

- *Capturing overlooked service revenues*   Lost, overlooked, or incorrectly captured charges plague virtually every industry that generates revenues from billing labor and materials. From doctors to field service workers, considerable revenue is lost due to under-billing. One way to grow revenues is to ensure that no billable charges are lost or overlooked, and that invoices are as accurate as possible. Arming professionals with mobile devices and wireless applications allows them to more precisely record work performed, time spent, and materials used. Capturing this information in real time at the point of delivery records charges that might otherwise be overlooked if entered from memory at the end of the day.

- *Cross-selling products and/or services*   Each customer contact point in a company's sales and delivery process represents an opportunity to cross-sell and up-sell accessories, upgrades, and additional products and services. Wireless applications can provide non-sales professionals, such as field service employees, the tools to capitalize on these opportunities. Access to good customer information can alert employees to potential sales opportunities such as the presence of aging equipment, imminently expiring warranties, or availability of a desired accessory.

- *Premium service as a selling tool* Premium service offers a competitive advantage and a means of collecting additional revenues. Customers are often willing to pay more for better service through premium pricing on products or by purchasing additional services. By allowing your company to improve responsiveness and enhance customer service, wireless applications can enable you to use superior service as a selling point to drive more sales.
- *Reduce cycle time* Wireless technology enables companies to move billing activities to the point of service delivery. Allowing field service workers to calculate, print, and deliver an invoice on the spot at the end of a service call cuts the cycle time to receive payment. Quicker payments allow your company to reduce float and take advantage of its capital sooner.

2.2.4 Enhance Customer Satisfaction

Satisfied customers are valuable business assets. They are loyal to your company, repeat buyers, and sources of referrals and recommendations. In most cases, the cost to maintain and sell to an existing customer is far lower than the cost to acquire a new one. Wireless technology is tailor-made to enhance customer satisfaction. By enabling mobility, it gives workers the ability to perform services remotely, closer and more convenient to the customer. Its communication and coordination capabilities help ensure that the right resource shows up at the right time. Its information exchange capabilities allow greater personalization and enhance responsiveness and delivery quality. Specific customer service benefits offered by wireless technology include:

- *Stronger customer relationships* Mobility allows salespeople and field service workers to spend more face-to-face time with customers, thereby building stronger relationships. Access to customer data, order and service histories, and other information enables salespeople or field service workers to further personalize their interactions to match the needs and interests of the customer.
- *Improved responsiveness* Customers prefer companies who are responsive to their needs. More productive and efficient technicians increase throughput, allowing more customers to be serviced sooner. The efficiencies offered by wireless technology can help a technician reach the customer more quickly while being better prepared to service the request. Wireless technology enables technicians to notify customers of estimated arrival times and delays, and

allows dispatchers to adjust schedules in real time. Access to on-line information helps a salesperson check back orders, research pricing, resolve problems, and handle requests on the spot.

- *More consistent, predictable service*   Customers don't like surprises. They want consistent, high quality service. Wireless applications give call takers and technicians access to repair histories, equipment inventory, and parts availability; wireless access to maps and directions keep technicians on schedule; and mobile devices let technicians keep customers abreast of scheduling issues. Wireless access to product manuals, diagrams, and repair instructions can guide technicians through repairs, thereby ensuring more consistent service.

- *Increased single-visit completion rates*   Downtime is costly to customers, placing a premium on the ability of the technician to complete all work in a single visit. Various factors influence this ability—the expertise of the technician, access to repair histories, parts availability, and access to product specifications. Wireless technology can help in all of these areas. Scheduling systems can ensure that assigned technicians have the expertise needed to perform the work. Mobile devices and wireless applications give technicians access to customer service history, equipment repair records, product diagrams and manuals, parts data, and other information needed to complete their work in a single visit.

2.2.5 Improve Safety

Wireless technology offers many possible tools for enhancing personal and worker safety. Enhancing personal safety is one of the top reasons cited for the purchase of personal mobile phones and is a major driver of the burgeoning telematics market. Telematics systems, such as GM's OnStar service, allow drivers to get assistance quickly in case of accidents or breakdowns. These systems automatically recognize problems, such as the deployment of a car's airbags, and rely on location technology to direct emergency personnel to the scene. Through their ability to track and locate mobile devices, wireless applications allow companies to stay in touch with workers venturing into truly remote locations—mines, mountains, space—often under hazardous conditions. Telephone companies in the northeast U.S. equip workers in mountainous regions with GPS tracking chips and wireless devices so that they can monitor worker safety and dispatch help quickly if needed.

Safety improvements are not limited to communications and location applications. The use of wireless monitoring and tracking devices can reduce the need for workers to enter hazardous environments to collect data or check equipment. These devices can also warn of potential equipment problems or situations that could become hazardous if not corrected. By capturing more accurate data, wireless applications can make downstream decision-making more accurate and avoid potential safety issues. For example, pharmacists are more likely to dispense the correct drugs when they can rely on electronically transmitted prescriptions rather than illegible handwritten ones. Combining the drug prescription application with drug interaction and dosage checking further improves the safety of the resulting prescription.

2.2.6 Strengthen Company Image

A less tangible, but potent benefit of wireless technology, is its marketing cachet. Everyone from consumers to executives are enamored with mobile phones, PDAs, and other wireless devices. This attraction is useful for building company image and targeting key market demographics.

- *Position as a technology leader* Companies positioning themselves as technology leaders will deploy wireless technology to enhance their image. For example, Fidelity Investments has been a leader in deploying wireless applications to their customers. Their Fidelity Anywhere[SM] service allows customers to monitor their portfolios and make trades using their preferred wireless device. Fidelity also offers Alerts, which sends messages to customers notifying them about stock prices changes, margin calls, and other market activities. Fidelity even supports trading through GM's OnStar Virtual Advisor telematics service. Although used by only a small percentage of their customers, these services have tremendous marketing appeal and position Fidelity to capitalize on growing consumer acceptance of wireless technology. Even applications with marginal practical use are effective in suggesting better things to come. Web searches disclose numerous sites offering free services for individuals signing on through a PDA or other wireless device.

- *Reach desired clientele* Wireless services and advertising campaigns can attract the attention of individuals in specific market demographics. For example, by offering wireless access to the Internet in its coffee shops, Starbucks hopes to attract mobile professionals and fill its seats during off peak hours. Other companies, such as Cadbury in the U.K., take advantage of the high level of wireless usage among teenagers to offer special promotions aimed at them.

- *Build support among executives* Wireless devices are equally effective for internal marketing purposes. Many executives like to show off sleek new tools and were early adopters of mobile phones and PDAs. Providing executives with wireless access to e-mail or other internal applications builds support for wireless deployments in other areas of the company.

2.2.7 Sharpen Competitiveness

Looking at the bigger picture, all of the benefits listed previously enhance competitiveness to some degree. For example, gaining process improvements in service delivery may yield pricing and customer satisfaction advantages over competitors. Through these benefits, even tactical wireless technology implementations bring some level of competitive advantage. However, your company can enhance its competitive advantage by thinking strategically when designing and selecting wireless solutions. For instance, wireless technology can be the catalyst for a strategy to differentiate your company as the service leader in its market segment. To become the service leader, your company must commit to improving the effectiveness of its service delivery, and must be able to measure its performance and proactively adjust service levels. By addressing key factors such as personalization, more effective scheduling, and higher single-visit completion rates, wireless applications are tailor-made to support rapid and cost-effective service improvement. While the technology offers the ability to gain competitive advantage, capitalizing on this advantage requires recognizing and exploiting the opportunity before your competition. First mover advantage is considerable, allowing a company to gain proficiency and market position before facing serious competition. The time and effort to work out people, process, and technology issues will pose a considerable barrier to entry for late adopters. Enhancing service is far from the only way that wireless technology can offer strategic competitive advantage. Other important examples include introducing new wireless product or service offerings, unleashing attention-getting wireless marketing campaigns, and differentiating existing products/services through wireless-enabled capabilities. For instance, using wireless technology to empower its insurance adjusters and flexibly set premiums brought Progressive Insurance numerous competitive advantages in positioning, cost of operation, and quality of service.

2.3 Approach _____

Potential opportunities to gain significant advantage from wireless technology abound throughout your organization. The challenge is recognizing the opportunities that will provide the greatest value to your company. As a starting point, it is tempting to take an incremental approach by wirelessly enabling an existing wired task. This approach is technology driven. You find an existing wired task, such as connecting PCs to the corporate network or retrieving e-mails, that is supported by a wireless counterpart. Incremental wireless applications are simple to recognize, low risk, easy to implement, and easy to assimilate. While they can provide an attractive return on investment (ROI), with some notable exceptions, they tend to provide only incremental benefits over their wired counterparts and offer less opportunity to build knowledge about wireless technology. The alternative approach is to start with a business need and determine whether wireless technology offers a novel means of addressing that need. This approach requires more initial analysis. Wireless technology does not support every business need; it doesn't necessarily offer an improvement over other methods. But when applied creatively to the right situation, it offers the opportunity to obtain breakthrough improvements by reshaping the way a business activity or set of activities is performed. Also, starting with a defined, and valued, business need ensures a higher ROI than can be obtained from the incremental approach. Since the technology is applied to a task chosen by need rather than technology fit, implementation can be more challenging. These challenges are more than balanced, however, by the benefits of solving a valuable problem, gaining a greater understanding of wireless technology, and finishing the effort with a showcase application. This approach is process-based.

2.3.1 It's All About Processes!

Wireless technology is simply a tool. It provides its value by how it is used. A new tool gives value by outperforming its predecessor on a given task, by bringing automation to a manual task, or by enabling a task that could not be done before. If the task at hand has little value, applying a new tool provides marginal additional benefit. Conversely, if the task is extremely valuable, even an incremental improvement from a tool provides significant benefit.

Processes and workflows organize our tasks. They describe the steps by which we do our work. These steps may be formal, well documented, and followed closely by all employees working with a specific function. They may also be informal, taught by example or word of mouth, and loosely followed. A major process may have formal and informal components. For example, a company may follow a formal sales process with specific steps for identifying prospects through closing business. While the company expects its sales personnel to follow the major steps and track key information throughout the process, individual salespeople have considerable leeway in how they perform the tasks within those steps. Technology can be applied to both formal and informal processes. Providing a salesperson with a PDA that offers access to corporate information systems may automate specific steps in the sales process while also providing personal productivity tools, such as appointment lists and call logs, which serve that salesperson's informal workflow.

Looking across a process as a whole identifies opportunities for improvement. To maximize benefits, start by determining the desired outcome, then look for actions that can support the outcome. Those actions may enhance the effectiveness of the task or may remove roadblocks that are currently inhibiting the desired outcome. For example, the desired outcome of sales cycle improvements may be a higher close rate during a sales call. Closing sales during the call eliminates the risk of a competitor derailing the sale in the time between the sales call and contract close, resulting in a significant increase in revenues from the additional sales. One action that would support this outcome would be the ability to generate and deliver a proposal during the sales call to capitalize on the momentum generated by the meeting. To deliver that proposal, the salesperson would need, among other things, access to current pricing and delivery information and the ability to print the proposal on site. At this point, wireless technology becomes an attractive option for providing the needed capabilities. Chapter 4 expands this concept of using processes as a source for high value opportunities for wireless technology.

Once a wireless technology becomes a potential solution, the process drives the specific technology decisions. Rather than trying to force a particular wireless solution to fit a given situation, the needs of the process dictate the selection of the appropriate components, immediately reducing an otherwise overwhelming set of options down to a manageable number. Furthermore, the final solution is a much better fit than is possible with a "technology seeking an opportunity" approach. To support our sales example, the wireless solution must be able to operate in the customer location, have access to corporate data, and allow the printing of finished proposals. While there are multiple ways of addressing these requirements, they already preclude a large number of technology options. Chapter 5 describes the approach for translating process needs into specific technology implementations.

2.3.2 Example: Improving Field Service

The best way to understand the advantages of taking a business process-oriented approach to wireless technology is to explore an example.

The situation   Figure 2.3 shows a simplified version of the workflow within a typical field service organization. A customer requesting service calls the company's customer service center for assistance. If a telephone service agent cannot resolve the issue over the telephone, the call is logged, classified, and assigned to a field service rep. A work order is printed out and is picked up the next morning in the field service office. The field service rep plans his or her day and travels to the customer site to perform the necessary work. At the customer site, the problem is diagnosed, and if the necessary parts are available within the service truck's inventory, the rep performs the repair. At some point, hopefully immediately after completing the work, the field service rep manually logs his or her actions, time, and parts used. The completed work order documentation is dropped off when the field service rep returns to the field office at the end of the day. The next morning, a data entry clerk in the customer service center enters the work order into the company's field service system

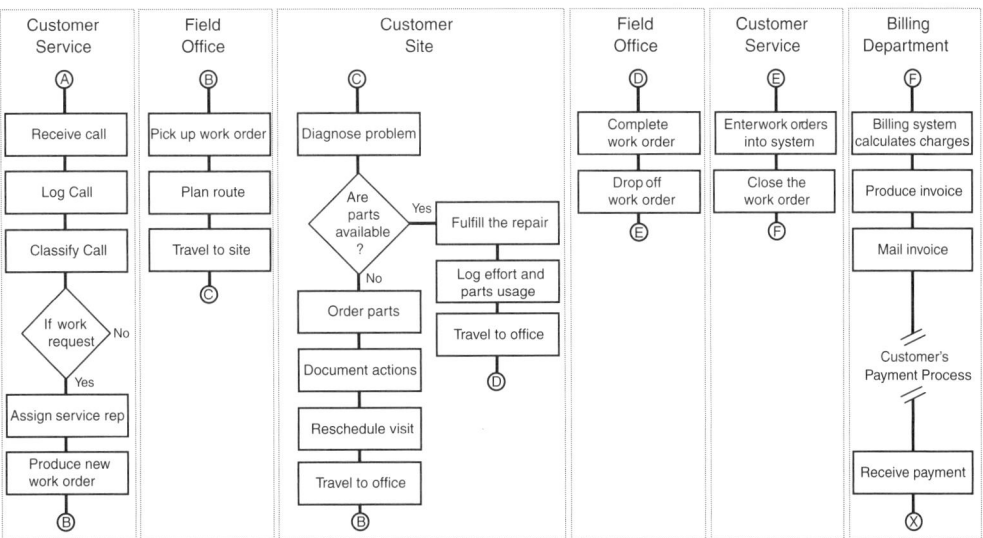

FIGURE 2.3
Simplified Field Service Workflow

and marks the order as closed, making it available for billing. At the next billing cycle, the billing system calculates the charges and produces an invoice to be mailed to the customer. The customer receives and pays the invoice following their usual payment practices and the company receives its payment for services rendered.

   The business issue   Companies around the world use variations of this basic field service process. While this process offers many opportunities for improvements enabled by wireless technology, let's address an issue that is near and dear to every Chief Financial Officer—receiving payment sooner. One major issue is the inefficiency of the billing portion of the process. This process relies on manually recorded data that is often documented at the end of a hectic day, which often results in errors and missed charges. Second, a data entry clerk must enter this manual documentation into the company's computer systems, requiring extra effort and creating another opportunity for error. Third, the lag time between service delivery and receipt of payment can be significant. A call taken on a Monday is completed on Tuesday, but the results are not entered into the computer system until Wednesday. An invoice won't be generated until the next billing cycle. If the work order is received at the beginning of a monthly cycle, up to 30 days may elapse before the invoice is mailed to the customer.

A wireless process improvement   If we could produce a bill for the customer on the spot at the completion of a work assignment, we could remove much of the lag from the billing cycle. We are still at the mercy of the customer's payment practices, but we have eliminated our own delays and the time needed for the invoice to arrive by mail. To accomplish this goal, we need to connect the field service rep to the company's field service and billing systems. This need can be addressed with wireless technology. One option is to provide each field service rep with a PDA, a wireless modem, and a portable printer. The field service rep downloads their work orders at the start of the day (or alternately, could have orders transmitted to them throughout the day) and travels to the customer. At the end of the service call, the field service rep enters his or her documentation into the PDA. This information is transmitted to the company's systems, which close the order, calculate the charges, and return the information needed to produce the invoice. The field service rep prints the invoice and hands it to the customer—thus completing that portion of the transaction. This improved version of the process is shown in Figure 2.4. Notice the desirable side-effects produced by our changes. The new process eliminates the need for a data

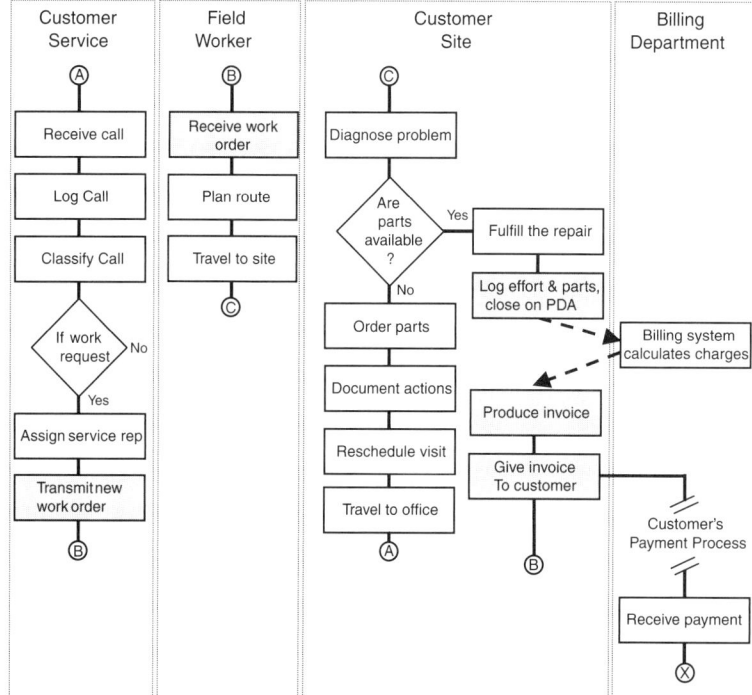

FIGURE 2.4
Improved Field Service Workflow

entry clerk to enter the completed work orders and it eliminates the effort required to generate and mail invoices in the billing department. The field service rep no longer has to return to the field office to pick up and drop off paperwork. The two major areas for billing errors have also been eliminated. The field service rep completes his or her documentation immediately at the completion of the assignment, and there is no need to re-enter data manually.

Final observations   This simple example shows only a few of the many opportunities to improve the field service process, and the illustrated solution represents one of many options for solving our need. For instance, if the field service rep was further equipped to take payment via credit card, the billing cycle would become negligible. We could use a location-based application to simplify the field service rep's routing and directions. Additional business benefits could be gained by increasing the number of service calls resolved on the first visit. To accomplish this goal, we could extend our solution to include wireless access to diagnostic information to improve problem solving and more robust parts inventory management to increase the likelihood of the right parts being available for each service call.

2.3.3 Process Opportunities

As the rest of this book will illustrate, the opportunities for wireless process improvements are endless; they are limited only by our ability to recognize a need and creatively apply the available technology components. Any process that involves mobility in any of its steps is a possible candidate, whether that mobility involves travel or simply freeing a line worker from the constraints of a physical cable. When researching business needs that could be served by wireless technology, consider these possibilities.

- *Automate manual components of an existing process*   As shown in our field service example, any process that involves the manual collection of data on paper forms by a mobile worker is a potential candidate for automation. Automating these steps increases the speed and accuracy of data collection and eliminates the cost, time, and error rates of re-entering that data into electronic form. Possibilities appear in every industry, including a retail store's inventory forms, restaurant inspector's reports, doctor's prescriptions, and salesperson's sales call logs.

- *Bring mobility to a stationary activity*   Mobility can be limited by data access, physical constraints, or the difficulty of moving an electronically wired activity to a new location. Whether a professional is chained to a desk by the need to access computer-based information or a package sorter is tied to her position by her scanner's umbilical cord, wireless technology can set them free.

Activities that require wired connections, such as point-of-sale terminals, become flexible through wireless access. This flexibility allows activities such as check-in at hotels and hospitals and check-out in stores to be moved as needed.

- *Streamline a process*   Free exchange of data, along with elimination of paper forms and physical connections allows process steps to be bypassed or eliminated. As illustrated in Figures 2.3 and 2.4, numerous steps can be eliminated to gain speed, increase process throughput, and reduce cost.

- *Launch a new process*   While improving existing processes can be highly valuable, breakthrough advantages can be gained by using wireless technology to conduct business in ways that were previously impossible. A new process could take advantage of wireless access to, or by a consumer, such as NTT DoCoMo allowing consumers to bill items from vending machines to their mobile phones. Or, like GM's OnStar, the process could pioneer a new range of services.

- *Support a larger initiative*   In addition to reshaping processes, wireless technology can be used to gain greater value from existing investments in Enterprise Resource Planning (ERP), Customer Relationship Management (CRM), and other major initiatives. For example, wireless access to CRM systems increases the accuracy of CRM data and allows mobile workers to gain the benefits of the CRM system in their daily activities.

2.4 Why Start Now?

When should your company get started on its wireless technology explorations? In some industries, such as logistics or package delivery services, your company is significantly behind unless it is actively deploying wireless applications. In other industries, companies are still sitting on the sidelines waiting for the technology to mature. This attitude is understandable if we take an outside view of the whole wireless industry; the technology appears to be evolving too quickly and too many standards appear to be in flux. From this standpoint, it seems prudent to wait until the technology becomes stable and mainstream before starting. Going beneath the surface, however, reveals that this conservative stance, while avoiding the trials of early adoption, can be quite costly over the long run. Significant business benefits are available from today's technology and the experience gained from current deployments will provide a major competitive advantage over later entrants. While future technology will extend the value and usefulness of wireless technology, waiting allows these benefits to slip away. Starting your company's wireless explorations now brings the following benefits.

2.4.1 Gain Immediate Business Benefits

As described throughout this chapter, wireless technology already offers benefits that are as diverse and wide ranging as the potential uses of the technology. From cost savings to productivity improvements and enhanced job satisfaction, wireless applications have a strong track record of providing tangible, quantifiable business benefits. Even in situations where future enhancements in coverage and bandwidth will enable wider and more feature-rich deployments, applications built with the current generation of technology can more than justify their investments. The advantages of mobility and wireless access to information are well established, and the next chapter provides many examples of currently implemented wireless applications that are providing significant returns to their owners.

2.4.2 Gain Implementation Experience

Simple implementations, such as installing a WLAN for accessing the corporate intranet through PCs, are minimally different than using an equivalent wired solution. Building a full solution that changes the way people do their jobs or how they interact with customers, however, involves issues far beyond technology installation. Working out kinks and moving up the learning curve in any new technology is harder and takes longer than most organizations expect. Experimenting with wireless technology now lets your organization gain valuable experience working out people, process, and technology issues that will pay many dividends when designing and rolling out future applications. For example, one large healthcare group went through several design iterations before arriving at their final successful emergency room applications. Issues such as device fragility, display performance in extreme lighting conditions, and tolerance of data refresh rates did not surface until doctors began using the first prototypes. Ideas that were attractive in theory proved less than successful in practice. The lessons learned from the initial implementations led to significant design, workflow, and training changes, and provided an extensive body of knowledge and best practices for building new applications. The experience also identified opportunities for new wireless functions and applications that would not otherwise have been conceived.

2.4.3 Be Better Prepared for Future Capabilities

Building from existing experience and capabilities is always faster than starting from scratch. Changes are incremental rather than revolutionary and most underlying issues have already been solved. As new wireless capabilities arrive, an experienced organization will know immediately how best to deploy them. Employees are already accustomed to using wireless devices and applications and will be able to assimilate and gain benefits from additional capabilities sooner. Late adopters, although starting with more powerful versions of the technology, are forced to work their way through an even steeper initial learning curve. In a competitive environment, the first mover organization has considerable advantage over the late adopter.

How Others Are
Using Wireless

3

The concept of using wireless technologies might seem new, but it isn't. Companies of all types and sizes have relied on wireless technology for years, using it to support some of their most critical business processes. Wireless technology has allowed these companies to gain greater mobility, exchange data when and where needed, direct resources to the most useful and expedient locations, track and monitor people and assets, and simplify their own workflows.

Shipping and trucking companies developed some of the earliest wireless applications to help track valuable cargo, vehicles, and packages. Using wireless technology, they have been able to optimize the logistics of their far-flung operations, perfect their delivery capabilities, and reduce theft and damage. Government agencies like NASA and the Department of Defense have relied on satellite technologies for decades to track the movement of troops, weaponry and military assets, receive and broadcast data, and communicate with staffers over great distances. IBM, with its enormous field service organization, developed the ARDIS wireless service in conjunction with Motorola to give its field service workers access to IBM's central mainframe from remote customer locations and data centers. Simple wireless applications, like paging services, have been used for years by doctors and business executives.

In contrast to some of today's wireless applications, early enterprise wireless applications were primitive and limited in their capabilities. With no off-the-shelf wireless packages available, these applications were custom-developed and proprietary. Many operated on private networks, whose coverage was built out to suit the needs of the applications. Mobile devices tended to be large, clumsy, and constructed to serve a particular user community and application. Many applications featured two-way radio voice communications, and only the most elite supported any type of data collection or transmission.

By today's standards, these wireless applications might have been rudimentary, but for the companies that adopted them, wireless technology provided break-through capabilities and powerful competitive advantages. For these early adopters, wireless technology is now fundamental to conducting business. Fulfillment organizations, shipping companies, and distributors are reliant on wireless inventory management and logistics applications. McKesson HBOC, a distributor of pharmaceutical products, was able to reduce errors in customer orders by over 60% using wireless technology and applications. United Parcel Service of America Inc. (UPS), already at the forefront of the wireless brigade, is investing another $100 million to standardize and upgrade its wireless infrastructure throughout its worldwide distribution centers; UPS expects payback within 16 months. FedEx is spending a similar amount to modernize its existing wireless capabilities in its distribution centers and sorting facilities, and is beginning to install wireless capabilities in its aircraft fleet to facilitate the transfer of engineering and maintenance data. Even car rental companies from Avis to Hertz, already dependent on wireless technologies to process rental car returns, are seeking ways to extend their wireless capabilities to the check-in process to offer greater customer convenience, eliminate lines, and reduce paperwork. Large field service organizations, from IBM to Sears and Sun Microsystems, use a panoply of wireless technologies and applications to dispatch, schedule, and guide their armies of field service representatives, and to automate and streamline the delivery of service to customers.

What compelled these early adopters to turn to wireless technologies? While some were innovators willing to experiment with a novel technology, most were driven by pressing business circumstances, i.e., they were looking for a solution to a dilemma. When FedEx originally turned to wireless technologies, it was in desperate straits, having outgrown its ability to manage and communicate with its thousands of delivery personnel. The company's astronomical success, and high number of field employees operating nationwide, had overburdened its two-way radio channels. For FedEx, wireless technology was the *only* viable way to resolve a mounting business problem. No "out of the box" wireless application was available to make FedEx's problem magically disappear. FedEx had to innovate and master wireless technologies to overcome a life-threatening situation. Now a wireless diehard, FedEx is staying ahead of the learning curve and seeking to extract even greater returns from its next wave of uniquely adapted wireless solutions.

The purpose of this chapter is to present actual, functioning wireless applications that companies are relying on today to derive some type of advantage. For the reader, it is helpful to consider how others are using wireless technologies to glean ideas about where and how you can apply wireless in your own organization and to your own business issues. Often, it is easier to look at existing examples of applications and extrapolate a solution from there, coming up with applications that are suited to your particular situation. It is undoubtedly useful to look at what others in your industry are doing. But it is equally valuable to examine what companies in other industries are trying, as some of these applications could be modified to work in your business. Adapting the examples cited throughout this chapter to serve your specific business situations, strategic plans, and objectives is a good way to start with wireless technologies. By adapting solutions that others have already battle-tested, your odds of success are that much higher. But you should remember that coming up with a breakthrough opportunity for your company takes more than incremental thinking. Once you've gained experience and comfort with wireless technologies, consider stepping "out of the box" to develop solutions that can propel your company past its competitors.

This chapter surveys the current landscape of wireless applications. The following sections consider some of the high-level categories of wireless applications used by companies throughout various industries. Starting with the most ubiquitous wireless application—voice communications—the chapter continues by looking at data applications including enterprise, consumer, and telemetry categories. Real-life examples of wireless applications are summarized in tables for easy reference. The latter portion of the chapter offers more in-depth case studies to illustrate several types of working wireless applications.

3.1   The World of Wireless Applications _____

If you were put on the spot and asked to name a few examples of wireless applications, you might be stumped. Yet, wireless applications surround us. Although the flashier types of applications garner all of the press, many straightforward, less glamorous applications have been providing business and personal value for well over a decade. These wireless applications can be broken down into two major categories: voice and data.

3.1.1   Voice Applications

By far, the number one wireless application continues to be voice. Beginning with first generation, analog cellular networks and spurred on by the higher quality of second generation digital networks, wireless voice communications have grown steadily and continue to accelerate. Mobile telephone usage has penetrated the far reaches of the globe, with Europe and Japan leading the way. Reports conflict, but analyst groups peg cellular telephone subscribers at upwards of 700 million and climbing at the time of writing.

Voice is a natural fit for both present day, and future, wireless applications. The ability to speak to another person, no matter where the calling parties are located, offers people the kind of immediate access and instant gratification that is the bedrock of wireless technologies. Telephones are ubiquitous in the wired world, and our desire to use them is tempered only by our ability to locate a convenient apparatus and pay for the service. Being able to carry our own mobile telephones around with us completely eliminates one of the impediments to usage (locating a device). Being able to reach other people anytime, anywhere, is yet another driver for widespread wireless voice communication.

All of the major land-based telecommunications carriers offer wireless telephone services. Market penetration has skyrocketed with the improved connectivity and voice quality offered by digital networks. But pricing schemes in the U.S. continue to dampen usage, at least among consumers; subscribers must always pay for incoming calls, and roaming charges, applicable outside the home calling area, can be steep. But in many other countries, notably Japan and European nations, where the expense and hassle of obtaining wired telephone service is sometimes formidable and where wireless billing models are more "user-friendly," cell phone use is pervasive. In developing countries, where physical, land-line connectivity is simply not an option, mobile phones have made dramatic inroads.

On the device front, voice applications will evolve along with devices. As handheld units become ever more powerful, and as Internet-ready telephone handsets become more prevalent, wireless voice communications will continue to dominate these devices. The reasons are simple. Voice applications are ideally suited for small, sometimes miniscule, handheld devices with limited real estate and therefore limited textual capabilities. Voice applications play to the strength of these devices—communications—while not being constrained by their frustrating limitations, i.e., paltry display areas and unwieldy, tiny keypads.

3.1.2 Data Applications

To better understand some of the data applications enabled by wireless technologies, pause and think about the kinds of data applications used in the wired world today. We rely on traditional, wired applications and equipment to perform a range of tasks and operate a host of data applications. We use personal productivity tools to schedule meetings, arrange calendars, and manage contact information. We use word processing applications, spreadsheet applications, and custom-designed transactional systems. We access enterprise information via intranets and public information over the Internet. Our supply-chain management, inventory management, sales force automation, and enterprise resource planning systems operate on our wired networks. We may retrieve information, update information, submit transactions, collect data, and disseminate it. These tasks and applications are very different, but they all depend on the transmission, collection, management, and/or manipulation of data. What they have in common is that they are all data-based applications.

Wireless technologies introduce two nuances to this wired world. First, they allow us to perform the same tasks and access existing data-based applications, but in a new way—wirelessly. Second, they enable the creation of entirely new types of data-based applications, particularly ones that support a range of mobile activities and tasks.

- *New ways to access existing applications* Wireless technologies allow us, for the most part, to perform the types of work we have always done through our hardwired desktop components. To a worker situated in an office, the difference may not seem that great. But to workers and users that need to be mobile, the ability to move about freely and yet still access needed information is a giant leap forward.

  Wireless access to existing applications may not be "breakthrough" per se, but it can dramatically improve efficiency, productivity, and responsiveness, and even lower overhead. Wireless personal area networks (WPANs) and WLANs (described in more detail in Chapter 11) provide wireless access to intranets and enterprise systems without the hassle and expense of pulling cables through walls and floors. Students can use infrared networks to access course information rather than lug around catalogs and handouts. Users rely on Bluetooth-enabled devices to synchronize data and swap information wirelessly with other Bluetooth users. Employees use WLANs to access corporate intranets and enterprise data, just as they use their wired desktop components to accomplish the same things. Executives use Research in Motion (RIM) BlackBerry devices to send and receive e-mail no matter where they roam. A salesperson that receives a lead via e-mail while traveling has the ability to respond immediately and potentially close business that might otherwise be lost.

- *New kinds of data-based applications* Wireless technologies also allow the creation of new types of data applications to serve users, customers, and consumers. Some of these applications build upon existing capabilities, making them even better than before. UPS, for example, is combining several types of wireless network technologies, from Bluetooth to 802.11x WLANs, and deploying several types of special purpose devices, from ring scanners to belt-worn data collection terminals, to automate and standardize package management and tracking across all of its delivery centers. Southwest Gas, a utility firm in the southwest U.S., is using a WLAN and vehicle-mounted laptops to give its field technicians access to up-to-date gas line maps rather than having them carry around bulky, weighty paper maps and manuals. Now, technicians park their vehicles in company lots overnight where the WLAN automatically transmits accurate gas line maps directly to the vehicle-housed devices. At Avis, lot attendants use handheld devices to check-in customers and process car returns, capturing digital signatures in real time and transmitting signed documents directly to the company's enterprise systems.

The discussion of wireless data applications that follows will focus on three major groupings: enterprise applications, consumer-oriented applications, and telemetry (machine-to-machine) applications. Within these groupings, we will highlight various examples of wireless applications in use today. Note that many applications do not fall neatly within one segment. For example, many wireless sales-oriented applications help empower the sales force, and also have the effect of strengthening customer relationships through more responsive service. Brandow Automotive, a group of car dealerships in the Philadelphia area, has a wireless lead management application that allows salespeople to pursue leads more quickly and efficiently—a boon for time-strapped salespeople—while shortening overall response times for prospects and customers. Atlantic Envelope Company, a manufacturer of specialty envelopes, has a wireless application that allows salespeople to provide immediate responses to customer queries about orders, inventory, and historical purchases by looking up the information from their handheld devices. The application has also improved the effectiveness of the overall sales process by reducing the number of calls to the central customer service center.

3.2 Enterprise Applications

Wireless data applications are ideally suited for businesses and many public sector entities. Although not as prevalent as wireless voice communications, wireless data applications can provide demonstrable qualitative and quantitative returns on investment to their adopters.

The most successful wireless implementations focus on process first, and application second, as explained more fully in Chapter 2. Savvy organizations first consider how their existing processes, whether mobile or not, could benefit from more timely access to information, the ability to gather and transmit information from its point of origin in the field to a central location, or the ability to perform additional steps off site. As these processes become more robust, producing and handling data more efficiently, they often benefit related internal processes by eliminating redundant steps, reducing the need for paperwork and providing quicker access to field-generated data.

Using a range of private, public, regional, and national networks and a host of mobile devices from PDAs, to WAP-enabled telephones, to custom data collection and transmission devices, organizations have built an extraordinary portfolio of wireless data applications. These applications fall into the following broad categories.[1]

3.2.1 Customer Relationship Management (CRM)

The bulk of wireless applications have a primary or secondary purpose of improving customer relationships. For example, when the State of California wirelessly notifies residents of impending power blackouts, it is trying to strengthen relationships with its constituents. When McKesson HBOC implements a wireless inventory management application, a secondary effect of the solution is to reduce errors when packaging orders, thereby increasing the satisfaction of its customers.

The advantages sought by a wireless CRM implementation, just as for any type of CRM implementation, are: greater customer loyalty, higher customer retention rates, lower acquisition costs, more targeted marketing and sales, greater personalization, and expanded opportunities to cross-sell and up-sell products and services. Wireless CRM applications foster a range of activities and functions that play to these objectives. They give customers the ability to communicate more quickly and efficiently with hard-to-reach salespeople, field technicians, or support personnel. Similarly, these applications allow company workers to contact customers more readily, provide fast answers to queries, and supply additional levels of service beyond the norm. Sometimes, they enable workers to initiate, perform, and conclude transactions with customers from any remote site, whether at the customer's office, on a trade show floor, or while traveling from place to place.

Examples of wireless CRM applications abound. As previously mentioned, Brandow Automotive outfits its car salespeople with handheld devices and wireless applications so that they can immediately receive, process, and respond to leads entered through various on-line channels. (See Figure 3.1.) GM and Ford Motor Company offer telematics applications like OnStar and Wingcast, respectively, to

---

1. For additional information on these example applications, please refer to the citations found in Tables 3.1 and 3.2.

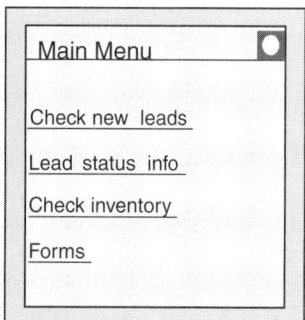

FIGURE 3.1
Lead Management Application

maintain relationships with car owners after the sale, by providing a new level of service from navigation to roadside assistance to stolen vehicle tracking. Rental car company, Hertz, uses wireless applications to check-in customers while they are in the complimentary shuttle bus, enroute to the rental car lot. Upon arrival, rental car stalls are illuminated with customer names, and keys are in the ignition, eliminating waiting in line at the counter.

The financial services industry has been actively promoting wireless CRM applications for years. Fidelity Investments, a pioneer in wireless customer services, offers Fidelity Alerts and Fidelity Anywhere[SM] to tighten relationships with high-volume trading customers. Through Fidelity Alerts, a customer can have portfolio and market activity messages sent to a wireless device. Fidelity Anywhere[SM] allows brokerage customers to use a variety of wireless devices, and even the OnStar telematics system, to monitor stock and 401(k) portfolios, check account balances and positions, obtain quotes, execute orders, check order status, trade stocks, and more. Brokerage houses Charles Schwab, Morgan Stanley Dean Witter, and Donaldson, Lufkin, and Jenrette offer similar services. Other financial services companies, from Bank of America to E*TRADE, allow customers to perform a variety of functions—checking account balances, paying bills, transferring funds, finding nearby ATMs, and getting real-time quotes, news, and research—from a slew of devices.

Cadbury, the chocolate maker, launched a highly-publicized wireless campaign to retain customers and attract new ones. Cadbury hid "codes" inside candy wrappers and encouraged buyers to send in their codes to be entered in prize drawings. The wireless catch? Buyers had to transmit the codes using wireless short message service to be eligible for entry in the drawings. In Japan, NTT DoCoMo in partnership with Coca-Cola (Japan) is launching a "smart" soda machine that will let consumers perform a variety of functions—from storing cash in their phone accounts to buying tickets—simply by passing a wireless phone over a sensor located on the vending machine.

The Venetian Hotel in Las Vegas, Nevada, is turning to wireless applications to expedite guest check-ins and maximize convenience. Guests can now check in from multiple locations in the hotel, including parking lots and lounge areas. (See Figure 3.2.) As mentioned above, the State of California turned to wireless to warn citizens of impending power blackouts, giving impacted individuals time to take action beforehand. Other companies use wireless to eliminate some of the inevitable hassles of doing business. McKesson HBOC realized that its improved wireless inventory management system would not only provide internal efficiencies and cost savings, but would also result in fewer incorrect orders, fewer aggravated customers, and fewer profit-eroding returns.

Progressive Insurance is also leveraging wireless to strengthen its customer relationships. Through its Immediate Response® claims service, Progressive provides its claims representatives with wireless applications so they can deliver in-person, 24x7 roadside service at accident sites. Claims representatives travel to claims sites

FIGURE 3.2
Hotel Check-in Application
Source: Symbol Technologies

in specialty vehicles equipped with the mobile and wireless technology needed to service a claim on site. The representatives are able to write estimates and even cut checks to cover repairs right at the scene of an accident. This on-site service helps Progressive reduce opportunities for fraud, and also keeps customers satisfied and loyal. The fact that Progressive spends less money settling claims means it can charge its customers lower premiums. Producers Lloyds Insurance Company of Texas, a crop insurance provider, is also using wireless applications and devices to give its agents access to real-time pricing data, allowing agents to file claims and authorize payments for farmers straight from the farm fields.

Cybex, an exercise equipment manufacturer, and Atlantic Envelope Company give their sales forces wireless applications to do things like track customer orders, look up inventory information, and check customer purchase histories. Dining establishments such as Starbucks offer patrons enticements like Internet and e-mail access to lengthen stays and increase purchases of food and beverages. Aureole restaurant in Las Vegas, Nevada offers diners wireless winebooks to review wine information, store favorite selections, and drum up repeat business. (See Figure 3.3.) Golf courses in the U.S. use the ProLink wireless application to attract golfers by offering course pointers and analysis during playtime and easy snack food ordering right from the golf cart. Moreover, the ProLink application uploads player statistics to a web site where players can spend time analyzing their performance and plotting improvement strategies.

Where else are wireless CRM applications making inroads? Livery taxis in New York City process credit card transactions wirelessly, eliminating the need for passengers to carry cash to pay fares. Airlines such as United Airlines, Northwest Airlines, and American Airlines allow passengers to sign up for e-mail alerts to mobile devices. These alerts provide notification of flight status and gate assignments so

FIGURE 3.3
Winebook Application

that customers are able to make any necessary adjustments to their travel plans before arriving at the airport. Delivery and transportation companies from UPS to FedEx allow customers to track packages from source to destination by using a combination of wireless technologies, networks, devices, and applications. Metra, greater Chicago's rail transit system, can pinpoint the location of trains to deliver real-time schedule information to waiting passengers, mitigating the frustration caused by delays. Florida Power & Light, a utility company in Florida, uses "smart" meters to allow its employees to read meters wirelessly from outside a building, eliminating on-site visits and enhancing customer convenience.

3.2.2 Field Service

Field service organizations are uniquely suited to benefit from wireless applications. Field staffers are constantly on the go, and historically have not had access to the tools or types and quality of information available to their office-bound counterparts. These disadvantages have created inefficiencies in the field service process as a whole, and make it particularly amenable to wireless solutions. Things like handheld devices and wireless applications can allow field services workers to access information precisely when they need it; deliver more responsive and targeted services to customers by knowing more about repair histories and parts availability; and perform more process steps in the field from closing work orders to logging repairs and invoicing customers. These advances streamline the entire field service process, and eliminate some of the redundancies and paperwork associated with back-end tasks such as billing and collections.

Although every field service process has its own nuances, here are some common ways in which organizations are using wireless applications to make improvements. To better balance workloads and deal with urgent requests, dispatchers communicate with field workers through wireless applications, directing them to the highest priority jobs. GPS location devices allow dispatchers to know where reps are at all times, and permit them to adjust workloads on the fly to accommodate more customers and service calls. Field reps use their mobile devices to obtain maps and directions to customer sites, cutting down on travel times and delays. On site, technicians use wireless applications to review repair histories that might speed the resolution of the current problem. Technicians look up product specifications and schematics to provide quicker and more accurate service, and use device-resident applications to guide them through the repair process. Wireless parts management applications allow technicians to order parts and update stocking records, increasing the odds that they will have the right parts available to make a repair. At the end of the visit, technicians can update work orders, record labor and parts used at the point of service, capture customer signatures, create invoices on the spot, and even accept payment.

Sears Roebuck's HomeCentral repair service has been relying on wireless technology for over a decade. Sears' field technicians communicate wirelessly with dispatchers using notebook computers, and place orders for parts from their wireless devices. Field technicians can also generate and transmit bills on the spot and accept payment, eliminating paperwork and reducing billing cycles. First Service Networks (FSN), a national contractor providing plumbing and heating system repairs for retail chains, works with several thousand local repair firms across the country. FSN relies on a wireless application to dispatch customer requests, along with complete customer histories, to a suitable local contractor. Repairpeople use their devices to record arrival times, review equipment repair history, log information about the repair, and forward billing information straight to a central server. Similarly, Honeywell uses wireless applications to connect building system technicians with dispatchers, perform more work remotely, eliminate paperwork and process steps, and speed collections. As previously mentioned, IBM was a pioneer in wireless field service applications, and participated with Motorola in defining and rolling out a wireless network, ARDIS, that connected field technicians with IBM's central mainframe. Sun Microsystems is also rolling out a wireless field service application to streamline its service delivery.

3.2.3 Locating and Tracking

Wireless data applications are often used to track the whereabouts and well being of employees and physical assets. These applications either use GPS-enabled devices and satellite networks, or some combination of radio frequency tags and transceivers to communicate location information from remote assets or devices to central computer systems.

Where employees perform work remotely under hazardous conditions, companies rely on wireless applications to ensure their safety. Nynex, a former telecommunications company in the northeastern U.S., equipped its fleet of repair trucks with GPS locators and wireless devices so that it could easily locate and dispatch aid to at-risk employees working in sparsely populated areas. In the North Sea, companies stay in touch with workers on remote oil drilling platforms using various wireless technologies.

Companies also use wireless applications to monitor the location and movement of valuable tangible assets, both large and small. Companies track these assets to guard against damage and theft, or simply to ascertain the whereabouts and anticipated arrival of assets in transit. Delivery firms like FedEx, UPS, and Airborne Freight Corp. are investing in wireless technologies to provide customers with real-time, in-transit tracking and delivery information. Associated Food Stores, a food distribution center, uses wireless applications to manage its inventory and track and locate food products to ensure that only fresh goods are sent to food stores. Chicago's Metra rail transit system uses wireless applications to pinpoint the location of trains and inform passengers of anticipated arrival times and delays. J&B Wholesale Distributors uses a wireless delivery application to track arrival and departure information for inventory in its trucks, allowing the company to record discrepancies between delivered goods and purchase orders. Penske Logistics uses a similar wireless application to track deliveries and relay critical information, such as delays or route changes, via satellite to company headquarters. Anderson Trucking Services uses a GPS-enabled wireless application to let customers track the whereabouts of their loads by locating and following the movement of the truck that contains the goods.

Progressive Insurance offers a unique program in some markets to its drivers, adjusting their monthly premiums based on their actual driving habits. Progressive installs wireless applications and GPS devices on customer cars to monitor how, when, and where the cars are driven. Data is transmitted to Progressive's computer systems so that premium adjustments can be made.

Retail establishments, including many banks, are investigating the use of location-based services to direct customers to nearby ATM machines or branch offices. Other service providers, like OnStar and Wingcast, use wireless location and tracking technologies to provide maps, directions, and other navigation assistance, find broken down vehicles and recover stolen cars.

3.2.4 Customer Service

Many companies have designed wireless applications specifically to provide some type of self-service to their customer base. These special purpose applications provide customers with select information and allow them to perform a subset of self-service functions like those available through the company's web site. The financial services industry has developed some of the earliest and most publicized examples of customer-facing wireless applications. Fidelity Investments' Fidelity Anywhere[SM], Charles Schwab & Co.'s PocketBroker[SM], and Morgan Stanley Dean Witter's TradeRunner[SM] are wireless applications offering a range of information and capabilities including price alerts, account balances, order executions, order status, and more. (See Figure 3.4.) Car rental agencies like Thrifty and Dollar Rent A Car allow customers to make reservations from their wireless devices. United Airlines and American Airlines are building out WLANs in airport lounges and gates to give travelers Internet connectivity and other travel-related services. Ticket agencies such as TicketMaster, sports teams, and other event organizers have designed wireless applications that allow patrons to reserve seating at events. FTD.com, an on-line florist, offers a wireless flower-ordering service. Other web retailers have also adapted their ordering capabilities so that they can accept orders from many types of wireless devices.

FIGURE 3.4
Brokerage Application

3.2.5 Sales Force Empowerment

Many wireless applications also have the effect of putting more power in the hands of salespeople in the field. Unlike weighty sales force automation systems, these wireless applications offer "point" capabilities that can greatly improve the productivity of a salesperson and his or her ability to respond to prospects and leads, and customer queries, quickly and efficiently. Mobile applications and wireless access give salespeople the ability to look up answers to customer questions on the spot, submit orders and close business on site, better manage sales leads, and improve their sales forecasts through timely submission of account updates. (See Figure 3.5.)

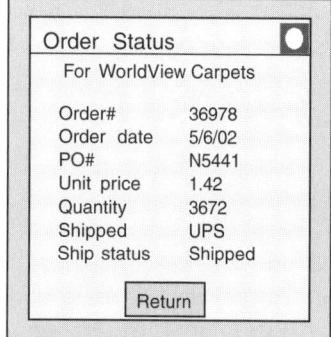

FIGURE 3.5
Sales Force Application

At car dealerships, salespeople are doing a better job at managing leads and closing business by relying on wireless applications to transmit customer queries virtually instantaneously. At Brandow Automotive, salespeople equipped with RIM BlackBerry devices receive leads immediately after queries are placed on the company's web site or a portal, reducing response times from several hours to under 30 minutes and increasing the percentage of closed deals from 10% to 15%.

Atlantic Envelope Company is using a custom wireless application that allows salespeople to check orders, inventories, and purchase histories while in the field, saving them the time and effort of calling the customer service center to obtain the same information. Celanese Chemicals, a chemical manufacturing company, relies on an "out of the box" wireless application to hook its salespeople to its web-based SAP/R3 system to obtain real-time product availability, pricing, and order status data. Exercise equipment maker, Cybex, added a simple wireless interface to its PeopleSoft ERP system to give salespeople access to customer order status from the road. AT&T Wireless Services' (Southeast Region) sales force uses a wireless application to access a customer activation database from the field in real time and run credit checks, activate products, and complete other requirements while they are sitting with the customer.

3.2.6 Information Access

Simply giving workers immediate access to information while mobile can improve their productivity and efficiency in carrying out their daily work. While this type of information access and retrieval capability does not fall neatly into any particular application category, it can provide sufficient workflow improvements and savings to justify its implementation. Giving executives and other key decision-makers access to their e-mail via a RIM BlackBerry device may seem like a straightforward and simplistic wireless application at first blush, but the results can be dramatic. Wireless access to e-mail can allow these executives to provide timely responses and act quickly on critical information, actions that can help increase the bottom line or outflank competitors.

A wireless application that gives the Illinois State Police access to criminal justice data from the field has the ultimate affect of improving public safety, but it also helps reduce calls to and from call centers and eliminates paperwork through its report submission capabilities. Similarly, providing emergency room doctors with wireless access to patient registration information and test results helps doctors deliver more accurate and quicker healthcare services. But it also eliminates the extra steps and time involved in walking to view a central "whiteboard" of patient information, waiting for test results to be hand carried from the labs, calling labs for test results, and shuffling papers to and fro.

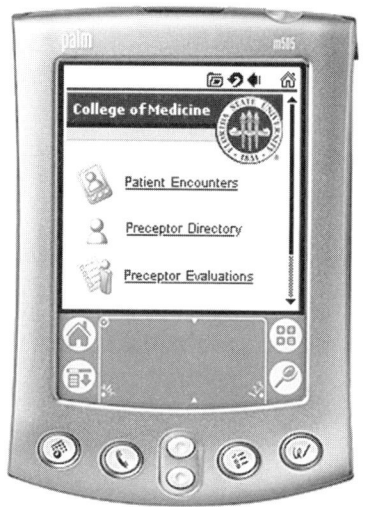

FIGURE 3.6
Education Applications
Source: ArcStream Solutions

Providing wireless access to information serves to keep constituents updated, and save costs too. Colleges and universities across the U.S., including Stanford University School of Medicine and the University of South Dakota, give students wireless access to course information and faculty members. (See Figure 3.6.) Students no longer need to lug around course materials, or return to their wired dormitory rooms for access. Instead they can look up just what they need from select locations around campus. In a similar fashion, Southwest Gas uses wireless applications to transfer updated gas line maps to laptops mounted in service vehicles rather than require its technicians to carry around huge books with paper maps that are quickly outdated.

3.2.7 Data Collection

A variety of wireless applications assist mobile workers in collecting and transmitting data straight from the field to central computer systems. These applications improve the efficiency of the data collection process by eliminating paper forms and the burden of re-entering the data back at headquarters. Data accuracy is also improved when data is collected at its source; errors are easily introduced when data must be re-keyed from often-illegible handwritten forms. Retail establishments, in particular, are heavy users of data collection applications and devices to help them

FIGURE 3.7
Data Collection Application
Source: Symbol Technologies

control and manage inventory. Bar code scanners are common fixtures in retail stores of all types, to check out purchased items and to track stock on the shelves. (See Figure 3.7.) Delivery companies like FedEx and UPS also rely on data collection applications to track packages as they make their way through distribution centers and into the hands of customers. (See Figure 3.8.)

At international airline sites, the United States Postal Service (USPS) equips its field personnel with handheld devices and wireless applications to monitor the condition of physical mail receptacles. USPS workers produce incident reports at the scene to catalog damage to receptacles, rather than write and submit handwritten reports that might take weeks to process. The airlines can take corrective action right away and the USPS is better able to track vendor performance and enforce penalties.

FIGURE 3.8
Data Collection Application
Source: Symbol Technologies

Inspectors at Michelin North America, the tire manufacturer, also use wireless applications to capture inspection data on site instead of filling out paper forms that are re-keyed by data entry clerks. The inspection data feeds directly into the product design process.

In Florida, Miami Dade County adopted a wireless system that allows building contractors to access inspection records mere minutes after they are generated rather than the customary wait of two days. Police officers in Flint, Michigan use wireless applications to write, submit, and correct incident reports from their patrol cars. The Parking Authority of New Jersey allows its officers to communicate wirelessly with the central office to research warrants and prepare tickets. Resident psychiatrists at the University of California, Davis, Medical Center, collect and wirelessly transmit patient data between the devices of doctors working different shifts. At SUNY Downstate Medical Center, medical students can record and collect information about patient encounters as shown in Figure 3.9. Progressive Insurance collects information about drivers' driving habits to track "how, when, and where" they travel to adjust monthly premiums accordingly. Aureole restaurant's wireless winebook application and Pro-Link's golf management application collect and save data about an individual's preferences or playtime statistics as a way to encourage repeat visits.

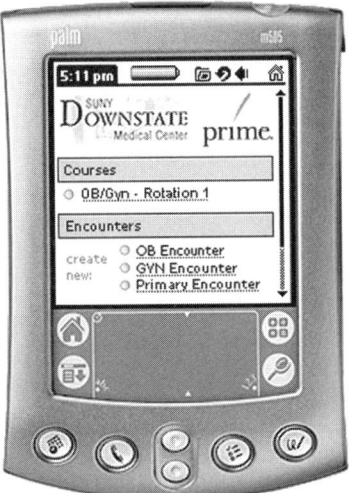

FIGURE 3.9
SUNY Downstate Medical Center Wireless Application
Source: ArcStream Solutions

3.2.8 Inventory Control and Quality Assurance

Manufacturing and distribution sites are constrained environments when it comes to deploying technical solutions. The physical characteristics of the sites, the presence of moving equipment, safety regulations, the need for sterility, or periodic reconfiguration of production lines precludes more permanent, wired solutions. In addition, manufacturing and distribution sites must manage and control inventory, and perform tests and quality assurance on finished products, functions that can also be aided through selective application of wireless technology. Tracking, locating, and accounting for inventory is aided by wireless data collection techniques, as mentioned above, and also by deploying wireless applications throughout the production process, as noted below. (See Figure 3.10.)

To overcome these environmental limitations, many manufacturing and distribution organizations are turning to wireless technologies. At Xerox Europe, the company uses wireless testing centers to check the performance of office products as they leave the production line. Installing wired cables was infeasible since Xerox's manufacturing environment must be "clean" to avoid damaging, electro-static discharges. Now, office products are connected via a wireless network to a central server that runs automated tests and diagnostics on the equipment before it leaves the plant.

FIGURE 3.10
Inventory Control Application
Source: Symbol Technologies

Avon Products relies on a wireless network in its manufacturing facility for inventory control. The wireless network enables Avon to generate bar code labels right at the production lines so that products can be tracked as they leave the site for shipment to a distribution center. Wireless-enabled printers are installed at the end of each production line, and request and receive bar code data over the air from a central server. The printer is able to generate the correct label, which is affixed to the product. By using a wireless solution, Avon was able to avoid the headaches of rewiring the print stations as production lines changed, allow line operators to remain mobile, avoid the physical limitations of the building and sidestep safety concerns over exposed cabling.

3.2.9 Safety

Location and tracking wireless applications can be used to enhance the safety of people and assets as mentioned above. Telematics applications such as OnStar and Wingcast allow drivers to notify a service center of vehicle problems or break downs, and permit the service center to dispatch a tow truck or repair truck to provide roadside assistance. Public safety organizations such as police forces and investigative agencies have sophisticated wireless applications to help them perform work in the field. The Illinois State Police, for example, have outfitted their force with vehicle terminals that connect officers in the field with criminal justice systems. The officers can perform on-the-spot warrant inquiries, check on stolen property, communicate silently with other police cars, print reports, and transmit and receive images. The State of California relays alerts of impending power blackout to residents as a service, but also to help ensure the safety of individuals during a blackout.

Doctors use wireless electronic prescription applications to improve the health and well-being of their patients. These applications automatically check for drug interactions and potential side effects and alert the physician before the prescription is sent. They also relieve pharmacists of the burden of deciphering sometimes illegible handwriting, a contributing factor in the distribution of incorrect medications. At hospitals, nurses and attendants use wireless applications to admit patients, record patient histories, monitor health conditions, and maintain electronic charts. (See Figure 3.11.) At CareGroup Healthcare, emergency room doctors have wireless access to patient information, including laboratory, radiology, and EKG data, to help them make faster and more informed decisions regarding patient care.

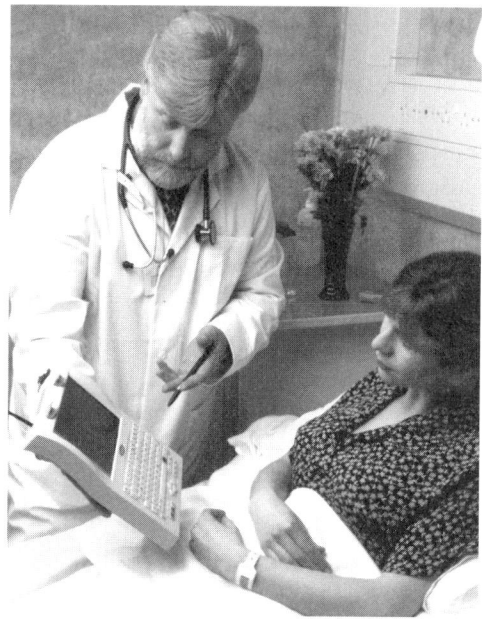

FIGURE 3.11
Safety Application
Source: Symbol Technologies

3.3 Consumer Applications _____

Many companies are directly targeting mobile consumers with specialized wireless applications. In areas where mobile, Internet-enabled telephone usage is high (especially Europe and Japan), the wireless craze has become something of a cultural phenomenon. Among young devotees, it is de rigeur to be equipped with one or more wireless devices. When compared to the strides made with business applications, however, consumer applications appear mostly frivolous. While some of the predicted mobile commerce (m-commerce) consumer applications are exciting, the anticipated deluge of annoying advertising and marketing may well do them in.

Drawn from examples around the world, listed below are some of the more popular consumer wireless applications.

3.3.1 E-mail

An obvious target for consumers, the ability to send and retrieve e-mail messages is the second most popular wireless application after voice calling. From special-purpose devices like the RIM BlackBerry to Internet-enabled phones, e-mail is one of the leading reasons why consumers turn to wireless.

3.3.2 Short Message Service (SMS)

SMS is a two-way text message service for mobile phones similar to many of the "instant message" chat services found on the Internet. People use the keypads of their devices to dash off short, text messages to friends or business associates. Popular in Europe for quite a while, U.S. providers also offer SMS services. In the U.S., SMS is generally an extra charge on top of network service fees. Large carriers in the U.S. are ironing out technology differences to provide cross-network SMS service.

3.3.3 Locator Services

One of the most widely anticipated wireless consumer offerings are location services. Through GPS-enabled devices and satellite networks, the ability to locate users precisely is already available, and telematics applications like OnStar and Wingcast rely on these features. Soon, in response to federal E911 mandates, telecommunications carriers will be able to pinpoint the location of users relying on a combination of technologies implemented at their land-based stations and user-owned, location-enabled device handsets. The ability to locate a user exactly and to apply contextual information to the location, promises to be a real boon for wireless services. In Korea, a user's cell phone location is used to recommend nearby restaurants. In Japan, parents can verify their children's whereabouts via their devices. Banks in the U.S. are working on applications that will direct customers to nearby ATMs based on their present locations. Wireless device users can also look up nearby businesses, movie theaters, and transportation facilities.

3.3.4 Telematics Services

Telematics applications, as mentioned above, rely on location-based technology like GPS to offer services to drivers such as roadside assistance, directions, and even stolen vehicle tracking. (See Figure 3.12.) Car manufacturers are expanding their telematics offerings, however, to include a range of additional services and conveniences such as Internet and e-mail access, digital video, traffic reports, music, and map updates. Mercedes-Benz is working on a telematics sedan that can receive short bursts of high-volume data as it passes transceivers located at strategic roadside locations. GM is investigating ways to extend its OnStar service to offer consumers a warning system that alerts them to vehicle repair problems.

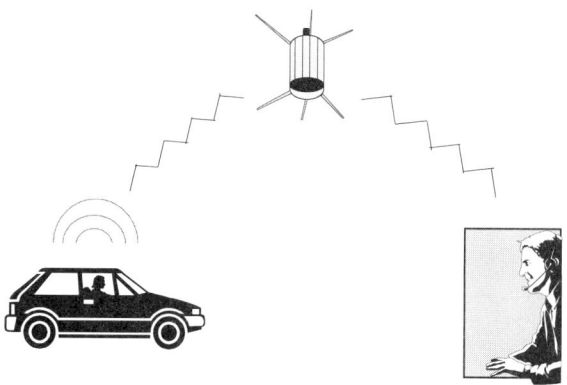

FIGURE 3.12
Telematics Application

3.3.5 Amusements

Many consumer wireless applications exist just to provide fun. Simple games are available for handhelds. Teenagers regularly customize their cell phones by down-loading special icons and personalized ring tones. Horoscope and astrology features are the rage in Japan and Italy. Karaoke, complete with scrolling lyrics and MIDI music, is appearing on handhelds in Japan.

3.3.6 Commerce

Despite the hype, consumer commerce applications are few and far between. Perhaps the most widespread type of wireless commerce application is associated with credit card processing devices that enable financial applications to be conducted from more mobile locations. Consumers can check-in and return rental cars, concluding transactions right in the parking lot with the attendant thanks to wireless applications and devices that capture signatures and process credit card transactions. This type of wireless application is also used in field service situations, in taxicabs, livery, and transportation services, and in restaurants where waiters can process credit card transactions at tableside.

In countries other than the U.S., mobile commerce is a bit more experimental. As mentioned above, in Japan, consumers can use their cell phones to conduct a range of transactions at smart vending machines. In Finland, rather than feed coins into vending machines and parking meters, people use their cell phones to pay, with charges appearing on their telephone statements.

3.3.7 Internet Access

Wireless Internet service providers and telecommunications carriers offer Internet access to subscribers. With current network bandwidth limitations, only rudimentary web browsing is capable on handheld devices. Although 3G networks promise greater data transmission speeds, Internet experiences on handhelds will continue to be constrained for the foreseeable future due to inherent device limitations (small display sizes, tiny keypads, low battery life) and a lack of web site content tailored for wireless users.

3.4 Telemetry or Asset Monitoring Applications ___

Monitoring the location and status of valuable physical assets or inventory is a burgeoning wireless application area. Many of these applications rely on telemetry—one machine speaking to another—to perform their functions, and are used where it would be impossible or inconvenient to locate a person to perform the monitoring task. For example, oil rigs located in hostile environments such as the North Sea are equipped with wireless monitoring devices and applications that allow land-based operators to check their status from the safety of shore. Non-profit organizations use similar applications to monitor the health of fragile coral reefs. Packaged Ice, a company that supplies ice-making equipment to retail stores, incorporates wireless applications and tiny satellite dishes atop store-based ice-making equipment to monitor air temperature, ice production, and door status, enabling the company to alert stores quickly whenever problems arise. Utilities like Florida Power & Light use "smart" electricity meters that can transmit usage information to specially equipped vans as they drive by.

Manufacturers and distributors are also exploiting telemetry applications, combining wireless technologies with either bar code ID scanning or radio frequency identification (RFID) tags to more tightly manage their production facilities, inventories, and supply chains. Ford uses a combination of RFID tags and wireless applications to detect and arrange pick up of empty parts containers on the manufacturing floor. McKesson HBOC uses a wireless inventory management system in its warehouses to reduce errors in customer shipments and thereby avoid returns. Georgia-Pacific, a manufacturer of plastic boxes used to deliver produce and perishable items to food stores, embeds tags within the boxes to track data like shipment and delivery dates, cleanliness (date and time that the box was washed), and a variety of environmental conditions such as temperature. The system is accessible not only by Georgia-Pacific but by retailers, transportation and shipping companies, and even the produce growers. American Airlines also installs tags on its cargo assets like freight carts and dollies so that it can track their locations around the airport and speed cargo and mail transfers. The Swiss Federal Railway also uses a telemetry-based application to track passenger and freight cars, and monitor and record maintenance history.

FedEx is experimenting with a wireless keyless entry and ignition system. Couriers wear a wristband containing transponders that, when placed near a reader on one of the four doors of the delivery truck, will unlock the door for up to six seconds to permit entry. The wireless application eliminates the hassle of searching for keys, and FedEx can quickly generate new codes if a wristband is lost instead of having to re-key the vehicle.

Consumers are also using telemetry applications, often without fully realizing it. Many gas pumps now offer "speed passes" that allow a person to "wave" a tag in front of a transceiver on the pump to authorize the transaction and activate the equipment. Many toll booths are now equipped with wireless readers that allow cars outfitted with a "smart pass" to drive straight through while recording toll charges against the vehicle. In Japan, NTT DoCoMo in partnership with Coca-Cola (Japan) is launching a "smart" soda machine that will let users perform a variety of functions—from storing cash in their phone accounts to buying tickets—simply by passing a wireless phone over a sensor located on the vending machine.

3.5 Wireless Application Examples _____

As you can easily tell from the examples cited above, wireless applications not only surround us, they abound. To illustrate this point, the tables below highlight some of the types of wireless applications that organizations across industries are deploying to great success. The first chart shown in Table 3.1 lists applications by industry type. As mentioned at the start of this chapter, surveying what others in your industry are doing with wireless can provide tremendous insight and help to jumpstart your own efforts. The second chart shown in Table 3.2 lists applications by their characteristics. Even though a company may not be in your industry, the type of application that they are using—CRM, data collection, inventory control—may be adaptable to your company's situation.

TABLE 3.1
Applications by Industry

Continued

Industry	Company	Application Description	Application Characteristics	Source
Auto	Brandow Automotive; Nissan	Two-way wireless messaging for salespeople to manage online leads	SFE, CRM	"The Wireless Work Force Pays Off," Robert L. Scheier, M-Business Daily, October 3, 2001.
	General Motors; Ford	GPS-enabled telematics apps (OnStar and Wingcast) to provide maps, directions to drivers; stolen vehicle tracking; roadside assistance	CRM, location and tracking, safety	"Open Platform Paves Way for Location-Based Wireless Services," Jay Wrolstad, Wireless NewsFactor, November 26, 2001.
	BMW Dealership (Greenwich, CT)	Car salespeople use wireless app to gather customer data, check inventory, register customers for follow-up and issue reports	CRM, SFE	"BMW Dealer Invests in Wireless System to Coax Sales," Dan McDonough, Wireless NewsFactor.com, October 5, 2001.
	Fiat Credit France	Wireless app allows salespeople to make offers on the showroom floor, access information and print contracts	SFE, CRM	www.palm.com
	Ford	Combination of RF tags and wireless apps allows detection and pickup of empty parts containers on assembly line	Asset monitoring; parts replenishment	"Leading the Way," Alan Radding, ComputerWorld ROI, September/October 2001.
	Thrifty Car Rental	Wireless reservation requests	CRM, customer service	"Leading the Way," Alan Radding, ComputerWorld ROI, September/October 2001.
	Avis	Wireless app speeds check-ins and returns by capturing digital signatures; connects support, maintenance, repairs and training	CRM	"Broadband system will capture digital signatures, speed customer service, repairs," Bob Brewin, ComputerWorld, July 24, 2000.
	Hertz	Wireless app checks in customers enroute in rental car shuttle bus; at the lot, rental stalls are illuminated with customer names and keys are in the ignition eliminating waiting in line	CRM	"Zamba Solutions Helps Hertz Put Its Customers in the Driver's Seat," Business Wire, February 23, 2000.
Building & Construction	First Service Networks	Wireless field service app for dispatching local subcontractors, accessing equipment histories, and submitting bills to central office	Field service	"The Wireless Work Force Pays Off," Robert L. Scheier, M-Business Daily, October 3, 2001.
Chemicals/Energy	Celanese Chemicals	Sales force wirelessly accesses customer/ERP data	SFE, CRM	"Leading the Way," Alan Radding, ComputerWorld ROI, September/October 2001.
Education	Stanford University School of Medicine	Wireless campus access to course schedules, drug reference and disease guide, weekly quiz modules, patient record-keeping app and curricula materials	Information access	"Future Doctors Go Digital," David Lustig, Field Force Automation, September 2001.
	University of South Dakota	Wireless campus access to maps, schedules and curricula information	Information access	"PalmPilot Devices for First-Year Students," Phillip J. Britt, Field Force Automation, October 2001.
Financial Services	Fidelity Investments; Charles Schwab; E*TRADE	Wireless apps to check account status; make trades; receive alerts	CRM, customer service, information access	"Leading the Way," Alan Radding, ComputerWorld ROI, September/October 2001.
	Hull Trading	Wireless app and pen devices give traders on exchange floor access to pricing information for offers and bids, and records transactions as they occur	Information access	www.proxim.com
	Bank of America	Wireless banking and brokerage services to view account balances and histories, transfer funds, check stock quotes and maintain watch lists	CRM, customer service, information access	www.bankofamerica.com
Food & Beverage	Coca-Cola & NTT DoCoMo	"Smart" soda vending machine uses wireless apps that allow consumers to purchase sodas, concert tickets and other items by debiting mobile phone accounts; dispenses maps and coupons; provides information access to events.	CRM, telemetry, commerce, information access	"Smarter" Coke Machine to Link with Wireless Web," Dan McDonough, Wireless NewsFactor.com, August 8, 2001.
	Cadbury	Prize drawing promotion; hide printed codes inside candy wrappers and have purchasers wirelessly transmit codes to be entered into drawing	CRM	"Eat yourself a win," Mark Ward, BBC News Online, May 31, 2001.
	Associated Food Stores	Food distribution center uses wireless app to track, locate and monitors food products	Telemetry, location and tracking, inventory control	"Wireless Tags Help Grocers Deliver Fresh Food," Jay Wrolstad, October 2, 2001.

TABLE 3.1
Applications by Industry

Industry	Company	Application Description	Application Characteristics	Source
Government	Illinois State Police	Wireless app to connect police in the field to criminal justice systems	Information access, data collection, safety	"Calling All Cars," Steve Barth, Field Force Automation, October 2001.
	Miami Dade County, FL	Building inspectors fill out and upload reports wirelessly to give contractors immediate access	CRM, data collection	"Leading the Way," Alan Radding, ComputerWorld ROI, September/October 2001.
	Daytona Beach, FL	Police force uses a wireless app to perform background checks on suspects by accessing a national warrants database, and to submit reports from the field	Information access, data collection, safety	"2002 Mobile Master Awards for Enterprise Deployments," Howard Baldwin, mbusiness, January 2002.
	Flint, MI	Police force use wireless app to write, submit and correct reports from patrol cars	Data collection	"The Wireless Work Force Pays Off," Robert L. Scheier, M-Business Daily, October, 2001.
	Parking Authority of New Jersey	Parking officers communicate wirelessly with central office to research warrants and prepare tickets	Data collection, information access	"New Wireless System Helps Cops Write Parking Tickets," Brian McDonough, Wireless NewsFactor.com, August 8, 2001.
	State of California	Wireless notification system sends alerts to subscribers of impending power blackouts and traffic conditions	CRM, safety	"Wireless CRM: Strings Attached," Marc Songini, ComputerWorld, November 5, 2001.
Healthcare	Various doctors' practices	Wireless voice and data apps check prescriptions against formularies and insurance lists and transmit them electronically to pharmacies	Safety	"Wireless Industry Writes Prescriptions for MDs," J. B. Houck, Wireless NewsFactor, February 6, 2001.
	CareGroup Healthcare	Wireless bedside registration of ER patients; ER doctors have wireless access to just-in-time "white board" with patient registration, lab, radiology and EKG data	Data collection, information access, safety	"Upwardly Mobile: A Wireless Primer," Ian S. Hayes, Software Magazine, August/September 2001.
	University of California, Davis, Medical Center	Resident psychiatrists collect and wirelessly transmit patient data between devices used by doctors on different shifts	Data collection	"Taking the Leap," Matt Hamblen, ComputerWorld ROI, September/October 2001.
Hospitality	Venetian Hotel	Wireless check-in of guests from multiple entrances and locations	CRM	"Las Vegas Hotel to Try Wireless Check-in," Matt Hamblen, ComputerWorld, May 24, 2001.
Insurance	Progressive Casualty Insurance	Mobile claims app puts adjusters at accident sites to prepare estimates and cut checks	CRM, information access, data collection	www.progressive.com
	Progressive Casualty Insurance	GPS-enabled wireless app tracks "how, when and where" driving to adjust monthly premium rates	CRM, location and tracking	www.progressive.com
Manufacturing	Packaged Ice	Incorporated wireless app and tiny satellite dish on store-based ice-making equipment to monitor air temperature, ice production and door status, and alert company of problems	CRM, telemetry	"Kicking Back," Mary Brandel, ComputerWorld ROI, September/October 2001.
	Cybex	Wireless sales force app to track customer orders	CRM, SFE	"Growing pains slow wireless CRM rollouts," Ann Bednarz, Network World, November 5, 2001.
	Atlantic Envelope Company	Wireless sales force app to check orders, inventories and customer purchase histories	CRM	"Wireless CRM yet to kick in," Gina Fraone, eWEEK, June 24, 2001.
	Georgia-Pacific	Wireless app to track reusable plastic containers used to deliver fruit and perishables to grocery stores.	Asset monitoring	"RFIDs: More Versatile Than Bar Codes," Jennifer M. Sakurai, Field Force Automation, July 2001.
	Avon	Wireless app provides cordless bar code printing capability at production lines	Inventory control, location and tracking	www.proxim.com
	Xerox	Wireless app permits cordless testing of office products at stations on the production floor	Inventory control, QA	www.proxim.com
	Honeywell	Wireless app, devices and networks connect field technicians and dispatchers, eliminate paper work, reduce process steps and speed collections	Field service, CRM	"Honeywell Embraces Field Service Automation," Erika Morphy, www.CRMDaily.com, December 7, 2001.

Continued

TABLE 3.1
Applications by Industry

Industry	Company	Application Description	Application Characteristics	Source
	General Electric Medical Systems	Outfit U.S. field service engineers with smart phones and wireless app to access diagnostic information on malfunctioning equipment, order parts and track shipments	Field service, CRM	"2002 Mobile Master Awards for Enterprise Deployments," Howard Baldwin, mbusiness, January 2002.
Oil & Gas	Southwest Gas	Wireless app updates and transfers gas line maps to laptops mounted in service vehicles while parked in lot overnight	Information access	"Southwest Gas Pumps Data Through Wireless LAN," Kimberly Hall, Wireless NewsFactor.com, August 30, 2001.
Pharmaceutical	McKesson HBOC	Wireless app to track and manage warehouse inventory	Inventory control, location and tracking	"Reworking Wireless," Joshua Kwan, Mercury News, October 21, 2001.
	Pfizer	Wireless apps to monitor and manage manufacturing process, optimize equipment usage and track inventory	Inventory control, location and tracking	"Leading the Way," Alan Radding, ComputerWorld ROI, September/October 2001.
Restaurants	Starbucks	Wireless Internet access offered to patrons of retail coffee shops	CRM, information access	"Starbucks takes wireless leap," Bob Brewin, ComputerWorld, January 8, 2001.
	Aureole	Wireless winebook lets patrons view related wine information and compile history of selections	CRM, information access, data collection	"Wireless Winelist," Stephanie Overby, CIO, November 15, 2001.
Retail	Sears Roebuck	Field service workers use wireless notebooks to create and close work orders, order parts, generate bills and accept payment	Field service	"The Wireless Work Force Pays Off," Robert L. Scheier, M-Business Daily, October 3, 2001.
Sports & Leisure	ProLink	Wireless golf app with course pointers, snack food ordering and uploading of player statistics to web site	CRM, location and tracking, data collection	www.goprolink.com
	Geocaching	Treasure hunting game using GPS-enabled devices	Location and tracking	www.geocaching.com
Telecommunications	AT&T Wireless Services (Southeast Region)	Wireless sales force app runs credit checks, activates products and submits applications	CRM, SFE	www.sierrawireless.com
Transportation	United Airlines, American Airlines, Northwest Airlines	Wireless alerts with flight information, changes and confirmations	CRM	"American Airlines offering text, voice notification service," Katherine Young, DallasNews.com, February 7, 2001.
	American Airlines	Wireless supply chain monitoring system to track cargo handling assets (freight carts and dollies) at the airport	Asset monitoring	"American Airlines Deploys Wireless System to Monitor Cargo," Jay Wrolstad, CRMDaily.com, October 12, 2001.
	NYC livery taxis	Wireless credit card terminals allow passengers to pay for fares without using cash	CRM, customer service	"Plastic Payment to Add Safety, Convenience to NY Taxis," Brian McDonough, Wireless NewsFactor.com, August 9, 2001.
	Metra	Wireless app pinpoints location of trains to deliver real-time schedule info to passengers	CRM, location and tracking	"Using Satellite for Train Tracking," Jon Hilkevitch, Wireless NewsFactor, November 21, 2001.
Trucking & Shipping	UPS; FedEx	Wireless apps that span warehouse and delivery functions and track packages in transit	CRM, data collection, customer service	"UPS to deploy Bluetooth, wireless LAN network," Bob Brewin, ComputerWorld, July 23, 2001.
	J&B Wholesale Distributors	Wireless delivery app to track arrival and departure info and onboard inventory; record discrepancies between POs and invoices, and collect customer signatures.	CRM, location and tracking	"Trucking with wireless," Ephraim Schwartz, InfoWorld, November 9, 2001.
	Grenley-Stewart	Wireless truck fueling app authorizes fuel delivery, signals pump mechanism, totals costs and uploads data for customer viewing	CRM	"Mobile Commerce," Bob Brewin, ComputerWorld, October 23, 2000.
	Penske Logistics	Onboard wireless app tracks deliveries and relays critical info in real time	CRM, location and tracking, data collection	"Penske outfits fleet with wireless terminals," Bob Brewin, ComputerWorld, June 11, 2001.
	Anderson Trucking Services	GPS-enabled wireless app lets customers track their loads in real time by location of truck	CRM, location and tracking, customer service	"BroadVision Satellite Tracking Helps Truckers Deliver," J.B. Houck, www.CRMDaily.com, March 28, 2001.

Continued

Industry	Company	Application Description	Application Characteristics	Source
Utilities	Northeast Utilities	Wireless app supplies info on power outages and trouble spots to speed dispatch and repairs	Field service, CRM	"The Wireless Work Force Pays Off," Robert L. Scheier, M-Business Daily, October 3, 2001.
	Florida Power & Light	"Smart" meters allow wireless readings of electricity usage by passing vans	CRM, telemetry	"The Wisdom of Starting Small," Susannah Patton, CIO, March 15, 2001.

SFE: Sales Force Empowerment
CRM: Customer Relationship Management

TABLE 3.1

Applications by Industry

TABLE 3.2
Applications by Type

Application	Company	Application Description	Industry	Source
Asset monitoring	Georgia-Pacific	Wireless app to track reusable plastic containers used to deliver fruit and perishables to grocery stores.	Manufacturing	"RFIDs More Versatile Than Bar Codes," Jennifer M. Sakurai, Field Force Automation, July 2001.
Asset monitoring	American Airlines	Wireless supply chain monitoring system to track cargo handling assets (freight carts and dollies) at the airport	Transportation	"American Airlines Deploys Wireless System to Monitor Cargo," Jay Wrolstad, CRMDaily.com, October 12, 2001.
Asset monitoring; parts replenishment	Ford	Combination of RF tags and wireless apps allows detection and pickup of empty parts containers on assembly line	Auto	"Leading the Way," Alan Radding, ComputerWorld ROI, September/October 2001.
CRM	Avis	Wireless app speeds check-ins and returns by capturing digital signatures; connects support, maintenance, repairs and training	Auto	"Broadband system will capture digital signatures, speed customer service, repairs," Bob Brewin, ComputerWorld, July 24, 2000.
CRM	Hertz	Wireless app checks in customers enroute in rental car shuttle bus; at the lot, rental stalls are illuminated with customer names and keys are in the ignition eliminating waiting in line	Auto	"Zamba Solutions Helps Hertz Put Its Customers in the Driver's Seat," Business Wire, February 23, 2000.
CRM	Cadbury	Prize drawing promotion; hide printed codes inside candy wrappers and have purchasers wirelessly transmit codes to be entered into drawing	Food & Beverage	"Eat yourself a win," Mark Ward, BBC News Online, May 31, 2001.
CRM	Venetian Hotel	Wireless check-in of guests from multiple entrances and locations	Hospitality	"Las Vegas Hotel to Try Wireless Check-in," Matt Hamblen, ComputerWorld, May 24, 2001.
CRM	Atlantic Envelope Company	Wireless sales force app to check orders, inventories and customer purchase histories	Manufacturing	"Wireless CRM yet to kick in," Gina Fraone, eWEEK, June 24, 2001.
CRM	United Airlines, American Airlines, Northwest Airlines	Wireless alerts with flight information, changes and confirmations	Transportation	"American Airlines offering text, voice notification service," Katherine Young, DallasNews.com, February 7, 2001.
CRM	Grenley-Stewart	Wireless truck fueling app authorizes fuel delivery, signals pump mechanism, totals costs and uploads data for customer viewing	Trucking & Shipping	"Mobile Commerce," Bob Brewin, ComputerWorld, October 23, 2000.
CRM, customer service	Thrifty Car Rental	Wireless reservation requests	Auto	"Leading the Way," Alan Radding, ComputerWorld ROI, September/October 2001.
CRM, customer service	NYC livery taxis	Wireless credit card terminals allow passengers to pay for fares without using cash	Transportation	"Plastic Payment to Add Safety, Convenience to NY Taxis," Brian McDonough, Wireless NewsFactor.com, August 9, 2001.
CRM, customer service, information access	Fidelity Investments; Charles Schwab; E*TRADE	Wireless apps to check account status; make trades; receive alerts	Financial Services	"Leading the Way," Alan Radding, ComputerWorld ROI, September/October 2001.
CRM, customer service, information access	Bank of America	Wireless banking and brokerage services to view account balances and histories, transfer funds, check stock quotes and maintain watch lists	Financial Services	www.bankofamerica.com
CRM, data collection	Miami Dade County, FL	Building inspectors fill out and upload reports wirelessly to give contractors immediate access	Government	"Leading the Way," Alan Radding, ComputerWorld ROI, September/October 2001.
CRM, data collection, customer service	UPS; FedEx	Wireless apps that span warehouse and delivery functions and track packages in transit	Trucking & Shipping	"UPS to deploy Bluetooth, wireless LAN network," Bob Brewin, ComputerWorld, July 23, 2001.
CRM, information access	Starbucks	Wireless Internet access offered to patrons of retail coffee shops	Restaurants	"Starbucks takes wireless leap," Bob Brewin, ComputerWorld, January 8, 2001.
CRM, information access, data collection	Progressive Insurance	Mobile claims app puts adjusters at accident sites to prepare estimates and cut checks	Insurance	www.progressive.com

Continued

Application Characteristics	Company	Application Description	Industry	Source
CRM, information access, data collection	Aureole	Wireless winebook lets patrons view related wine information and compile history of selections	Restaurants	"Wireless Winelist," Stephanie Overby, CIO, November 15, 2001.
CRM, location and tracking	Progressive Insurance	GPS-enabled wireless app tracks "how, when and where" driving to adjust monthly premium rates	Insurance	www.progressive.com
CRM, location and tracking	Metra	Wireless app pinpoints location of trains to deliver real-time schedule info to passengers	Transportation	"Using Satellite for Train Tracking," Jon Hilkevitch, Wireless NewsFactor, November 21, 2001.
CRM, location and tracking	J&B Wholesale Distributors	Wireless delivery app to track arrival and departure info and onboard inventory; record discrepancies between POs and invoices, and collect customer signatures.	Trucking & Shipping	"Trucking with wireless," Ephraim Schwartz, InfoWorld, November 9, 2001.
CRM, location and tracking, customer service	Anderson Trucking Services	GPS-enabled wireless app lets customers track their loads in real time by location of truck	Trucking & Shipping	"BroadVision Satellite Tracking Helps Truckers Deliver," J.B. Houck, www.CRMDaily.com, March 28, 2001.
CRM, location and tracking, data collection	ProLink	Wireless golf app with course pointers, snack food ordering and uploading of player statistics to web site	Sports & Leisure	www.goprolink.com
CRM, location and tracking, data collection	Penske Logistics	Onboard wireless app tracks deliveries and relays critical info in real time	Trucking & Shipping	"Penske outfits fleet with wireless terminals," Bob Brewin, ComputerWorld, June 11, 2001.
CRM, location and tracking, safety	General Motors; Ford	GPS-enabled telematics apps (OnStar and Wingcast) to provide maps, directions to drivers; stolen vehicle tracking; roadside assistance	Auto	"Open Platform Paves Way for Location-Based Wireless Services," Jay Wrolstad, Wireless NewsFactor, November 26, 2001.
CRM, safety	State of California	Wireless notification system sends alerts to subscribers of impending power blackouts and traffic conditions	Government	"Wireless CRM: Strings Attached," Marc Songini, ComputerWorld, November 5, 2001.
CRM, SFE	BMW Dealership (Greenwich, CT)	Car salespeople use wireless app to gather customer data, check inventory, register customers for follow-up and issue reports	Auto	"BMW Dealer Invests in Wireless System to Coax Sales," Dan McDonough, Wireless NewsFactor.com, October 5, 2001.
CRM, SFE	Cybex	Wireless sales force app to track customer orders	Manufacturing	"Growing pains slow wireless CRM rollouts," Ann Bednarz, Network World, November 5, 2001.
CRM, SFE	AT&T Wireless Services (Southeast Region)	Wireless sales force app runs credit checks, activates products and submits applications	Telecommunications	www.sierrawireless.com
CRM, telemetry	Packaged Ice	Incorporated wireless app and tiny satellite dish on store-based ice-making equipment to monitor air temperature, ice production and door status, and alert company of problems	Manufacturing	"Kicking Back," Mary Brandel, ComputerWorld ROI, September/October 2001.
CRM, telemetry	Florida Power & Light	'Smart' meters allow wireless readings of electricity usage by passing vans	Utilities	"The Wisdom of Starting Small," Susannah Patton, CIO, November 15, 2001.
CRM, telemetry, commerce, information access	Coca-Cola & NTT DoCoMo	"Smart" soda vending machine uses wireless apps that allow consumers to purchase sodas, concert tickets and other items by debiting mobile phone accounts; dispenses maps and coupons; provides information access to events.	Food & Beverage	"Smarter Coke Machine to Link with Wireless Web," Dan McDonough, Wireless NewsFactor.com, August 8, 2001.
Data collection	Flint, MI	Police force use wireless app to write, submit and correct reports from patrol cars	Government	"The Wireless Work Force Pays Off," Robert L. Scheier, M-Business Daily, October, 2001.
Data collection	University of California, Davis, Medical Center	Resident psychiatrists collect and wirelessly transmit patient data between devices used by doctors on different shifts	Healthcare	"Taking the Leap," Matt Hamblen, ComputerWorld ROI, September/October 2001.

TABLE 3.2

Applications by Type

Continued

Application Characteristics	Company	Application Description	Industry	Source
Data collection, information access	Parking Authority of New Jersey	Parking officers communicate wirelessly with central office to research warrants and prepare tickets	Government	"New Wireless System Helps Cops Write Parking Tickets," Brian McDonough, Wireless NewsFactor.com, August 8, 2001.
Data collection, information access, safety	CareGroup Healthcare	Wireless bedside registration of ER patients; ER doctors have wireless access to just-in-time "white board" with patient registration, lab, radiology and EKG data	Healthcare	"Upwardly Mobile: A Wireless Primer," Ian S. Hayes, Software Magazine, August/September 2001.
Field service	First Service Networks	Wireless field service app for dispatching local subcontractors, accessing equipment histories, and submitting bills to central office	Building & Construction	"The Wireless Work Force Pays Off," Robert L. Scheier, M-Business Daily, October 3, 2001.
Field service	Sears Roebuck	Field service workers use wireless notebooks to create and close work orders, order parts, generate bills and accept payment	Retail	"The Wireless Work Force Pays Off," Robert L. Scheier, M-Business Daily, October 3, 2001.
Field service, CRM	Honeywell	Wireless app, devices and networks connect field technicians and dispatchers, eliminate paper work, reduce process steps and speed collections	Manufacturing	"Honeywell Embraces Field Service Automation," Erika Morphy, www.CRMDaily.com, December 7, 2001.
Field service, CRM	General Electric Medical Systems	Outfit U.S. field service engineers with smart phones and wireless app to access diagnostic information on malfunctioning equipment, order parts and track shipments	Manufacturing	"2002 Mobile Master Awards for Enterprise Deployments," Howard Baldwin, mbusiness, January 2002.
Field service, CRM	Northeast Utilities	Wireless app supplies info on power outages and trouble spots to speed dispatch and repairs	Utilities	"The Wireless Work Force Pays Off," Robert L. Scheier, M-Business Daily, October 3, 2001.
Information access	Stanford University School of Medicine	Wireless campus access to course schedules, drug reference and disease guide, weekly quiz modules, patient record-keeping app and curricula materials	Education	"Future Doctors Go Digital," David Lustig, Field Force Automation, September 2001.
Information access	University of South Dakota	Wireless campus access to maps, schedules and curricula information	Education	"PalmPilot Devices for First-Year Students," Philip J. Britt, Field Force Automation, October 2001.
Information access	Hull Trading	Wireless app and pen devices give traders on exchange floor access to pricing information for offers and bids, and records transactions as they occur	Financial Services	www.proxim.com
Information access	Southwest Gas	Wireless app updates and transfers gas line maps to laptops mounted in service vehicles while parked in lot overnight	Oil & Gas	"Southwest Gas Pumps Data Through Wireless LAN," Kimberly Hall, Wireless NewsFactor.com, August 30, 2001.
Information access, data collection, safety	Illinois State Police	Wireless app to connect police in the field to criminal justice systems	Government	"Calling All Cars," Steve Barth, Field Force Automation, October 2001.
Information access, data collection, safety	Daytona Beach, FL	Police force uses a wireless app to perform background checks on suspects by accessing a national warrants database, and to submit reports from the field	Government	"2002 Mobile Master Awards for Enterprise Deployments," Howard Baldwin, mbusiness, January 2002.
Inventory control, location and tracking	Avon	Wireless app provides cordless bar code printing capability at production lines	Manufacturing	www.proxim.com
Inventory control, location and tracking	McKesson HBOC	Wireless app to track and manage warehouse inventory	Pharmaceutical	"Reworking Wireless," Joshua Kwan, Mercury News, October 21, 2001.
Inventory control, location and tracking	Pfizer	Wireless apps to monitor and manage manufacturing process, optimize equipment usage and track inventory	Pharmaceutical	"Leading the Way," Alan Radding, ComputerWorld ROI, September/October 2001.
Inventory control, QA	Xerox	Wireless app permits cordless testing of office products at stations on the production floor	Manufacturing	www.proxim.com

Continued

TABLE 3.2

Applications by Type

Application Characteristics	Company	Application Description	Industry	Source
Location and tracking	Geocaching	Treasure hunting game using GPS-enabled devices	Sports & Leisure	www.geocaching.com
Safety	Various doctors' practices	Wireless voice and data apps check prescriptions against formularies and insurance lists and transmit them electronically to pharmacies	Healthcare	"Wireless Industry Writes Prescriptions for MDs," J. B. Houck, Wireless NewsFactor, February 6, 2001.
SFE, CRM	Celanese Chemicals	Sales force wirelessly accesses customer/ERP data	Chemicals/Energy	"Leading the Way," Alan Radding, ComputerWorld ROI, September/October 2001.
SFE, CRM	Brandow Automotive; Nissan	Two-way wireless messaging for salespeople to manage online leads	Auto	"The Wireless Work Force Pays Off," Robert L. Scheier, M-Business Daily, October 3, 2001.
SFE, CRM	Fiat Credit France	Wireless app allows salespeople to make offers on the showroom floor, access information and print contracts	Auto	www.palm.com
Telemetry, location and tracking, inventory control	Associated Food Stores	Food distribution center uses wireless app to track, locate and monitors food products	Food & Beverage	"Wireless Tags Help Grocers Deliver Fresh Food," Jay Wrolstad, October 2, 2001.
CRM: Customer Relationship Management				
SFE: Sales Force Empowerment				

TABLE 3.2

Applications by Type

72

3.6 Wireless Application Case Studies _____

As the preceding sections demonstrate, many companies are actively and success-fully using wireless applications to make a difference in their organizations. In this section, we take a more in-depth look at several companies and their specific wire-less applications to examine how they approached the task of harnessing wireless technologies and the benefits received. These forward-thinking companies are truly innovative—willing to adopt new technologies, and adapt them to their individual situations. They deserve a tremendous amount of recognition for forging the way and for demonstrating to all that wireless technologies are real and worthwhile.

The case studies in this section cover: Atlantic Envelope Company, a manufac-turer of specialty envelopes; Honeywell's Automation & Control Solutions division, which services building systems; and Penske Logistics, a transportation and logistics services company.

3.6.1 Case Study: Atlantic Envelope Company

Case Study: Atlantic Envelope Company
Industry: Manufacturing
Wireless Application Type: Customer Service; Sales Force
                            Empowerment

Atlantic Envelope Company (AECO), a division of National Services Industries, is a leading manufacturer of custom envelopes specializing in direct mail and high-end pieces for businesses. Headquartered in Atlanta, Georgia, the company has nine manufacturing facilities in the U.S., employs approximately 1,200 people, and takes in roughly $240 million in revenues. AECO serves predominantly regional and national customers in the energy, finance, transportation, direct mail, and package delivery markets using a direct sales team. For more information about the company, visit *www.atlanticenvelope.com*.

Industry   Manufacturing

Background   A strong believer in customer service, AECO knew that to sharpen its competitive edge it had to give its salespeople tools to be more respon-sive to their customers. During sales calls, sales reps responded to customer queries about order status, inventory, and invoices by calling the customer service depart-ment for information. Sales reps would have preferred having this information at their fingertips, but there was simply too much to keep track of given the high level of account activity. Calling the customer service department multiple times a day, while not the most efficient process, was the best way to obtain current information.

For their part, customers were increasingly looking to reach their AECO sales reps via e-mail. With their hectic travel schedules, however, sales reps seldom had access to e-mail from the field, delaying their ability to read and respond to messages. To bridge the gap, sales reps used AECO's customer service department as a go-between, asking the service reps to retrieve and read pending e-mails over the phone.

Although AECO was open to tools that would empower the sales force to communicate more quickly with customers and provide requested information more readily, it was hesitant to accept the cost, risk, and disruption of a full-blown sales force automation system. Nor were wireless technologies in particular on its radar screen. But when Ed Ringer, the CIO at AECO, brought in a RIM BlackBerry and showed it to the VP of Sales and Marketing, Kerry Reedy, things started to click. Recognizing the potential applicability of the technology to his sales staff's needs, Mr. Reedy decided to investigate the device and talk to wireless experts about what it would take to craft an application to put information at the sales reps' fingertips.

Application Description   To meet its needs, AECO created a wireless customer service application. Sales reps throughout the U.S. use the application to perform a range of functions. The most important functions are aimed at responding to customer inquiries. But the sales reps can also use the application for remote e-mail access, to obtain directions, and to receive news and information about their accounts.

AECO's wireless implementation has four components:

- Sending and receiving e-mail
- Internet access
- PC synchronization, if desired
- A custom AECO application

By far, e-mail is the most popular component. Because the device has an "always on" network connection, e-mails are delivered to the device in real time, which means that sales reps can respond more quickly. Sales reps receive e-mails from customers, and also from AECO sites alerting them to inventory situations such as low stock, recent shipping activity, and other account-specific information. The sales reps also use e-mail to communicate with other AECO employees. Internet access and PC synchronization are available to the sales reps to use as they wish.

AECO's custom wireless application has two major capabilities. First, it allows a sales rep to answer customer questions in real time at the customer site by retrieving current information from a central server. Second, it permits the sales rep to submit requests and changes affecting orders direct to a manufacturing facility or other designated spot.

The first capability—information retrieval—works extremely well because AECO purposefully limits the amount of information at the sales reps' disposal. From the outset, the company's methodology was to identify the top ten questions that customers were likely to ask pertaining to orders, inventory, and invoices, and design the application around these questions. Because the universe of information is limited in this way, the interface, screens, and functions are kept simple. Sales reps can look up order status information, such as the date an order is scheduled to ship, or the date of the next scheduled manufacturing run. They can look up order history and invoice information such as the specifications for an item, the last time a customer ordered a particular item, the date it shipped, the order quantity, the PO number, the unit price, and the invoice number. Sales reps can do all of these things through a few screens and a few clicks. Typing is intentionally minimized.

The second capability—submitting requests—allows sales reps to send predefined requests affecting orders to the manufacturing facilities. Again, these items are limited to those most commonly requested by customers. Through a set of preformatted, standard templates with required input fields, sales reps can: request an order estimate, request delivery improvement for an order, request a warehouse/inventory release, and submit a change request for an order. These requests are sent and acknowledged via e-mail, and sales reps are free to print out hard copies of the requests if they choose.

AECO Benefits

- Greater efficiency through standardized processes for accessing information, sending alerts, and submitting requests
- Greater customer satisfaction through quicker response to customer queries and concerns
- Stronger competitive advantage, resulting in greater revenues, by cultivating an image of responsiveness and becoming the "supplier of choice" for its customers
- Cost savings from eliminating pagers carried by sales reps and reducing phone calls to customer service desk

Sales Rep Benefits

- Enhanced personal productivity through e-mail and contact access
- Greater productivity and efficiency by eliminating calls to the customer service desk
- Greater autonomy in obtaining current account information
- More knowledgeable about important account activity and inventory status
- More responsive to customers resulting in greater numbers of closed sales

Customer Benefits

- Fewer phone calls to track down a sales rep or get an answer to a question
- Greater access to sales reps through e-mail
- Quicker responses to questions and concerns
- Instant information on stock, orders, shipping, accounting, and specifications
- Easier to place orders

Technology   AECO's solution is composed of four components: network, devices, application software, and information architecture.

- *Network*   AECO purchases wireless data network services from Cingular Wireless. In customer locations, coverage exceeds 90%. In sales reps' home areas, coverage dips below 90%.
- *Devices*   AECO selected the RIM BlackBerry 957 after evaluating a range of devices. AECO felt that the BlackBerry was the most user-friendly device. It allowed for typing and could be used with a single hand, did not have a cryptic handwriting scheme, and did not require a stylus. The BlackBerry does require "all thumbs" typing, but AECO reports that sales reps become proficient within a week of using the device. The BlackBerry was also preferred for its "always on" operation and the simplicity of retrieving e-mails.
- *Application*   Wireless OnRamp developed device-resident and server-resident software. On the device are simple forms and templates for retrieving information and submitting requests. On the server is functionality to process information requests, interpret forms and templates, retrieve any necessary information and forward it to the device. Security is accomplished by authenticating users via an ID and a password.
- *Information Architecture*   AECO draws the data needed for the wireless application from several back-end systems. This data is stored on a mobile application server, and is refreshed twice a day, a timeframe that is sufficient to provide accurate account information to the field.

The Project   AECO launched the project with the assistance of two outside vendors. Wireless OnRamp was engaged to develop the wireless application software. Cingular Wireless was selected to deliver network services. AECO's IS organization was tasked with preparing the internal infrastructure to support the application.

The project proceeded in four phases:

- *Strategy and Planning Phase* to define the application, select the device, and determine the necessary infrastructure components. To scope the functionality of the application and keep things simple, AECO focused on the top ten questions that a customer might ask and the top four requests that a sales rep might submit from the field to a manufacturing site. Forms and templates were chosen to make navigation simple, minimize input, and reduce the amount of data transmitted.

- *Development Phase* where Wireless OnRamp developed the device and server-resident software, working with Research in Motion and Cingular Wireless to ensure that device and network hardware and software worked together properly. AECO's IS staff put the mobile application server components into place and developed routines to gather data from back-end systems and make it available to the wireless application.

- *Pilot Phase* where a handful of sales reps in the Atlanta area were given the devices and application to try out in real-life account situations. During this phase, AECO and Wireless OnRamp worked with the sales reps to fine-tune the application. The pilot was wildly successful, and the pilot team refused to give up their devices.

- *Implementation Phase* where the wireless application and devices were rolled out to the entire sales field of 85 reps. AECO's IS organization was responsible for prepping the devices and loading them with application software, contact lists, passwords, and other data. The launch occurred during AECO's annual sales meetings where the devices were distributed to the sales team and hands-on training was simultaneously conducted. Training consisted of a speaker leading the audience through the features and functions of the application, and ten trainers circulating throughout the room assisting the sales reps in working with the devices.

In production since September 2001, AECO's help desk now supports the wireless application. Few difficulties or problems have arisen. In the short term, AECO is looking at software that will allow it to centrally distribute updates to devices in the field, and software that will allow the reps to view e-mail attachments.

Design Challenges   AECO's major design challenges were to limit the scope of the application to keep responses speedy, simplify the interface to maximize usage, and keep the amount of transmitted data small to accommodate low network bandwidth.

The scope of the application is limited, as described above, to respond to the top ten questions that customers usually ask (order status, history, inventory, and invoices), and the four most common requests they make pertaining to their orders (change an order, estimate an order, release inventory, improve delivery). These limitations serve a dual purpose: ensuring that the application performs speedily and simplifying the back-end data integration task. Because AECO does not have a CRM system, it has to collect the data needed by the application from various sources. By limiting inquiries to order and invoice information, the task of gathering the data is greatly eased. With twice-daily refreshes of the data deemed sufficient, AECO further simplifies the integration chore.

The features, functions, and templates of the application are also kept simple to ease training requirements and encourage usage. A cumbersome application would lead sales reps to circumvent the application and resort to calling the customer service department. Data entry—clicking and typing—are minimized through the use of pre-formatted screens and templates. Navigation is never deeper than a screen or two. Input is checked for required fields before it is sent off to the server so that turnaround time is quick.

Finally, by employing the restrictions described above, the application necessarily limits the quantity of data transmitted over the network, keeping response times very quick.

Vendors

Wireless OnRamp, for the custom application software.

Research in Motion, for the RIM BlackBerry devices.

Cingular Wireless, for the wireless data network services.

Lessons Learned at AECO   Keep it simple. Try to rely on basic forms and templates where possible. Provide some kind of "hook," like e-mail, to get people excited and encourage usage. Think in terms of your users' perspectives. Don't waste time trying to quantify ROI when the objective is to gain a competitive advantage.

Acknowledgments   The author thanks Kerry Reedy, Vice President of Sales and Marketing at AECO, for his assistance in preparing this case study.

3.6.2 Case Study: Honeywell

Case Study: Honeywell
Industry: Manufacturing
Wireless Application Type: Field Service

Honeywell's Automation & Control Solutions Service (ACS Service) division provides field services—emergency repair and preventive maintenance—for a range of commercial building systems including heating, cooling, and security. Typical customer sites include university campuses, hospitals, large corporate buildings, office complexes, government facilities, and schools. ACS Service is staffed with approximately 1,400 technicians providing services throughout North America. For more information about ACS Service and its services, visit *www.honeywell.com*.

Industry   Field service within manufacturing.

Background   By nature, the field service business is highly transaction-oriented. A technician is dispatched to a customer site, provides some type of service, closes out the job, performs record keeping, and then moves on to the next job site. ACS Service had spent considerable effort over the years to enhance the efficiency and productivity of its field service business, and had automated back-office functions by installing a service management system. Moreover, as a Six Sigma *Plus* adherent, ACS Service had already optimized its service delivery processes. Like the Six Sigma methodology first developed by General Electric, Honeywell's proprietary Six Sigma *Plus* version seeks to improve the quality of processes, products, and services by reducing defects and variations. Applying Six Sigma *Plus* principles, ACS Service had defined, measured, analyzed, improved, and controlled its service delivery processes to improve quality, reduce costs, and eliminate defects.

With an optimized field service unit already in place, ACS Service decided to aim even higher—to deliver a level of service that would not only satisfy customers, but also exceed their expectations. Accordingly, ACS Service identified three strategic business goals that it would have to meet:

- Improve customer responsiveness
- Improve the quality of service rendered
- Provide customers with better and more timely access to service information

To guide its efforts, ACS Service developed a Strategic Roadmap, a collection of short- and long-term strategies and plans to help it meet these larger goals. Initiatives included in the roadmap had to meet two stringent conditions: they could not increase operating costs, and had to improve business performance. To obtain funding, an initiative also had to pass a rigorous cost justification process that called for unanimous agreement on anticipated benefits and cost savings.

Early in 2001, an initiative was identified that managed to pass this rigorous screening process. Dubbed FAST (Field Automation Service Technology), it would harness wireless technology to automate dispatching, ease record keeping burdens, and improve information dissemination within ACS Service and to customers, objectives that were aligned with ACS Service's larger business goals. ACS Service had investigated wireless technologies earlier, and believed that they were finally mature enough to support a deployment of the magnitude required by FAST.

Rolling out FAST across North America was an ambitious undertaking, but ACS Service had two factors working in its favor. First, ACS Service's Six Sigma *Plus* investments had created a solid foundation of processes so it did not have to start from scratch. Second, because ACS Service was a firm believer in the Pareto principle (the 80/20 Rule), it was willing to adjust its processes if it found a solution meeting at least 80% of its requirements. This flexibility would help ACS Service avoid the time and expense of developing a custom solution, and achieve an even quicker victory with FAST.

Application Description   ACS Service did not want to develop a custom solution, preferring instead to license a commercial product. Potential products had to meet four general requirements:

- Automate technician dispatch and improve communications between the dispatch center and technicians in the field
- Allow technicians to look up customer service histories and repair information on site
- Automate and streamline technicians' and back-office record keeping chores
- Permit near real-time transmission of service information from work sites to back-end systems to give customers an up-to-date and accurate snapshot of work orders and service status

ACS Service eventually selected the FX Mobile™ product from FieldCentrix. The product met the requirements outlined above, and required virtually no customization other than the creation of an equipment catalog. ACS Service integrated the FX Mobile™ application with two of its back-end systems: the service management system that performs dispatching and collects service metrics and repair information for each customer; and the financial system that captures labor and expense data for customer invoicing.

ACS Service's wireless field service application has three major capabilities:

- *Messaging*   Through its real-time messaging capabilities, the application sends dispatch instructions for both emergency service and routine preventive maintenance directly to the technician's handheld unit. The technician can also message the dispatcher while enroute, at the job site or upon departure. At the site, the technician can message other technicians to seek help with a service issue.
- *Service Information Lookup*   From the field, technicians can use the application to access service information contained within ACS Service's back-end service management system including the particular service history for a customer or site, or repair history for a piece of equipment.
- *Data Capture and Service Report Completion*   Rather than rely on paper reports, technicians can fill out and submit service reports electronically via their handheld devices. Checklists are used to ensure that pertinent information—repair description, labor and parts used, etc.—is captured to update service management records and to invoice customers or charge against contracts.

To allow customers to review current service requests and status, information collected by the wireless application is transmitted in near real time via wide area and satellite wireless networks back to ACS Service's servers. Service information is propagated in ACS Service's internal systems, and made available to customers via an on-line extranet, the ServicePortal.

ACS Service Benefits

- Improves the ability to communicate with, locate, and manage field resources
- Enhances customer satisfaction through improved responsiveness and quality of service
- Enhances efficiency through standardized processes for communications, information access, service report completion, and transmission
- Improves productivity and efficiency by replacing paperwork with electronic reports, eliminating re-keying of data, and cutting the steps required between service completion and invoicing from 17 to 3
- Results in shorter billing cycles and improved cash flow by streamlining the collection of labor and expense information from the field

Field Technician Benefits

- Facilitates more timely communications through messaging capabilities
- Offers greater productivity by collecting data electronically and eliminating paperwork
- Improves efficiency through on site access to service histories and the ability to message other technicians for help

Customer Benefits

- Provides quicker response to service requests through enhanced dispatching and messaging
- Provides higher quality service rendered by a more knowledgeable technician with on site access to service and repair histories
- Provides near real-time access to complete information about work requests and service information through the ServicePortal

Technology   ACS Service's solution is composed of four components: network, devices, application software, and information architecture.

- *Network*   ACS Service relies on two types of wireless networks to accomplish real-time messaging and data transmission. ACS Service purchases data services from Cingular Wireless over its wide area wireless network. If Cingular coverage is unavailable at a work site, then ACS Service reverts to satellite network coverage provided by Wireless Matrix. Technicians' vehicles are equipped with satellite transmitters, and when a handheld unit is docked in the vehicle, any queued messages are sent via a satellite network. The wireless application takes care of queuing and re-sending messages when coverage is unavailable; the technician does not need to take any action.

- *Devices*   ACS Service uses Fex21 handheld terminals from Itronix. To aid in the selection of a device, ACS Service developed a matrix of necessary capabilities. As each device was evaluated, from WAP phone to PDA to PC, ACS Service filled out the matrix for that device type. Following the 80/20 Rule, ACS Service was willing to go with a unit that provided at least 80% of the functionality it needed. At the end of the evaluation period, the completed matrix clearly indicated the unit with the most overlapping functionality.

- *Application*   ACS Service licenses a commercial software product called FX Mobile™ from FieldCentrix. The only "customization" required was the creation of an equipment catalog. ACS Service relied on internal IS resources to integrate the FX Mobile™ application with its back-end service management and financial systems.

- *Information Architecture*   The wireless application relies on information contained within ACS Service's existing service management system. This service management system has a well-defined information architecture that supports a variety of service data including customer and site histories, equipment repair records, etc. Technicians can access any of this information using their handheld devices, and can also update service records through the reports that they create in the field. In addition, labor and expense information captured in the field is fed into back-end financial systems for billing purposes. All data is transmitted to and from the field in real time.

Project Approach   Part of ACS Service's Strategic Roadmap, the FAST initiative was conceived by a team consisting mostly of field technicians, with one IS and one management representative. ACS Service went through a lengthy cost justification process to demonstrate that the project would not increase operating costs yet would provide concrete business benefits. As mentioned earlier, team participants had to agree unanimously on the project benefits and costs before they could be included in the business case.

ACS Service believed that with its solid processes, standards, and back-end systems already in place, it could follow a compressed project development lifecycle. Furthermore, ACS Service wanted to capitalize on the team's enthusiasm for the project and felt that a drawn-out implementation would squelch their motivation. At the same time, ACS Service knew that for this aggressive approach to succeed, it had to confront project risks head on. Using a tool called FMEA (Failure Modes Effect

Analysis), ACS Service attempted to identify every possible issue that could go wrong with the project, prioritized these risks, and developed appropriate mitigation strategies for the most important items.

After spending four months defining system requirements and one month evaluating and selecting software, ACS Service was ready to field test a prototype system in a 30-day "proof of concept" trial. Based on the success of this test period, ACS Service would receive funding for the FAST initiative. The prototype system enabled ACS Service to test the application software, devices, networks, processes, and workflows that would comprise its solution. Actual back-end integration with the service management and financial systems was deferred, and instead, back-office staff "simulated" the integration by manually entering data and intercepting requests for information.

Needless to say, the proof-of-concept stage was successful, the FAST initiative received funding, and ACS Service proceeded to develop the back-end portion of the system. When the back-end development phase was complete, ACS Service spent two weeks testing the system and then went straight into deployment.

Two other aspects of ACS Service's project were instrumental in its success. First, ACS Service had a solid communications plan in place from the outset of the initiative, designed to educate affected employees and raise awareness. Field and administrative staff were prepared for the new system well in advance of its deployment, and knew how it would affect their workflow and tasks, and their responsibilities in the roll out.

Second, ACS Service put together a robust training plan that called for both individual and group training. ACS Service issued a training CD with a series of lessons so that technicians could train independently. At the end of each lesson, the program would print out a certification indicating that the technician had undergone and passed the training. Once certified, the technician was eligible to attend a live, one-day training session where handheld units were distributed. Upon leaving the session, technicians were ready to start using the device and application.

The second phase of the project was a roll out in Quebec, Canada, complete with French language interfaces. The third phase was a roll out in Europe, with appropriate language interfaces.

Design Challenges   If ACS Service were starting from scratch, the design challenges of defining and standardizing field service processes and workflows, and automating back-end systems would have been monumental. By using Six Sigma *Plus*, ACS Service had the necessary building blocks in place, and was able to jump-start its efforts and sidestep many of the design challenges (immature processes, conflicting standards, lack of technologies) that kill projects at the outset. Furthermore, by being receptive to a commercial solution and by following the 80/20 Rule, ACS Service dramatically increased the likelihood that it would find a workable solution. Lastly, ACS Service's risk mitigation strategies and use of the FMEA tool, meant that it was prepared for, and had appropriate responses to, design challenges and project risks that did occur.

Vendors

FieldCentrix, for the FX Mobile™ application software

Itronix, for the Fex21 handheld terminals

Cingular Wireless, for wide area wireless data network services

Wireless Matrix, for satellite network services

Other vendors included Bell Mobility, Telus, Sasktell, MTS, and Alliant.

Lessons Learned   Building a solid foundation of defined, standardized pro-cesses is invaluable. Expect to put your best people on the wireless project, because the technology, while improving, still isn't "plug and play." Invest in communica-tion, which is the key to project success.

Acknowledgments   The author thanks Santosh Patel, ACS Service's Director of Service Operations for North America, and Shawna Todd, Director of Communi-cations at ACS Service, for their invaluable assistance in preparing this case study.

3.6.3 Case Study: Penske Logistics

Case Study: Penske Logistics
Industry: Logistics and Transportation
Wireless Application Type: National Tracking and Location

Penske Logistics (Penske), a subsidiary of Penske Truck Leasing, provides third-party logistics services to its customers. These services fall into three categories—transpor-tation management, distribution center management, and integrated logistics where Penske either serves as the Lead Logistics Provider or provides synchronized inbound services. Based in Cleveland, Ohio, Penske operates in North America, South America, and Europe. For more information about Penske and its services, visit *www.penskelogistics.com*.

Industry   Logistics

Background   Penske is committed to delivering high quality logistics services to its customers and has historically used technology to help meet this goal. To more tightly manage its operations, Penske's IS group had developed a logistics management system (LMS) integrated with its data warehouse. Using the information contained in the data warehouse, Penske had provided its customers with various reports to give them insight into Penske's performance and the performance of their supply chain partners.

Leveraging logistics information has become one of Penske's greatest strengths. By collecting and disseminating useful information, Penske allows its customers to measure, evaluate, and tune their supply chains. As customers increasingly perceive Penske as a purveyor of critical information, their expectations grow. They judge Penske's performance based on its ability to provide timely logistics information as well as its ability to provide logistics services. These customer expectations prompted Penske to look at ways to improve its data collection and management capabilities, and to wireless technologies in particular. Not only did wireless technologies look like the right mechanism to deliver more timely, higher quality data to customers, they also promised to improve the efficiency of Penske's processes and streamline operations.

Application Description   Penske implemented three different but related applications. These applications differ in terms of their functionality, and the devices and networks used, but they all interact with Penske's LMS system, providing data to or using data from that system. Penske's internal IS staff performed all integration between the LMS system and these applications.

Before looking at the three applications, consider how Penske's processes work. At the start of each day, Penske's LMS application calculates skeletal information—the planned routes and schedules for its fleet of drivers and trucks. During the day, as more detailed information associated with these routes and schedules is collected, it is fed into Penske's computing systems to populate the LMS data stores and update the plans. In the morning, drivers arrive at a central Penske dispatch center to pick up their route and schedule instructions and begin the day's run. Drivers then make stops along the various routes, pick up or deliver freight, and bring shipments to Penske's cross docks for processing. As these events occur, information is collected, transmitted, and integrated with the LMS system to record how the plan executed.

To streamline these processes and collect meaningful logistics information, Penske developed three applications:

- The first application allows drivers in their trucks and dispatchers at Penske centers to exchange information. This task is accomplished using a QUAL-COMM OmniTRACS device mounted in the cab of a truck. During the day, the application feeds route information to the driver and captures time-sensitive information such as location, arrival times, departure times, and downtime—which is integrated into LMS. The application also allows dispatcher and driver to communicate route and schedule changes and directions, and exchange freeform messages about events demanding attention. All of this information is transmitted in near real time via satellite networks. Once integrated into the LMS system, dispatchers can track the actual progress of each truck and view automatic recalculations of estimated arrival times for upcoming stops. In this way, dispatchers are able to manage by exception and respond to projected deviations proactively. Penske also uses OmniTRACS to automatically collect and transmit vehicle operating information used for fleet management and to monitor efficient driver practices.

- The second application is designed to track, locate, and manage shipments as they arrive at and leave Penske's cross docks. A cross dock is similar to an airport, where passengers and luggage arrive and must be re-routed and loaded onto connecting flights for departure. At the cross dock, Penske processes incoming shipments and re-packages them optimally, based upon their destination, into outgoing trucks. Unloading, tracking, and managing shipments that arrive at the incoming docks, and correctly loading them onto outbound trucks, is a detailed, painstaking process. Penske installed a WLAN within the defined space of its cross dock facilities, and outfitted its dock workers with handheld scanners from Symbol Technologies so that they can scan, track, and match items as they travel through the cross dock facility.

- The third application is mobile rather than wireless. The driver uses this application to collect detailed shipment information that is fed into the LMS system intermittently during the day. At the start of the day, the driver is given a handheld PDA loaded with delivery and pickup information obtained by synching the device via a docking cradle with the server-based LMS system. As the driver makes his deliveries or pickups, he enters information about the goods (number of pieces, parts, weights, etc.) processed, typically by electronically capturing data from a bar code label. The driver may also use the device to enter information about services rendered, and, optionally, to capture the recipient's signature and print a receipt. When the driver returns to the dispatch center, the data accumulated on the device is uploaded to the LMS system using a docking cradle and synching software. Penske chose not to transmit this data in real time because customers were not requesting immediate access to this level of detail and the cost of transmitting such data is high. Penske continues to monitor the coverage and cost of alternative wireless networks (such as 3G cellular) and expects to implement a wireless mobile application when the ROI is acceptable.

Penske also gives its customers access to the logistics information contained in its data warehouse via an on-line portal with a variety of reports.

Penske Benefits

- Improves the ability of the dispatchers to communicate with, locate, and manage drivers resulting in more optimal routes and schedules and greater productivity for drivers

- Enhances efficiency at cross docks by automating the tracking, locating and management of shipments, freight, and parcels

- Enhances the productivity of drivers and streamlines internal processes by capturing detailed pickup and delivery information electronically rather than on paper and eliminating the need for drivers and dispatchers to manually conduct time consuming "check calls" throughout the day

- Improves service delivery and heightens customer satisfaction by optimizing routes and schedules, and tightly managing assets at cross dock facilities to avoid shipping errors

- Raises customer satisfaction through the collection and presentation of timely and meaningful logistics information

Customer Benefits   In addition to reaping benefits from the improvements made to Penske's services as outlined above, customers also receive the following benefits.

- Improves oversight and planning through access to near real-time logistics information

- Provides greater visibility into the performance of supply-chain partners allowing customers to identify best-performing partners

- Offers fuller, more accurate access to performance data, making it easier for customers to monitor service levels and make adjustments

Technology   Penske's solution is composed of four components: network, devices, application software, and information architecture.

- *Network*   Penske uses two types of wireless network services. For real-time transmission of information from trucks to dispatch centers, Penske purchases satellite network services from QUALCOMM. By using a satellite network rather than a wide-area cellular network, Penske ensures that its drivers will have coverage virtually everywhere they travel. For tracking and locating assets at cross docks, Penske uses WLAN technology supplied by Symbol Technologies.

- *Devices*   Penske relies on several types of devices. In-vehicle units are QUALCOMM OmniTRACS terminals with integrated satellite communications. Intermec and Symbol Technologies supply the handheld devices, complete with label reading and WLAN capabilities, used at the cross docks. The third type of device, used by drivers to capture detailed pickup and delivery data, is a handheld PDA from Symbol Technologies. The PDA allows drivers to collect data electronically from labels, enter data, capture customer signatures, and print receipts.

- *Application*   Application code is distributed between the devices and Penske's back-end servers. Penske developed the proprietary application code, including interfaces between the wireless applications and its LMS system.

- *Information Architecture*   As mentioned previously, Penske's LMS system supplies data to the various wireless devices and applications. In turn, the data collected and transmitted by these devices is incorporated within Penske's repository of LMS data. Data on arrivals and departures is time-sensitive information, and is communicated to dispatch centers in real time so dispatchers can react to pressing issues. Detailed shipment data not requiring immediate attention is uploaded intermittently throughout the day as it becomes available. Information originating on the cross docks from scanning incoming and outgoing packages is available immediately to Penske's LMS system. All LMS data is uploaded nightly into Penske's data warehouse, although Penske is exploring the feasibility of uploading data at various points during the day.

The Project   Penske's wireless applications have been operational for some time. Penske's IS staff was responsible for the bulk of the implementation work, from selecting devices and networks, to coding proprietary interfaces, to deploying the solution. All of this work was done in close cooperation with the selected device and network vendors.

Penske is leveraging the confidence and expertise gained through its past initiatives and looking for additional ways to harness wireless technologies to bring greater efficiency and productivity to its processes. A number of possibilities are being explored, from smarter in-vehicle terminals to wider use of WLANs, next generation cellular, and dual-mode satellite/terrestrial networks.

Design Challenges   In designing its solution, Penske had several challenges to address, including:

- The ability to transmit information from trucks to dispatch centers in real time, no matter where drivers were located. Since Penske could not ensure that drivers would always be within the coverage areas of existing wide area cellular networks, it opted for the more ubiquitous coverage of satellite networks. To keep its satellite transmission costs down, Penske carefully partitioned and limited the amount of data traveling over the satellite network.

- One of Penske's goals was to improve and automate its control, end-to-end, over the logistics information generated by its operations. Because the processes involved in producing that information are so different, the challenge was to develop a multi-dimensional solution that would work in each type of venue—within trucks traveling anywhere, at the cross docks, and at pickup and delivery sites. The result was a three-pronged solution relying on three different device types, three different networks, and three different types of applications.

- Penske also had to ensure that it could deliver all of the collected information to its customers in an acceptable timeframe. Penske partitioned the universe of data that it was collecting, so that it could transmit the most critical, time-sensitive information—arrival and departure times—in near real time to its dispatch centers. Any anomalies or problems are noted instantly, giving dispatchers time to respond appropriately and notify customers if warranted. More routine information, such as detailed pickup and delivery information, are designated for intermittent uploading to the LMS system. Asset tracking information originating at the cross docks is instantly available to users of the LMS system, and improves the productivity of overall operations at those facilities. Finally, the accumulated LMS data is loaded into Penske's larger data warehouse once a day, overnight. This timeframe is acceptable to customers, but Penske is exploring the possibility of more frequent uploads at various points during the day, to give customers even quicker access to logistics data.

Vendors

QUALCOMM, for in-vehicle units and satellite network services

Symbol Technologies, for handheld PDAs and WLAN equipment at cross docks

Intermec and Symbol Technologies, for handheld devices used by workers at cross docks to scan and track assets

Lessons Learned at Penske   Research the technology components of your solution thoroughly. Expect to spend significant time and effort to get off-the-shelf software operating the way you need it to in your own environment. The value of wireless technology is derived from the process changes it enables.

Acknowledgments   The author thanks Kevin Lamanna, technical sales support analyst of Penske Logistics, for his kind assistance in preparing this case study.

3.7 Wireless Application Resources

Reports of new and interesting wireless applications appear frequently in a variety of publications, some of which are listed below. In addition, please visit *www.just-enoughwireless.com* for updates on wireless applications and implementations. Other resources include:

www.wirelessnewsfactor.com	for current wireless news and applications
www.mobilecomputing.com	for current wireless news and applications
www.crmdaily.com	for news stories about wireless CRM applications
www.aimglobal.org/technologies/RFID	for information about telemetry and radio frequency technology applications
www.computerworld.com	select the "Mobile/Wireless" Knowledge Center for news on latest wireless developments
www.mbusinessdaily.com	magazine on the mobile economy

Recognizing an Opportunity

Selecting the right wireless solution for your business, or even deciding if a wireless solution is appropriate in the first place, is a daunting challenge for an area as broad, complex, and rapidly evolving as wireless technology. Where do you begin? Do you start with a specific business challenge and search through the seemingly endless array of current, near-term, and long-term wireless solutions to find one that fits? Or do you first select a promising wireless solution and look for a business challenge that can be addressed by that specific technology?

With wireless technology, there are no simple answers. On one hand, starting with a business challenge seems impossible to do. The amount of research appears endless. But without a thorough knowledge of wireless technology, you risk missing valuable opportunities or defining your business problem in a way that precludes many more beneficial solutions. For example, unless you are familiar with location-based technology, you may not recognize an opportunity to reduce shipping losses by wirelessly tracking containers.

On the other hand, starting with a pre-selected technology brings its own set of challenges. For example, if your technicians favor Bluetooth as their network technology, they will be tempted to apply it to situations where another approach would be far better suited. Making the wrong choice may severely limit future growth options for your application. Starting with an application in mind is also risky if it is selected without sufficient research. Wireless access to e-mail through a RIM Black-Berry device may be the perfect solution for your needs. Or it may wind up as another unwanted gadget burdening your field employees if coverage is lacking where needed. With the rate at which wireless technologies are evolving, an insurmountable constraint today may be gone tomorrow.

Wireless application selection and design is also greatly influenced by environmental and usage factors. Unlike IT applications that are primarily accessed in fixed, controlled environments, many wireless applications will be used in less than optimal conditions. Displays may be incomprehensible under excessively bright or dark lighting, keyboards may be unusable by individuals wearing gloves, or users may not be able to tolerate the data refresh rates imposed by a given network solution. Other factors, such as the volume and confidentiality of the data to be exchanged, the geographic location where the application will be used, and the length of battery life all play into the desirability and feasibility of a solution.

To date, few true "out of the box" solutions are available for companies seeking wireless solutions. While more prepackaged solutions will appear as the technology matures, most wireless applications of any sophistication will be custom-built using hardware and software components from a variety of sources. Companies will choose these components for their technology compatibility and fit to the specific needs of the business problem they are meant to solve. Thus, more than is usual for IT projects, business requirements drive all aspects of a wireless solution down to very specific technical decisions. Ironically, a business-driven approach to wireless technology actually simplifies solution identification and design.

For all of their variety and complexity, the technology components of wireless solutions are very amenable to a top-down, business-driven selection approach. Components from wireless networks to client devices and data exchange software can be categorized by capabilities, constraints, and compatibility. Mapping business requirements to technical capabilities immediately narrows the range of choices to a manageable number. Initial feasibility becomes easy to research; if appropriate (and compatible) options cannot be found, a wireless solution is not feasible. If acceptable choices exist, further research and definition can continue within reasonable bounds.

This chapter will help you determine if a wireless solution is appropriate for your business objectives and, if so, translate those business objectives into a set of requirements for a wireless application. It advocates a simple top-down approach for identifying and capturing the business requirements for wireless solutions. It describes how to recognize business opportunities where wireless technology may be useful (see Figure 4.1). It explains the process for moving from a business objective to an implementation strategy and introduces the "Five W's" approach to capture functional requirements in a form amenable to wireless solution design.

4.1   Is Wireless Technology a Potential Solution? ____

As exciting as its capabilities are, wireless technology cannot solve all problems, nor is it applicable in all situations. Existing technologies may already be sufficient or may be simpler to implement and support than a wireless solution. The wireless capabilities needed to justify a solution may not be available yet. Or, although wireless technology may be the perfect solution, the problem may not be worth solving. Some wireless applications are obvious and their value easy to recognize, such as telephones for mobile communications and pagers to reach doctors. Others, such as golf cart-mounted wireless devices, require greater creativity to conceive. Providing wireless access to the Internet and creating mobile phone applications for consumers may be flashy, but many highly valuable opportunities exist in unglamorous areas such as meter reading, shipment tracking, and assembly line automation.

The wireless market is evolving. Many highly touted capabilities are not yet available or fail to work as advertised. Lack of technical maturity means competing and incompatible standards, few prepackaged solutions, and a heavy emphasis on custom-created solutions. Application designers are still learning and wireless user design is a rapidly developing art. But new capabilities appear each day. Even at this early stage, there are many opportunities to gain significant business advantage. The success of a company's wireless initiatives depends on its ability to find opportunities that provide business benefits and gain experience now, while laying the groundwork to take advantage of future capabilities.

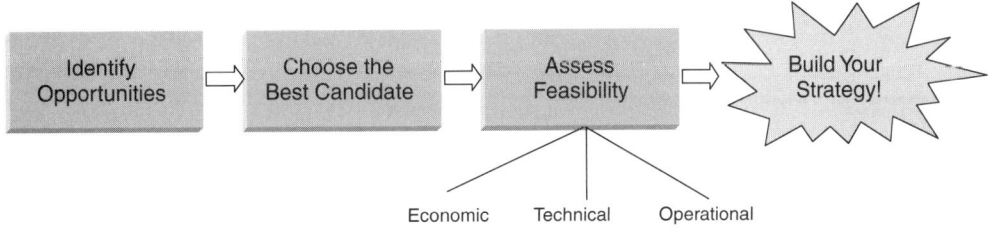

FIGURE 4.1
Finding the Right Opportunity

4.1.1   Identifying an Opportunity

Opportunities to extract significant business value from wireless technology are everywhere in most organizations. While some of these opportunities are obvious, many more will take a little research and creativity to identify. Ironically, some of these opportunities involve activities and tasks that we perform on a daily basis. We are so used to our current way of working that we fail to see them! While the obvious opportunities can bring good returns on our investments, it is the more creative uses of wireless technology that will lead to long-term competitive advantage. For this reason, it is worth investing the effort to brainstorm about potential wireless applications before leaping into the most readily apparent opportunities.

Ideas for wireless solutions can come from many sources. The more you read and learn about the topic, the more ideas will develop. The sources below offer starting points for your wireless considerations.

- *Known Technology Opportunity*   Wireless devices, such as mobile phones, pagers, and PDAs surround us. Applications such as voice communication, mobile dispatch, and wireless alerts are commonplace. If you are not already using these tools widely in your organization, consider them a good place to begin your wireless efforts. Although obvious, these applications are easy to deploy and offer known business value. Wireless access to e-mail is another example of a known opportunity that can provide tremendous productivity gains.

- *Examples of Existing Wireless Applications*   Examples of existing wireless applications are an excellent source of ideas, especially if you can extrapolate an application from another industry to serve a unique need in your company. Feasibility is not an issue; these applications have already been implemented using current technology. Chapter 3 describes many examples of wireless solutions, and additional examples can be found at the *www.justenoughwireless.com* web site that supports this book. Trade press articles, e-mail newsletters, and industry conferences are good sources of the latest application case studies. Watching other companies in your industry, and especially your competitors, is another source of ideas, but you are already relegated to playing catch up if you depend on it as your primary source.

- *Seeking Process Improvements*   As advocated in Chapter 2, focusing on business processes provides an excellent source of valuable opportunities. Processes that are already partly or wholly mobile are obvious targets of opportunity, and can benefit most easily from judicious use of wireless applications. But many other process opportunities exist. Processes that are inhibited from going mobile (or that suffer when they do) due to informational tethers are good prospective candidates for wireless solutions. And entirely new approaches to providing a service or executing a business function can be unleashed through wireless access to information.

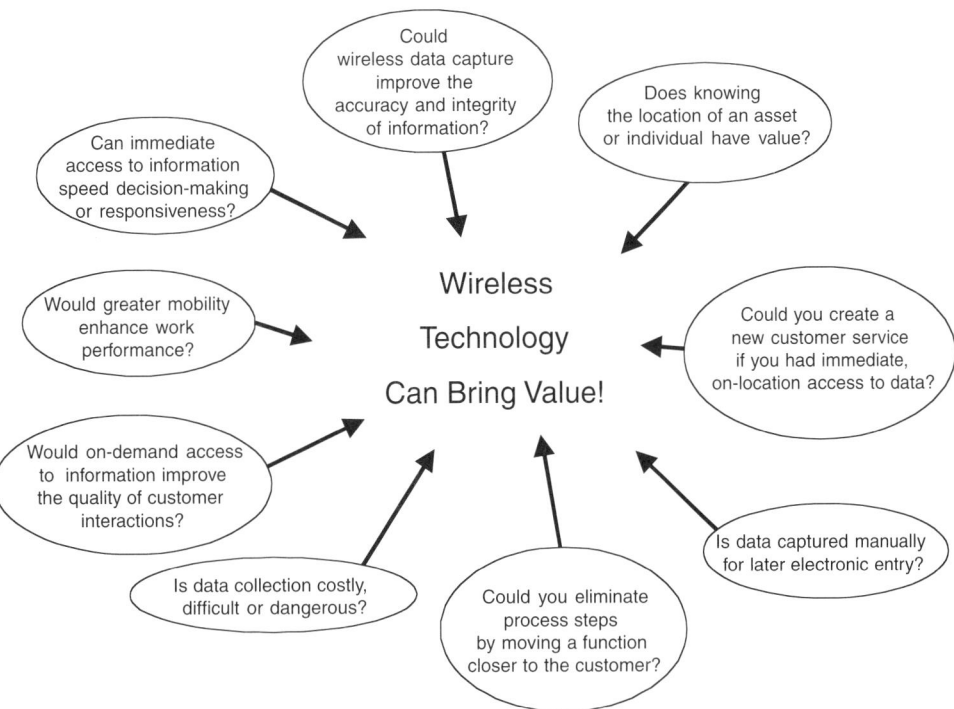

FIGURE 4.2
Possible Opportunities

Wireless technology should always be explored as a possible option whenever your company engages in business process improvements. As shown in Figure 4.2, consider whether the wireless capabilities described in Chapter 2 could benefit any part of the process. The following questions highlight possible process improvement opportunities.

- *Can Immediate Access to Information Speed Decision-Making or Responsiveness?* Wireless solutions are ideally suited for facilitating immediate action on high-value information. For example, stock and auction alerts allow quick response to market fluctuations and instant access to laboratory test results can help emergency room physicians save lives.

- *Does Knowing the Location of an Asset or Individual Have Value?* Wireless location and tracking capabilities enable companies to monitor the whereabouts of shipments, equipment, and employees. Potential applications include loss prevention, more efficient routing of deliveries, accurate notification of arrival times, better inventory control, and enhanced safety for remote workers.

- *Is Data Captured Manually for Later Electronic Entry?* Any process that involves a mobile worker collecting data on paper forms is a potential candidate for wireless automation. Automation possibilities include: laboratory notebooks, prescription writing, inspectors' reports, and trip logs.

- *Would Greater Mobility Enhance Work Performance?*   Mobility allows work to be performed where and when needed. Example applications include: wireless approval of purchase requisitions, mobile check-in within hospitals and hotels, and on-site auto accident claims adjustment.

- *Is Data Collection Costly, Difficult, or Dangerous?*   Wireless devices can lower the cost, effort, and/or risk of capturing and transmitting data from remote locations. Utilities use wireless meter readers to eliminate the cost and inefficiencies of home visits, oil companies track offshore equipment, and shippers monitor conditions within shipping containers.

- *Could Process Steps Be Eliminated by Moving a Function Closer to the Customer?*   For instance, allowing a field worker to calculate the bill and produce the invoice upon the completion of a service call simplifies the billing process, shortens collection cycle times, and increases customer satisfaction.

- *Could Immediate, On-Location Access to Data Allow the Creation of a New Customer Service?*   Offering new customer services enabled by wireless capabilities can enhance customer relationships, provide competitive differentiation, and increase profits. For example, airlines increase customer convenience through wireless notification of flight delays and changes. GM offers a host of services for drivers through its OnStar telematics program.

- *Could Wireless Data Capture Improve the Accuracy and Integrity of Information?*   Allowing doctors to submit prescriptions via a wireless device directly to pharmacies frees pharmacists from the risky task of deciphering illegible handwriting.

- *Would On-Demand Access to Information Improve the Quality of Customer Interactions?*   Providing repair workers with access to a customer's repair records prior to the service call via a handheld device enables them to diagnose and correct appliance problems more precisely.

- *Addressing Known Business Issues*   The wireless capabilities described above for process improvements can also be applied to other types of business issues. In these cases, wireless technology might serve as an enabler for a more direct solution. For example, a company may have the goal of improving its relationships with its suppliers. One component of this initiative is increasing the volume and quality of the personal interactions between the company's purchasing agents and its suppliers. Wireless technology can support the initiative by providing purchasing agents with services such as the ability to approve purchase orders while traveling and remote access to supplier information.

4.1.2 Choosing the Best Candidate

Given the full range of wireless capabilities, it is likely that you will have too many potential solution candidates rather than too few. How then do you select the best candidates? Is it better to choose the most obvious and easy to implement wireless application even if the level of return is low? Or should you aim for high returns despite higher risk? As always, it depends.

With more mature technologies, selections are often made based on standard business calculations such as ROI. While ROI and cost justification remain important for wireless technology, and are covered in Chapter 6, other factors may lead you to select one candidate over another. For example, a simple application, such as wireless e-mail access, may provide an easy and low-risk entree into the wireless world while offering enough business value to justify its adoption. Among your possible candidates, it may not offer the highest business benefit, but it can serve as a non-disruptive proof-of-concept and set the stage for more advanced efforts. Conversely, the business benefits of a more complex and technically risky application, such as wirelessly enabling insurance adjusters, may be sufficient to outweigh other concerns.

Consider the following factors as a starting point for selecting your candidate opportunities.

- *The Value of the Business Opportunity* How much money can the wireless solution save your company or bring in as new revenues? How quickly will the investment pay itself back?

- *The Extensibility of the Business Opportunity* Will your wireless solution serve as a base for additional high-value applications over time? For example, providing salespeople with wireless PDAs with the ability to research customer information can be the first step toward building a highly sophisticated sales force automation application. The long-term value of the solution may greatly overshadow its initial return.

- *Fit with Your Business* Does the application make sense in terms of your business or is it diversionary? Just because you can build it, doesn't necessarily mean you should. For example, unless you are a wireless software vendor, it makes no sense to build your own new wireless e-mail platform.

- *Competitive Value* A highly visible wireless application may have significant value as a competitive weapon. It may position your company as leading edge or provide another competitive knock-off.

- *Technology Complexity* Wireless solutions can become inordinately complex. Technology complexity is always a risk factor, but it becomes more important for immature technologies. Often, unexpected issues arise or components do not work well together. Wireless pioneers have been forced to design around significant hurdles. Avoid high-risk solutions unless their business benefits are substantial enough to justify the inevitable extra investments.

- *Extensibility of the Technical Solution*   Is the proposed solution a platform for additional applications or a potential technology dead-end? Providing PDAs as a platform for corporate e-mail may offer more future options than relying on a paging device. As wireless capabilities expand, can the solution take advantage of those capabilities or will it have to be replaced?

- *The Value of Experience*   Wireless technology is new to most companies. Implementing the first wireless applications will bring considerable knowledge in areas such as user interface design, security, system support, and development tools. Building a simple application or two before embarking on a more complex effort will help reduce project risks.

4.1.3 Assessing Feasibility

Feasibility is an important issue for a new and quickly evolving technology. Depending on the approach selected, a wireless solution may prove too costly or complex to implement, the particular combination of wireless components needed for your solution may not work together or may be untested, or a specific feature or area of coverage may not be available yet. A host of other implementation issues, technical, procedural, and political, can stop an otherwise perfect solution. While the ultimate feasibility of a project cannot be ascertained until deeper into its design and implementation phases, an initial assessment of feasibility allows you to discard questionable opportunities early, saving the time and cost of their investigation.

Assess the initial feasibility of a wireless opportunity by considering its economic, technical, and operational aspects. Although each of these categories is important individually, they are also interrelated. For instance, a project that promises considerable economic benefits may warrant taking on a greater level of technical risk as the benefits would justify the investments needed to solve technical issues as they arise.

- *Economic Feasibility*   Chapter 6 covers the cost justification of wireless solutions in more detail. At this point in the opportunity analysis, the goal is to judge quickly whether the benefits of the solution can justify its costs. To make this determination, a gross estimate of benefits and costs is sufficient. Assess the size of the potential benefits by considering:

  - How large are the potential direct benefits? (cost savings, new sources of revenue, increased sales, etc.)

  - How valuable are any secondary benefits? (marketing value, educational value, employee safety value, ability to support a larger initiative, etc.)

  - Is the solution necessary for competitive purposes?

While the answers to these questions will vary considerably by the type of opportunity under consideration, they should provide a rough idea of the worth of a wireless solution to your company. Even without a specific wireless implementation in mind, the following parameters will help you determine whether costs are likely to be small, medium, or large.

- How many people will use the solution? (size of problem)
- Will you have to provide them with mobile devices? (cost of devices)
- How large is the geographic area that will be covered by the solution? (cost of networking)
- Will the solution be unique to your company? (cost of development)

Obviously, the cost of the investment will be proportional to the number of people using the solution as well as the cost of any equipment that needs to be purchased. Similarly, the greater the level of customization, the more expensive the project.

- *Technical Feasibility*   At the technology level, feasibility for wireless solutions is a moving target. Capabilities are evolving quickly and new tools, devices, and services appear almost daily, removing constraints and creating new solution options. As is typical for technology, however, actual capabilities often lag behind claimed capabilities, particularly in development tools and the level of component integration. If the opportunity you are targeting can use currently available technologies and live within current device, bandwidth, and coverage constraints, then it is apt to be feasible. The same is true if your opportunity mirrors a solution already implemented elsewhere. As you begin to push these bounds, short-term feasibility declines. Unique solutions can have very high economic value, but will require custom programming and integrating components that may never have been integrated before. Similarly, if your solution will need capabilities that are not yet available, you have the choice of waiting for those capabilities, building the solution anyway to be ready when those capabilities materialize, or scaling back to work within current capabilities.

- *Operational Feasibility*   Economic feasibility determines whether a solution is worth using. Technical feasibility determines whether it is possible to build that solution. Operational feasibility determines whether the solution will actually be used. Too often, an otherwise perfect solution will fail when user requirements and preferences are not considered from the outset. For example, wireless access to customer information may appear to offer obvious and irresistible benefits to a mobile salesperson, but the solution may still fail if it requires carrying yet another device, or conflicts with the way the salesperson actually works. Security is another critical issue for operational feasibility. If data is so confidential that transmitting it to a mobile device would jeopardize its security, then a wireless solution is infeasible. A quick evaluation of operational feasibility requires a realistic review of the possible issues the solution may face and making a gut-level assessment of whether those issues are possible to address. Involving potential users and discussing the solution with them is an effective way to surface issues early in the process. You may find out that users are not even interested in having a wireless solution.

4.2 From Business Objective to Implementation Strategy

Wireless solutions depend greatly on the specifics of the problem they are trying to solve. From client devices to wireless application design and networks to information infrastructures, most wireless solutions integrate components separately chosen for their fit to a specific need. Issues such as who will use the application, how and where it will be used, and the type and volume of data that it will exchange affect the selection of individual components and have major implications for the solution's design and its supporting infrastructure. For example, although they are both methods for wireless data exchange, infrared networks and satellite networks have vastly different capabilities and are appropriate in very different situations.

The greatest benefit of a top-down business-driven approach is its ability to narrow technology choices to a manageable number. If your ultimate business objective is to add cash registers on-demand at any location in a department store, your hardware and software choices will fall within a relatively narrow set of options. These options are very different than the ones that can enable on-site billing by field service personnel. Each option brings its own set of needs, constraints, and issues that drive the subsequent levels of your solution's definition.

A simplified view of this approach is shown in Figure 4.3. This example is based on the field service process improvement first introduced in Chapter 2. Our business objective is to enable on-site billing of clients at the end of a service call. This objective imposes a host of **business requirements** on our prospective solution. We need a solution that will work at any client site. It must have access to our corporate billing application. We need to collect time spent, actions taken, and parts used. We need to be able to print an invoice on the spot. The list goes on. These business requirements drive **solution requirements** for the major components of our prospective solution. These components include our wireless network, client devices, application functionality, and corporate data sources. To meet our business requirements, we need a network solution that supports national coverage as well as data rather than voice transmission. Moreover, the network should be able to quickly

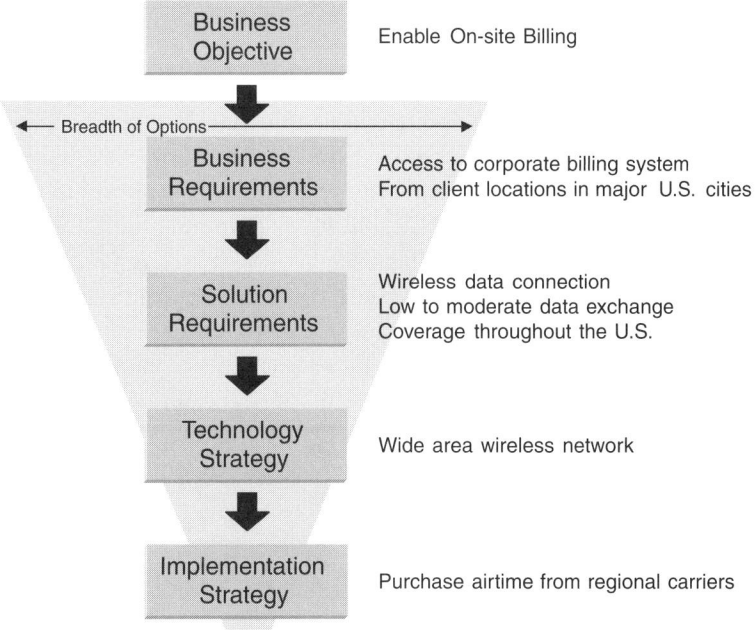

FIGURE 4.3
A Top-Down Approach to Wireless Strategy

handle low to moderate levels of data exchange. We can easily compare these requirements against the capabilities of our major network options and discover that our most promising **technical strategy** is a wide area wireless network. Similar decisions can be made for our other components. Once our strategy is chosen, it drives our **implementation strategy**. We already know, for example, that we will have to purchase airtime from regional carriers for our network and that we need to select development tools that support wide area wireless networks.

At each level, a set of choices is made and our breadth of options becomes more reasonable in size. By the time we reach our implementation strategy, we know enough to be dangerous. We have proven that it is at least conceptually possible to address our business requirement with a wireless solution and we have a "straw man" candidate ready for detailed analysis and feasibility assessment.

In practice, this process is more iterative than linear as our choices in each category interact. Our network options may not neatly match our application data requirements or our preferred choices for client devices. In many cases, we can resolve an incompatibility by moving up in our hierarchy and rethinking our approach within our now known and understood technology constraints. If our business requirements drive incompatible and irresolvable technology strategies, our solution is not technically feasible and we have to choose a non-wireless solution or turn our attentions elsewhere.

The core of this approach is developing a deep understanding of the business objectives and requirements that will drive our technology solution. Resist the urge to leap directly into technology implementation without first thoroughly understanding how, where, and by whom a wireless solution is going to be used. Environmental and usability issues are supremely important for wireless applications and can quickly ruin an otherwise functionally perfect solution. For example, if we are creating a wireless application for emergency room doctors, our solution must be able to withstand a harsh and fast paced operating environment, poor lighting conditions, and the working styles of harried physicians. Designs that may be acceptable in other situations are intolerable to a physician running between examination rooms. Information access must be quick and navigation efforts minimal. Subtleties, such as the ability to operate the device while wearing latex gloves, are easily forgotten at the requirements stage, but have significant impact on device selection and solution design. Failure to recognize and address any of these issues will result in a technically elegant, but unusable wireless solution.

4.3   The Five W's of Wireless Business Requirements

If your company has its own IT organization, you are likely familiar with the various methods of gathering business requirements for new application systems. With some nuances, these methods remain applicable to wireless solution design. First, we need to gather additional information about the solution's users to address issues not present in traditional IT systems. With wireless solutions used in many different environments and conditions, we no longer have the luxury of assuming that the user is sitting in an office at a stationary PC, that the office is temperature controlled and well-lit, that distractions and noise are minimal, or that a fast and consistently available network connection exists. Second, many important decisions about a wireless solution are driven by the demands of the application being provided, and require understanding characteristics such as where the solution will be used, the technical prowess of its users, and physical conditions under which it will operate. These requirements quickly set the parameters of the type of wireless solution that can be considered. Once this basic foundation has been established, standard business analysis approaches can be used to further refine the solution.

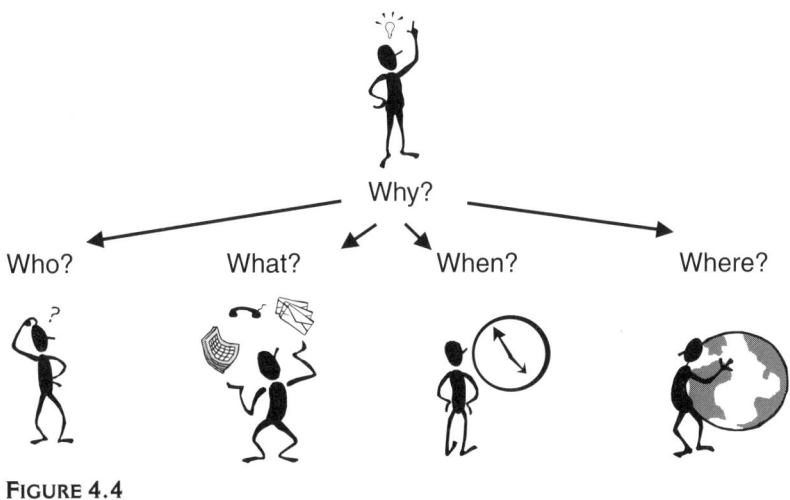

FIGURE 4.4
The Five W's

The Five W's approach described in this section is designed to address the nuances of capturing business requirements for a wireless solution. It organizes questions into five categories: Why, Who, What, When, and Where, to address the unique characteristics that shape device, network, information architecture, and application decisions.

- *Why* addresses the reasons for pursuing a wireless solution in the first place. It sets the mission of the project and helps to prioritize design decisions and resolve conflicts.

- *Who* gathers information about the intended users of the solution. For example, the device and application needs of a busy emergency room physician are very different in terms of information immediacy and ease of use than those of a quality assurance inspector using a forms-based system.

- *What* focuses on the functions and services the solution will provide. It examines the activities supported by the solution, investigates its information requirements, and considers user experience requirements. For instance, a customer information system supporting a field service worker may need to access information from a variety of customer and support databases. Access to this information has numerous implications from application integration and data security to transmission bandwidths.

- *When* examines requirements for data immediacy, latency, synchronization, and integrity. For example, a stockbroker may require instant notification about fluctuations in stock prices, while an order status application may require only periodic updates. Serving the stockbroker requires real-time processing and an always on connection, while the order status application can be supported by automatic hourly synching.

- *Where* captures information about the solution's geographic location, mobility, and environmental requirements. For instance, a system for registering hotel guests inside a hotel can use a short- or medium-range wireless LAN, but an application connecting mobile workers operating anywhere in North America will need to use a digital cellular or satellite network.

The questions covered in this section are contained in a questionnaire found in Appendix A. Use this questionnaire as a starting point for your efforts, and customize it as needed. The answers to the questions are the inputs for the design decisions covered in Chapter 5.

4.4 Why

The first step toward defining a successful wireless solution is understanding why you want that solution in the first place. While the other 4 W's capture solution requirements, "Why" provides the mission statement for the overall effort. It defines the ultimate objective of the solution and becomes the screen for all decision-making. Keep going back to this objective when defining a solution. Whenever a question of direction arises, pick the path that best supports the solution's goal. For instance, if the main goal is to increase the sales force's responsiveness to customers, don't compromise that goal to improve a secondary desire, such as performance tracking.

4.4.1 Why Care about "Why?"

Answer the why question by considering the main benefits that the proposed solution will provide. Is the goal to increase revenue, improve customer satisfaction, or reach new markets? If you have selected the solution candidate following a process such as the one described in Section 4.1, you will have already set your goal and have a feeling for the level of business benefits that can be achieved from the solution. This understanding is invaluable for the following purposes.

- *Set Priorities* The project's mission helps identify which activities are the most important and gives the project a clear sense of direction. It provides executives with a means of comparing the project with competing efforts when assigning scarce budget dollars and resources.

- *Make the Right Compromises* Technical, procedural, and business challenges will require compromises throughout the project. A clear understanding of the project objectives will help ensure that those compromises support rather than hinder the ultimate objective. For example, as described in Chapter 3, focusing on the top ten questions asked by customers allowed Atlantic Envelope Company to deploy a much simpler and more effective solution for its sales force than would have been possible had it tried for a 100% approach.

- *Avoid Diversions*   All projects encounter scope and feature creep. Rapidly evolving wireless capabilities will tempt project teams to add new features and capabilities throughout the effort. Providing users with a wireless device will attract other applications that can operate on that device. Additional features are fine if they do not divert effort or compromise the ultimate goal of the project.

4.4.2 Possible Goals and Their Implications

Chapters 2 and 3 describe many examples of wireless applications and their potential benefits. The list below summarizes some primary goals along with their implications for developing a wireless solution.

- *Generate Revenues*   Wireless technology can generate additional revenues by bringing new wireless services directly to customers, enabling a company to offer additional products and services, freeing workers for revenue-generating activities and making it easier to purchase goods and services. The possibilities for increasing revenues are limited only by the imagination of the solution designers. For example, golf courses using ProLink's wireless and GPS-enabled golf cart systems gain additional revenues by increasing the number of golfers that can use the course (by tracking the location of golfers), enhancing concession stand revenue by simplifying ordering (by letting players order from the cart using menu selection), and identifying prospects for personal training (through playing statistics gathered and saved to a web site).

- *Increase Efficiency*   Most wireless solutions will increase efficiency in some manner by providing mobility, eliminating redundant activities, or bypassing process steps. If improving efficiency is the primary goal of a project, a strong process improvement focus will provide the greatest benefits. UPS is seeking to make its distribution centers more efficient by deploying a new, standardized wireless system for scanning and tracking packages as they enter and leave the premises.

- *Improve Customer Service*   Wireless applications give workers the ability to perform services remotely, closer and more convenient to the customer. Better access to customer information is key to enhancing customer service. The more information about the customer available to the field worker, the more personalized the service, and the more satisfied the customer. Companies like Honeywell, Sun Microsystems, and Sears rely on wireless applications to improve technician dispatch, enable work order completion from the field, and even allow invoice generation on site.

- *Protect Assets*   Wireless tracking and monitoring applications can prevent theft and damage to expensive capital assets. These applications typically rely on wireless tags and/or monitoring devices over WANs to transmit small amounts of status information or requests for service. Packaged Ice outfits its ice-making equipment with wireless applications that transmit status information in real time to alert company representatives about problems and malfunctions.

- *Curb Losses*   Through a combination of tracking devices and wireless applications, companies can locate and monitor the condition of tangible goods. These applications rely on the same types of technology as described for protecting assets. Telematics applications track the location of a vehicle to minimize theft and damage.

- *Improve Safety*   Wireless applications help ensure personal safety through a combination of communication, location, and monitoring capabilities. Monitoring applications eliminate the cost and need to send workers to hazardous locations. Tracking technologies such as GPS allow companies to locate workers and dispatch help quickly if needed. A telecommunications company in the northeastern U.S. tracks and maintains contact with its employees that travel into remote or hostile areas to make repairs.

- *Expand Market Share/Reach New Markets*   Wireless applications can reach a growing number of consumers in industrialized countries, and wireless technology is the method of choice for bringing communications to developing countries. Applications that reach out to consumers must be simple to use and support the myriad of devices owned by these individuals. In Europe and Asia, the growth of wireless devices and communications exceeds traditional, land-based counterparts precisely because of their ease-of-use.

- *Improve Accuracy*   Inaccurate, incomplete, or outdated data can lead to many costly errors. By encouraging immediate and controlled data entry, wireless applications can more accurately capture and relay data. Access to corroborating information and device processing capabilities can further increase accuracy by validating data as it is entered. For example, an electronic prescription system eliminates errors due to illegible handwriting and can check dosage and potential drug interactions.

4.5 Who

The individuals that will use your wireless solution heavily influence its design, implementation, and support. Sensitivity to the needs and characteristics of your users will spell the difference between enthusiastic adoption of your application versus a dismal failure. Although solutions for mobile professionals garner most of the attention, every wireless solution has users. Even telemetry solutions, such as one that monitors and relays the conditions in an ice vending machine, is ultimately serving one or more users. Understanding the characteristics of these users, their preferences, and how they do their work is critical before attempting to select or design a solution. More so than other computer applications, the constraints imposed by wireless technology necessitate a user-friendly design. Despite its many advantages, a multi-step information search facility will not be used by a salesperson looking for a quick price quote. A doctor won't spend the time to learn a device-specific scripting style. Studying the traits, behaviors, and working styles of your actual users is essential to capture accurate solution requirements. Don't fall into the trap of assuming you understand your users' needs. Making assumptions will lead you to miss subtle, but important, details about your users.

4.5.1 Who Will Use the Solution?

An end user of a wireless solution may be a single individual, a group, department, division, or a defined community. In a telemetry solution, the "user" may be a machine or a piece of equipment. Several layers of users may exist. The primary users are those individuals who directly use the application to accomplish their job functions. These individuals may belong to a very narrow class, such as UPS package sorters, with well-defined characteristics and job requirements or they may be part of a large and diverse group, such as consumers, with few shared characteristics. In the former case, the solution design is very specific and can be optimized for maximum performance. In the latter case, the solution has to support the least common denominator in device capabilities and user prowess. Secondary users derive benefits from the primary users of the solution. For example, sales management executives gain information such as pipeline activity and sales forecasts from a wireless sales force automation application. Other indirect users are those that install and support the wireless solutions. They have a vested interest in issues such as ease-of-use and ease-of-deployment. Keep the roles of these separate constituents in perspective; if the solution does not meet the needs of the primary users, it will fail even if it successfully serves its other constituents.

4.5.2 How Do They Work?

To be effective, a wireless solution has to support the way users do their work. All too often, solutions that make sense logically fail from not fully understanding how their users work. The poor adoption of PC-based sales force automation applications is a prime example. Highly mobile salespeople do not want to lug around a heavy laptop. They perform many of their tasks quickly, in front of clients or during spare moments between sales calls. Salespeople have neither the time nor the patience to wait for a PC to boot up. Finding and connecting to a telephone line for communications adds to the inconvenience. Given these characteristics, it's not surprising that salespeople would hesitate to adopt a PC solution. Understanding the following areas will help your wireless solution avoid this trap.

- *How Frequently Do They Perform the Activities Under Consideration?* Would they use an application periodically to check the status of an order, retrieve and respond to e-mails, or enter small pieces of information? Or would they constantly use the application to monitor a situation or perform a highly paced and repetitive activity? Do they get to choose when to use the application (periodically logging on to get e-mail) or is the timing controlled by outside factors (receiving alerts of "perishable" information like stock prices)?

- *How Do They Usually Exchange Information?* Do users rely on paper forms, dictate notes for their administrative assistants or use a PC or PDA? What types of data entry mechanisms are they most familiar with? Do they use keyboards, Dictaphones, or pen-based tablets? Do they get most of their information from a computer screen, printed materials, videos, or voice communications?

- *What Is Their Working Style?* Would they use the application while walking, driving, standing, or sitting at a desk? For example, a building inspector may typically walk around writing notes on a clipboard, while a package sorter stands within his work area and a taxi driver works seated in a taxicab.

- *How Harried Are They?* A physician or emergency medical technician has to perform in situations where seconds can make a difference. A stocktrader is barraged by information from many sources. In contrast, a field repairperson works on one account at a time and can take a few minutes at the end of each task to power on a device and enter information.

- *What Types of Tools, Equipment, and Accessories Do They Carry for Their Jobs?* An executive or salesperson may have a briefcase, telephone, and appointment book. A service technician may carry a toolbox, pager, telephone, and clipboard. Will a wireless device add to their load or can it take the place of other equipment? If the solution is physically inconvenient, users will avoid it. How would users carry a device? Would they hold it in their hands, put it in a pocket or briefcase, wear it on a belt, or attach it to a work area? The way in which a device is carried has ramifications on features such as device size and ruggedness.

- *What Are the Conditions in Their Work Environment?* Environmental challenges include factors like temperature fluctuations, lighting conditions, noise levels, and distractions. These factors are covered in the Where section of this chapter.

4.5.3 How Comfortable Are They with Technology?

This trait determines adoption effort, ease-of-use, and training needs. Professionals used to computers and mobile phones will adapt quickly. Other groups may have little experience with technology. Consumer applications and other solutions built for diverse audiences must be able to handle a wide range of technical skills.

- *Do They Currently Use Technology?*   Are users already familiar with the use of computers? Have they used PDAs or do they have a mobile phone? What types of technology are most familiar to them? What are their biggest complaints about the technology that they use? Which features do they like the best? Will they need to synchronize their wireless solution with another computer-based solution?

- *What Types of Devices Do They Own or Use?*   Do users already own mobile devices that could support the desired solution? Leveraging an existing platform saves time and money. Conversely, choosing the best device for an application optimizes the solution and reduces development costs. If you are building an application that your customers will operate, you must support the full range of devices that they commonly use.

4.5.4 What Are the Personal Traits and Preferences of Users?

While the user community as a whole is likely to share common traits, individuals within the group will possess unique characteristics. The ability to customize and personalize a wireless solution to accommodate individual preferences can be a strong selling tool and encourage adoption. Examining some of the factors discussed above can identify opportunities for personalization. For instance, letting users select their own mobile devices is the highest form of personalization, while giving them options for their preferred mode of data input (voice or keypad) is another. Personalization can also be affected by language support, information display, and the presence of additional functionality.

4.5.5 What Social and Cultural Challenges Exist?

Social challenges abound in any human endeavor. Users may fear loss of privacy, resist workflow changes, or feel the use of a wireless device detracts from their image. These issues are most easily addressed if understood early in the analysis process. Some issues include:

- *Image*   Does a wireless solution add or harm their image? Some groups, like salespeople, often believe that having the latest tools and gadgets adds to their cachet. Others, like doctors may believe that using a mobile device will detract from their efforts to focus on patients.

- **Fear of Big Brother** *Wireless* applications, and especially location-based applications, bring a host of privacy concerns. Employers can theoretically contact employees anytime, anywhere, and workers may be reluctant to use a mobile device that includes a GPS locator to track their whereabouts.
- **Unions and Other Employee Groups**   Any solution that changes the way a job is performed or potentially eliminates or replaces current job positions is likely to need the agreement of unions and/or other employee groups. Involving these groups early in the process and understanding their concerns is likely to avoid problems and delays in implementation.

4.5.6 How Can You Support Them?

Knowing the characteristics of your users is also critical for understanding how best to support them. IT applications typically operate in offices, in a protected environment near sources of assistance. Conversely, wireless applications are often deployed far from sources of support, are used at odd hours, and are subjected to rough conditions. To help identify support requirements for your wireless solution, you should ask these questions.

- How many people will use the solution?
- When will they use the solution?
- Is it a business-hours only application or could it be used 24 hours a day, seven days a week?
- Is some window of unavailability acceptable for the solution?
- Is the solution simply "nice to have" or will the user be out of action if it fails? What are the costs of an outage? How long can an outage be tolerated? Are back-up solutions possible?
- How high is the risk of loss or damage?
- A solution used at a desk within an office building faces much less risk of loss or damage than one deployed on construction sites.
- How difficult is it to reach a user?
- How long would it take to ship a new device to a user? Can users easily receive (and install) software updates? Do users meet regularly at a central location where training can be offered and devices can be exchanged?

4.6   What _____

What does the user of the wireless solution need to accomplish? The purpose of this query is to gain a user-centric understanding of the types of functions and features that the solution must provide to meet its objectives. This understanding feeds directly into many of the solution's design requirements. The questions in this subsection consider the services the wireless application will have to support, the characteristics of the information it will exchange, how that information must be processed, and the ways in which the user will interact with the application. The answers drive the solution's application design, information architecture, device selection, and its presentation/user experience design. As shown in Chapter 3, wireless applications tend to fall into broad functional categories, and the applications in these categories share many attributes. An application for monitoring conditions within an ice-making machine has a different set of functions and features than a sales support application, but is similar to other monitoring applications. Gaining a thorough understanding of a solution's basic requirements early in the analysis process helps reduce design decisions and considerations.

To determine what your wireless solution needs to accomplish, first learn how the desired activities are currently performed. Note the specific activities performed by the user and ask what they need to know to perform their actions. Pay close attention to the nuances of how they work and the constraints that they face. Actual observation is the best way to avoid mistaken perceptions. For example, in designing a wireless clinical trial application, an analyst observing doctors and patients would quickly discover that they need to capture and record trial data, and communicate with each other regularly. On the patient side, they need the ability to record daily status information in diaries. Doctors need access to drug protocols and background information. Certain information, like drug side effects, must be communicated immediately, while the bulk of data can be submitted periodically. Only by making these observations would the analyst know to incorporate electronic diaries, e-mail, two-way messaging, form creation and submission, real-time download, and synching capabilities in the wireless solution.

While wireless solution design has its own nuances, the basic systems analysis methods used by IT organizations will work well for capturing wireless requirements. The sections below offer some questions to supplement your preferred system analysis approach.

4.6.1   What Does the User Need to Accomplish?

Whether wired or wireless, computer applications tend to be built from a basic set of functions. Some applications, such as pulling readings from a water meter, rely on a single function. More complex applications typically contain several functions. The primary objective of the solution may be satisfied through one function, but adding secondary functions may greatly expand the solution's overall value. For example, in the clinical trial application discussed above, data collection, communications, and look-up functionality were combined in one solution. This solution allowed doctors and patients to automate many of their activities.

The basic functions comprising most computer applications are listed below. Each of these functions has specific implications for the types of wireless solution selected.

- *Communicate*   Does the user need to communicate with other individuals? What is the purpose of the communication? What form of communication is needed—voice, text, or data? Does that communication need to happen in real time through an alert or can it occur periodically through synching?
- *Access Information*   Do users need to access information from another location? Do they need access to the Internet, corporate networks, or databases?
- *Collect Data*   Does the solution have to capture information in electronic form for later analysis and reporting or for use by other computer applications? Does the user need to fill out forms such as status reports and work orders? Is data scanned from bar code labels or collected using a special-purpose device?
- *Update*   Does the user need to update information held in another location or vice versa?
- *Transact*   What type of work does the user have to perform on the spot? Will the solution have to produce invoices, create proposals, calculate payments, submit orders, make reservations, issue tickets, or perform other types of specialized processing?
- *Monitor*   Does the user need to monitor the condition of assets or individuals in a remote location, such as the status of a patient, contents of a vending machine, or conditions on an oil rig?
- *Locate/Track*   Does the user need to locate or track a person, shipment, or piece of equipment?

4.6.2 What Does the User Need to Know?

Every wireless solution has information requirements. The solution may collect, display, and/or process multiple types of information, which has implications for device selection, information architecture, and network requirements.

- *What Forms of Information Are Used?*   Does the user currently rely on voice, text, graphics, or video information? Is this information confidential? How many different forms of information are used? Would other forms be more useful?
- *What Is the Source of the Information?*   Where is the desired information located? Is it readily available? Is it in multiple locations? How much effort will be needed to make the information usable for a wireless application? For instance, a CRM application may need to draw and combine information from multiple corporate sources.

- *How Much Information Has to Be Exchanged?* What volume of information has to be exchanged at any given time? For example, a simple paging network can be used to exchange short messages and alerts, but a more robust network is needed to support high-volume file downloads.

- *Where Will the Information Be Kept?* Does the user need access to both private and public information? What information is best kept with the user, and which is best kept in shared files?

4.6.3 How Will the User Interact with the Application?

Unless a wireless solution provides its users with significant benefits over their current methods of operation, it will never be used. Understanding the possible ways a user will interact with a wireless application helps to create a design that makes sense in the real world. The solution must allow users to easily access the functions and information they need in a format that matches the way they work. For example, providing a physician with instant access to patient x-rays on a handheld device has no value if the x-rays are indiscernible when displayed on a small screen.

- *What Is the Best Way to Interact with the User?* Does the user prefer keyboard, pen, touchscreen, or voice interaction with the application? How does the user currently capture data? How does the user currently enter data? What type of interactions can the user perform? For instance, someone driving a car cannot easily enter information through a keyboard.

- *How Does the User Find Desired Information?* How does the user currently search for and retrieve information? Can information retrieval be simplified through creative navigation or more efficient organization?

- *What Is the Best Way to Display the Information They Need?* Does the user require voice, textual, graphical, audio, or visual display? How is information currently delivered? Are more concise or efficient delivery methods available? Must information be specially processed to be useful? For example, if a stockbroker cares only when a particular stock moves above or below a price range, providing continuous updates on the entire stock market is superfluous and annoying.

- *What Types of Equipment Does the User Need to Support "On the Spot" Work?* Will the user need additional or special-purpose equipment to scan credit cards, print invoices, or perform other activities associated with the wireless application?

4.7 When

As the world becomes increasingly Internet-oriented, and as 24-hour-a-day, 7-day-a-week access to information becomes commonplace, it is easy to assume that every solution, including wireless solutions, will need to accommodate insatiable, instantaneous demands for information. And, as equipment prices decline and more powerful software products emerge, it becomes cheaper and more expedient to simply throw additional computing power at applications rather than design them to conserve resources and operate more efficiently. We become less concerned with whether people truly need 24x7 real-time access; it is easier to just build it into the solution in the first place.

With current generation wireless computing environments, we do not have the luxury of over-design. Providing 24x7, universal access to up-to-the-minute information is tough and sometimes impossible. Precisely because wireless technologies are constrained, you need to consider *when* users will need to have access to the wireless solution and to the underlying information. To understand these timing requirements fully, a two-part inquiry is needed. We need to ask:

1. When will the application be used?
2. When does information need to be made available?

4.7.1 When Will the Application Be Used?

Knowing when users will typically use the wireless solution helps to determine the timing of various other events and aids in capacity planning. The hours of use can determine the window that will be available for synching data, for example. If field inspectors use their mobile, forms-based application to collect data during daytime hours only, then data uploading and synching can occur overnight. Likewise, the hours of use will determine when call centers or support lines must be staffed with resources. The hours of operation of back-end or supporting systems, the timing of back-ups and other maintenance and support tasks are affected by when the wireless application will be used. Finally, application usage, especially concurrent usage, affects system load and capacity planning. To understand when the application will be used, consider:

- *What Hours of the Day Will Users Use the Application?* Will the application be used primarily during business hours? Daytime hours? Or at anytime of the day or night? Does the application have to be available on weekends or holidays?

- *In What Time Zones Are Users Located?* Will users be located in the same or multiple time zones? What type of collaboration must be supported?

- *Will the Application Be Used Concurrently or Intermittently?* How many users will use the application at the same time? Will they be doing the same or different things? Will the application be used randomly by different users? Or at predictable times by the same users? Will the application be used only periodically or at special events such as trade shows, registration periods, or high-volume shopping times?

4.7.2 When Does Information Need to Be Made Available?

People often tend to think that data must be one hundred percent current to be useful and reliable. In reality, up-to-the-second, accurate information is needed in only certain cases, usually where an irrevocable, life-altering, or financial decision is made based upon the data. The test results provided by a wireless application to an emergency room doctor must be completely current. Auction prices, bank account balances, and stock quotes must also be up-to-date for users to make reliable decisions. To place an order and promise a customer a shipping date, the inventory information used by a wireless order submission application must be accurate and current. In situations where data constantly fluctuates, such as wireless monitoring of temperature conditions inside a cargo container, it may be crucial to receive and analyze a constant stream of real-time information for the wireless application to provide any kind of value.

More typically, data is produced and refreshed at periodic intervals, and wireless applications use data that is "outdated" to a certain degree. When tracking a package using UPS's wireless application, a user may care that the package was loaded onto a truck at 9:30 am that morning, but is indifferent to the exact locations of the truck during the day. Oil and gas pipeline repairmen need current maps and drawings to perform their work, but a once-a-day refresh is sufficient to provide an accurate view. Propagating contact information to salespeople on a bi-weekly basis may be frequent enough to ensure that clients and co-workers can be easily reached while avoiding lengthy data transmissions.

Related to information availability and currency is the issue of integrity. Wireless applications that perform transactions or that exchange data with other applications may need the ability to recover gracefully and resume operations when transmissions are interrupted, as well as the ability to determine when data is corrupted or lost during transmission.

When considering time dependencies related to data, and issues of responsiveness, data quality, integrity and synchronization, explore these areas.

- *Immediacy* When information is needed, how quickly must it be delivered? Does information need to be "pushed" to a user as soon as it is available, or can it wait until the user asks for it? Can information be stored and sent at a later, more convenient time?

- *Latency*   How fresh and current does data have to be? Is data perishable? What is an acceptable "lag" time between data creation and dissemination? Is real-time access to information needed? How frequently must data be refreshed or repopulated for it to be reliable?

- *Synchronization*   Where is information created? Where does it reside? How closely do sets of data residing in different places have to match? What effort must be expended to synchronize data? Can data simply be overwritten or does it have to be compared, matched, and merged?

- *Integrity*   Is a transaction, and the data involved, so sensitive, valuable, or fleeting that it must be safeguarded against corruption, interruption, or loss? Can data be re-created, and at what effort or expense? Is it necessary to recover and resume interrupted transactions or can users start again from scratch without any adverse effects?

4.8 Where

When designing a mobile solution, the question of where that application will be used is a crucial factor affecting network and device selection, user interface design, distribution of data, security requirements, and user support practices. This broad question elicits two categories of information. The first focuses on the location of end users as well as their range of mobility. The second focuses on the physical conditions that will confront those users. When answering the questions posed within this section, consider:

- *Where Is the Activity Performed Now?*   Many wireless solutions bring technology into an already established work environment. Unless the purpose of the wireless solution is to move an activity to an entirely new set of locations and very different environmental conditions, the current work environment provides the baseline characteristics for answering the where questions. For example, bringing a wireless solution to emergency room doctors requires understanding the emergency room environment.

- *Where Would You Like the Activity to Be Performed?*   If the wireless solution will move an activity to a new location, the characteristics of that location have to be considered. Moving a desk bound function into the field will expose the activity to field rather than office conditions.

- *Where Might the Activity Be Performed in the Future?*   Once an activity becomes mobile, it may expand in scope and/or move to new locations. The original intention of a wireless heart monitor may be to check the status of

patien
patien
locatic

4.8.1

Even if yc
if the acti
performe
solution,

- *Ph*
  be
  a
  ve
  ca
  e
  c
  1

-

night expand to cover
hrough possible future
y limit later uses.

t solutions are required
ample, than if it may be
uirements for a wireless
mation.

here the activity is/will
e? Within a building? On
in a car or other motor
venient access to electri-
from these locations? For
usually has access to an
store to purchase replace-
y spend many hours away

eaker giving a presentation
es of a room. An engineer
s anywhere on a company
l centers. An executive may
bility ranges from relatively
ction to an office printer, to
container tracking.

solution selected, geographic
pes of network options avail-
vailability, coverage, and stan-
at is feasible in Europe may be
When describing geographic
untry and city locations, ask
or remote areas. Multiple geo-
ities. Be sure to note all of the
ust support. For example, one
were located in areas with good
eps tended to live in rural loca-
s handled by selecting a solution
a rep traveled to a covered zone.

4.8.2 Environment

The physical conditions surrounding the user of a wireless application are the second facet of the "where" question. Wireless technology allows IT applications to move outside of controlled office environments, placing a premium on designing solutions that will work in less than optimal working conditions. These working conditions affect how a user will use the application, the types of devices that are appropriate, and the design of user interfaces. Working conditions include environmental factors such as temperature, visual or auditory distractions, lighting, user posture, and health and safety hazards. For example, a wireless solution that will be used outdoors in extreme temperatures requires a very durable device, careful attention to battery life and perhaps an interface that supports users with gloved hands. Some of the environmental conditions to consider include:

- *Lighting*   Lighting conditions range from fairly constant, such as in an office environment, to poor, such as night use, to highly variable requiring the ability to handle both very bright and very dark conditions.

- *Temperature*   Wide fluctuations and extreme temperature conditions are tough on devices and electrical components and may require users to wear gloves or other protective gear that affects the ability to use keyboards and other types of interfaces.

- *Sound Level*   Loud sounds preclude voice applications and require designs that don't rely on auditory cues.

- *Dirt/Dust and Exposure to the Elements*   Rough outdoor conditions require especially durable and well-sealed devices as well as convenient support and repair procedures.

- *Vibrations/Shocks*   Vehicle or machinery-mounted equipment is susceptible to vibrations or shocks that can easily damage or destroy sensitive electronics. Specially designed devices and/or shock-resistant mounts are required.

- *Distractions*   Work environments such as a hospital emergency room or a financial trading floor are filled with visual and auditory distractions. User interfaces and data displays have to capture the attention of users and present information in a way that survives disruption.

- *Exposure to Hazards*   Will the device face environmental hazards such as toxic chemicals, exposure to germs, food spills, liquids, or other damaging materials? Will the device have to be cleaned, decontaminated, or sterilized?

- *Visual Appeal*   Solutions that are used in board rooms, in front of important clients, or in industries where image is especially important need to take visual appeal into account as part of their design.

- *Mobility*   Wireless solutions installed on moving vehicle such as cars, ships, and airplanes are exposed to varying environmental conditions. For instance, a car may drive through a tunnel that cuts off access to wide area wireless connections. Solutions built for highly mobile environments need to be very tolerant of interruptions.

4.9 Summary

Is a wireless solution right for your business requirements? This chapter has led you through a structured process to identify suitable candidates and evaluate their technical and business feasibility. Selecting the right wireless solution for your business problem does not have to be a daunting challenge. More so than for other IT application efforts, the business requirements of the target opportunity will drive many of the design decisions. Until more prepackaged solutions appear, most wireless applications of any sophistication will be custom-built and use hardware and software components from a variety of sources. These components will be chosen by a combination of technology compatibility and fit to the specific needs of the business problem they are solving.

To capture the solution's business requirements, this chapter offers the Five W's questions. This approach provides a business analysis method tailored to the unique needs of wireless solution design. Understanding why, who, what, when, and where questions about the solution's users and business requirements helps you to focus quickly on the narrow range of options that meet your needs. These business requirements are turned into technical solution requirements following the methods described in Chapter 5. Use the questionnaires included in Appendix A to support your requirements analysis efforts.

Defining a Solution

Turning a set of business requirements into a successful wireless solution is an exhilarating and challenging assignment. The technology is new and exciting, the results are very visible, and if the project is based on a strong business case, the impact on your company will be high. The most obvious parts of the exercise, such as selecting the type of wireless devices that will be deployed, have a "toy factor" appeal, making them appear fun and relatively straightforward. As you delve into the nuances of the selection, however, arriving at the right decision no longer seems so simple. The myriad options, issues, and considerations appear to grow exponentially. Worse yet, you will find that many of your desired choices are incompatible or require painful tradeoffs of functionality or features. From networks to devices to applications and implementation tools, there are simply too many complicated choices. Your seemingly straightforward solution has somehow turned into a jigsaw puzzle, composed of disparate pieces that must somehow fit into a complete picture. And, like a jigsaw puzzle, these pieces fit together only in a certain way. Put in one wrong piece and you won't be able to fit in the next right piece later. If approached in the wrong way, pulling together a complete wireless solution can be a daunting proposition even to experts.

Selecting the right components for a wireless solution requires navigating a complex and confusing maze of options and solution providers. The magnitude of capabilities, choices, and limitations of wireless components preclude the creation of a "one size fits all" wireless solution applicable to any business requirement. An architecture that works perfectly for one solution will be hopelessly inadequate for another. For example, the wireless solution used by American Airlines to track freight carts and dollies at the airports is vastly different from the one used by Fidelity Investments to allow investors to monitor stock prices and make trades.

Fortunately, a number of techniques can greatly simplify the conversion of business requirements into a well-defined wireless solution. Perhaps the most important trick is to reduce your range of options before becoming mired in the details of solution definition. Assembling a workable solution in a reasonable period of time is almost impossible if you must consider every potential device, network, application, and implementation option. The nature and constraints of your business requirements can be used to your advantage, however, to eliminate many of these options before you start, greatly reducing your research and evaluation efforts. For example, let's assume that we operate a food delivery service throughout the New England area and wish to provide our drivers with directions to drop-off locations and capabilities to accept on-site credit card payments for deliveries. As shown in Figure 5.1, these constraints significantly reduce our pool of options. We need only consider network options that support moderate data volumes from providers offering full coverage across New England. Short- and medium-range network options such as infrared, Bluetooth, and 802.11b don't apply, and the data volume requirements eliminate satellite networks. Since coverage varies by carrier, we'll have to pick a network service provider who covers New England. Similarly, our choice of devices is limited to those that can handle outdoor conditions, support credit card scanning, and work with our selected network. The implications of these decisions set parameters and refine options for other aspects of our solution. We have to choose (or develop) software that operates on the selected device; we need strong security to protect credit card information; and our training and support processes must be designed to fit this solution.

This chapter describes the process for turning business requirements into solution requirements. It uses the answers to the Why, Who, What, When, and Where questions from Chapter 4 to provide a framework for winnowing your wireless decisions into a manageable number. As shown in the example above, each business requirement imposes needs and constraints that create specific technical and operational requirements for the major components of our wireless solution. This chapter will explain how to develop specific requirements for devices, applications, data, and wireless networks. Comparing these requirements against the tables and other component-specific information in the second half of this book will enable you to quickly identify the wireless options that best apply to your needs.

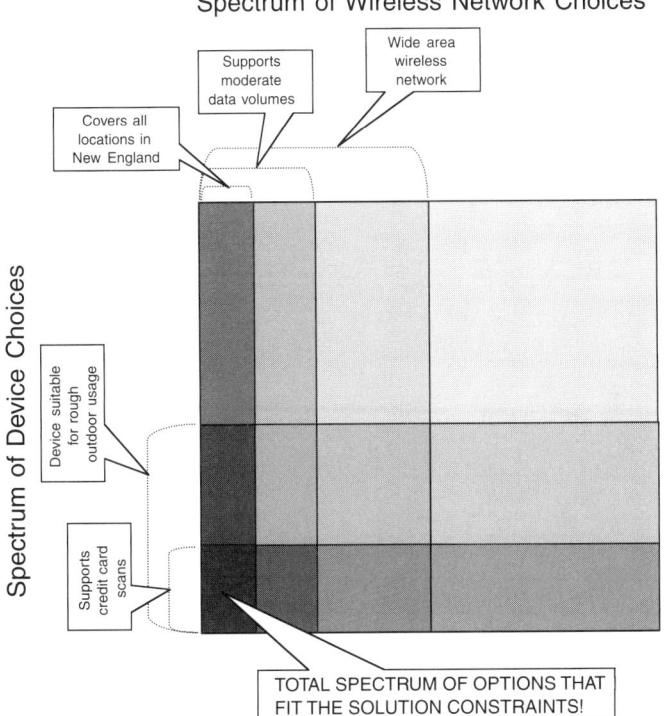

Spectrum of Wireless Network Choices

FIGURE 5.1
Shrinking the Solution Spectrum

5.1 Wireless Building Blocks _____

Before jumping into the mechanics of solution development, it is worth reviewing the basic building blocks that compose a complete wireless solution. While wireless solutions vary widely in characteristics, they all draw items from four categories of architectural components: client devices, wireless applications, information infrastructure, and wireless networks. These components are shown in Figure 5.2.

Client devices are the most visible component of a wireless solution. They are the physical platform for wireless applications and provide services such as voice communications, data capture and display, information processing, and location detection. These devices may be carried by users, mounted within shipping containers, or installed inside a car. Client devices include smart phones, pagers, PDAs, e-mail appliances, and special-purpose units for scanning, bar coding, and credit card reading.

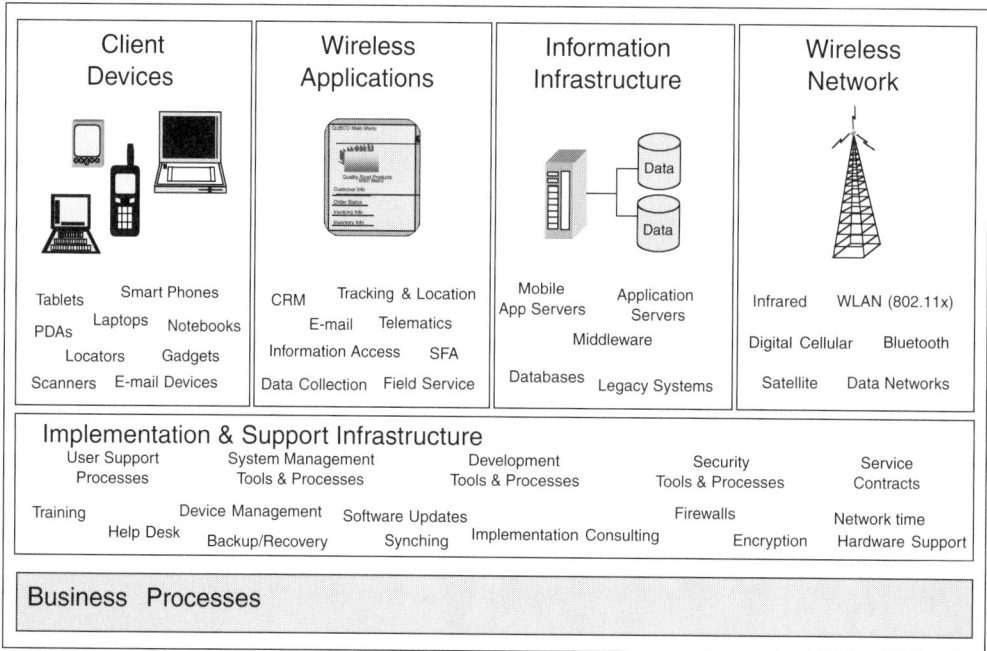

FIGURE 5.2
The Components of a Complete Wireless Solution

Wireless applications supply the business functionality behind the wireless solution. They can cover any need from personal productivity to safety and asset monitoring. Depending on the functionality required, these applications may be "off-the-shelf" packages, custom developed, or "re-purposed" from existing web applications.

The **information infrastructure** is the repository of knowledge incorporated within the wireless solution. Although these data components are invisible to most users, access to information is the "raison d'être" for most wireless solutions. This information may be environmental data captured on an oil rig for display at a monitoring station or it may be an amalgam of customer information drawn from a variety of corporate information systems and databases. The information infrastructure consists of the back-end applications, databases, voice systems, e-mail systems, middleware, and other components needed to support the information requirements of the chosen wireless application.

Wireless networks serve as the conduit, or transport mechanism, between devices or between devices and traditional wired networks (corporate networks, the Internet, etc.). These networks vary widely in cost, coverage, and transmission rates; they include options such as infrared, Bluetooth, WLAN, digital cellular, and satellite.

Together, these four components constitute the wireless solution's architecture. In the simplest case, this architecture consists of a single device type, using a single application and connected to a single network. However, many business solutions will be more complex, supporting multiple client devices, offering a variety of applications, and stitching together multiple networks to gain the desired level of coverage.

The solution's **Implementation and Support Infrastructure** provides the processes, tools, and resources used to create, operate, and support the wireless solution. This infrastructure ensures that users are trained, data is backed up, secured and synchronized, system and application software is kept up-to-date, devices remain functional, and networks operate efficiently. Although not part of the wireless architecture, the quality of this infrastructure is crucial for the success of the overall solution. As such, it merits as much consideration as the other wireless components when designing the solution.

Business Processes form the final component of a complete wireless solution. These are the processes that inspired the solution in the first place. Depending on the goals of the project, the wireless solution should enable your company to perform these processes faster, cheaper, and more efficiently than before. Gaining these benefits, however, requires redesigning and implementing new versions of processes that take advantage of the wireless solution. To capture the benefits of immediate, on-site invoicing offered by the field service example in Chapter 2, a company needs to change processes and job responsibilities in the customer service, field service, and billing organizations. Without these changes, work orders will still be entered manually in the company's systems by customer service, invoices will still be produced by the billing department, and the wireless device will simply end up as a new toy in the hands of the field service worker. While they are an integral part of a successful solution, business processes are usually outside the scope of responsibility of the technical team implementing and supporting the wireless solution. Implementing new business processes is its own project and requires knowledgeable resources backed by management commitment to the change.

5.2 Wireless Decision Process

Since wireless solutions depend greatly on the specifics of the problem they are solving, the process of selecting the right wireless solution is greatly simplified by focusing on the business requirements first. A given business requirement, such as the ability to deliver work orders to a user who spends many hours away from convenient power sources each day, imposes some critical technical constraints that have major implications for the entire solution design. The need for long battery life drives the selection of the client device. This selection may put constraints on the size and type of display and network connectivity options, which, in turn, affect the quantity of work order data transmitted, and its formatting and presentation. The resulting wireless architecture has implications for security, support processes, development tools, and service contracts with network and software providers.

This basic flow is true for every type of wireless solution, enabling us to offer a structured, top-down decision approach for turning business requirements into a design for a complete wireless solution. This approach proceeds in set levels from business requirements down to implementation strategy. It quickly narrows technology choices to a manageable number by focusing only on choices that are compatible with previous decisions. At each level, a set of choices is made and our breadth of options shrinks. If we follow the approach through to its conclusion, we have proven that it is conceptually possible to address our business requirement with a wireless solution and we have identified a "straw man" candidate for detailed analysis and feasibility assessment.

The diagram in Figure 5.3 illustrates the Wireless Decision Process. It uses a combination of business requirements gathered through the "Why, Who, What, When, and Where" questions from Chapter 4 and management considerations to drive require-

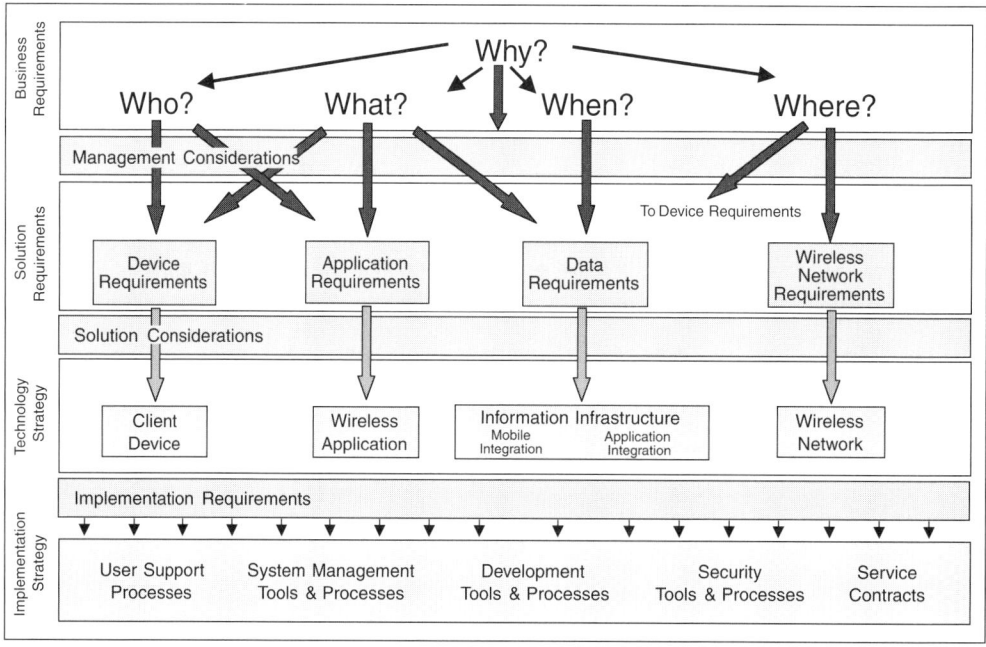

FIGURE 5.3
The Wireless Decision Process

ments for devices, applications, data, and wireless networks. When combined with technical considerations, such as the ability to secure the company's network, the solution requirements allow us to pick a specific set of components for our wireless architecture. The resulting technical strategy drives the implementation requirements for our solution, which defines the tools, processes, and resources for our Implementation and Support Infrastructure.

In practice, this process is not purely linear. Choices within each level interact. For example, our desired network option might not neatly match our application data requirements or our preferred client devices. To address these issues, we must cycle between levels, resolving an incompatibility by moving up in the hierarchy and rethinking our approach within our now known and understood technology constraints. We continue cycling in this manner until we have a solution that balances our desired capabilities with the available technologies. In the unlikely case that compromise is not possible, or our business requirements drive incompatible and irresolvable technology strategies, wireless technology is not a feasible solution for our problem.

The sections that follow provide an overview of the activities and deliverables within each layer of the Wireless Decision Process.

5.2.1 Business Requirements

This layer identifies the characteristics the wireless solution must have to successfully address its desired business objectives. It defines the goals of the project (**why**), identifies **who** is going to use the solution, **what** functions it must perform, **when** its information must be available, and **where** the solution will be deployed. Answers to these questions are business focused, leaving the technical details for the subsequent layers. For example, we need a solution for our field service representatives that will work at any client site. It must have access to information in our corporate billing application. We need to collect time spent, actions taken, and parts used. We need to be able to print an invoice on the spot, and so on. To implement this layer, gain a thorough understanding of the business objective being addressed and conduct a detailed analysis of the environment in which the solution will operate. Use standard business analysis techniques, such as interviews, assessments of current and desired processes and tools, and documentation reviews, guided by an assessment questionnaire that focuses on the nuances of a wireless and mobile solution. The final deliverable of this phase is a formal, documented set of business requirements. This document is the starting point for developing solution requirements, supporting cost justification efforts, and soliciting proposals from external solution providers.

5.2.2 Management Considerations

Wireless and mobile solutions raise a host of issues with implications for corporate policies, standards, and operations. For instance, allowing workers to perform important tasks at off-site locations may have audit impacts, affect union contracts, or expose the company to a new set of legal issues. Remedies to these issues, although not necessarily part of the original business requirements, will certainly affect the design and implementation of the final wireless solution. They also have ramifications on the supporting business processes. For example, the wireless solution may require a new set of audit policies and procedures.

Management considerations permeate all levels of the Wireless Decision Process, down to the details of the project's implementation and support infrastructure. Begin implementing this layer by reviewing the project's business requirements with company executives and appropriate individuals within the corporate legal, audit, human resources, and operational areas. Treat management considerations as a filter through which project requirements and design documents must pass. For more information on management considerations refer to Chapter 8.

5.2.3 Solution Requirements

The solution requirement layer translates the business requirements into technical requirements that will be used to select specific items within the four components of the wireless architecture. For example, a business requirement to provide a user with extended access to the wireless application in remote areas creates a technical requirement for long battery life. Many technical requirements flow directly from the answers to the business requirement questions. **Who** will use the solution sets requirements for the choice of device(s) and the design, function, and user interface of the application(s). Answers to the **What** question drives requirements for application functionality, data sources and access, and impacts device capability requirements. Major technical requirements are generally obvious and quickly determined, for instance, the mobility and location of an emergency room physician immediately rule out options such as laptops and satellite access. More detailed requirements will necessitate research into available options. Depending on the size and complexity of the wireless solution, this layer can be performed by an individual or a small team of solution designers. The major task is to translate business requirements into the parameters used to select specific components. For example, device selection parameters include size, weight, battery life, and display characteristics. The deliverable of this layer is a formal, documented set of technical requirements for the wireless solution.

5.2.4 Solution Considerations

Solution considerations are similar to the management considerations described above, but are technically oriented. Wireless solutions face a host of externally imposed technical constraints and design considerations. Unless the wireless solution is standalone, it has to work within the company's defined technical strategy and standards. For example, your company may prefer to standardize on a single vendor's products wherever possible or it may require a product "agnostic" approach that trades unique features for the ability to easily switch vendors and products at a future date. Other design considerations involve issues such as security, transaction integrity, and development environment standards. Like management considerations, solution considerations are a filter affecting all subsequent levels of the Wireless Decision Process. Implement this layer by reviewing the project's technical requirements with experts within your IT organization. These experts include IT strategy, security, database, network, and standards specialists. If the proposed solution requires integration with corporate applications, appropriate application specialists should be consulted. For more information on solution considerations, refer to Chapter 9.

5.2.5 Technology Strategy

The technology strategy layer produces the wireless architecture design. Using the technical requirements from above, the solution designer evaluates options and selects the appropriate items in each of the four wireless architecture categories. For instance, based on the solution's device requirements, the designer may recommend the use of a Palm or Pocket PC PDA as the preferred client device. Steps within this selection process can include researching option specifications, performing product evaluations, and visiting corporate users of the desired technology. While the decisions made in each architecture category flow directly from their respective requirements, choices must be compatible with each other. Incompatibilities can be resolved by changing one or both competing selections or by devising an appropriate workaround to be incorporated during implementation.

5.2.6 Implementation Requirements

This layer provides the requirements for the Implementation and Support Infrastructure component of the wireless solution. The characteristics of the wireless architecture automatically impose many requirements on the solution's implementation. If the architecture uses a digital cellular network, that portion of the implementation involves negotiating a service contract with a cellular provider. Conversely, an infrared network may be created from components purchased and installed by the solution's users. Selecting a particular wireless device means buying into that device's operating system, which in turn, drives the development tools used to create applications for that device. The choice of a particular wireless consulting firm as a partner may be based on its experience with the chosen technology. While the wireless architecture is the source of many implementation requirements, the other layers of the Wireless Decision Process provide additional input. If the wireless solution needs new audit procedures, those procedures must be designed and deployed as part of the implementation. The solution designer gathers implementation requirements at each step of the Wireless Decision Process and consolidates this information after completing the wireless architecture design.

5.2.7 Implementation Strategy

The final layer of the Wireless Decision Process turns the implementation requirements into an implementation strategy and project plan. The implementation strategy identifies all of the people, processes, and tools needed to implement and operate the wireless solution. For instance, the strategy might opt to perform development work with internal company resources or to hire a consulting partner. In addition to addressing the construction and roll-out the wireless solution, the strategy also covers training, user support processes, system management tools and processes, and the myriad of details needed to support the production use of the wireless solution. The level of effort for creating the implementation plan is proportional to the scale and complexity of the wireless solution. A simple solution, such as providing RIM Black-Berry devices to executives for e-mail access, will have a relatively simple implementation, although it will still involve training, support, and other roll-out and operational issues. In contrast, a large-scale solution, such as supporting UPS delivery drivers throughout the U.S., is a major endeavor and implementing each solution component can be a significant project requiring its own plan.

5.3 From Business Requirements to Technical Requirements

This section describes how to implement the Solution Requirements layer of the Wireless Decision Process. It offers guidance for translating the business requirements captured through the 5 W's approach from Chapter 4 into a set of technical requirements to facilitate component selection and architecture design. This process is primarily an exercise of mapping solution needs, described in business terms,

The building inspector's lighting needs

into technical parameters by wireless component category. For example, evaluating the physical environment where a building inspector performs his or her job identifies the ability to operate under poor lighting conditions as an important business requirement. Within the wireless architecture, lighting conditions are addressed by the characteristics of client device's display. The building inspector's lighting needs translate into a requirement for a backlit display within the Device Requirements section.

The table in Figure 5.4 shows the mapping of the 5 W's questions into the major selection criteria for the four wireless architecture categories.

The subsections below help you to identify quickly the major technical requirements for each wireless architecture category. Each subsection has a short "cheat sheet" questionnaire to assist you in setting high-level selection parameters. Use

Business Requirements	Technical Requirements			
	Devices	Applications	Data	Network
Why	Cost Extensibility 3rd Party Availability	Approach Source		Cost
Who	Voice Size/Weight/Portability Display Data Input Durability Battery Life Built-in Capabilities Internet Access E-mail Access	Functionality Device/Network Support Strategy User Experience Interface Security	Immediacy Format Security	Coverage Interoperability
What	Voice Display Data Input Processing Power Memory Battery Life Built-in Capabilities Bundled Software	Functionality Approach Source User Experience Interface Security	Data Source Type of Data Access Volume Format Latency Data Integrity Synching Requirements Security	Bandwidth Reliability Security
When			Latency Immediacy Synching Requirements	Latency
Where	Display Data Input Durability Battery Life	Interface Security		Coverage Interoperability

FIGURE 5.4
Five W's Table

these parameters to hone in on the component options that best fit your needs. While more thorough and detailed research is needed to create formal technical requirements documentation, use the "cheat sheet" approach as a shortcut for moving from requirements to design in Section 5.4.

5.3.1 Devices

The two biggest influences on device selection are the solution's users (**Who**) and the functions or applications (**What**) the solution will provide. Other important factors are the environments (**Where**) where the devices will be used and business considerations (**Why**) such as cost and extensibility. The relative importance of these business requirements to device selection is shown in Figure 5.5.

The characteristics of the application(s) will largely determine the device, since the device must have features to facilitate the operation of the application. If the application is highly specific or closely tied to a vertical process, then a special-purpose device may be needed. An inventory management application may call for a bar code scanning device, while a parts replenishment application may call for telemetry equipment. If the application is aimed at improving mobile worker productivity, then a more versatile, general-purpose device will likely suffice, and factors like user preferences, device portability, extensibility, and cost will greatly rule in or rule out certain device types. Since devices are the most personal component of your wireless solution, your future end users are likely to have strong

Device Requirements	Business Requirements			
	Why	Who	What	Where
	Management Considerations	User	Application	Environment
Voice		Primary	Primary	
Size/Weight/Portability		Primary		
Display		Primary	Tertiary	Secondary
Data Input		Primary	Secondary	Tertiary
Processing Power			Primary	
Memory			Primary	
Durability		Secondary		Primary
Battery Life		Secondary	Tertiary	Primary
Built-in Capabilities		Primary	Secondary	
Extensibility	Primary			
Bundled Software		Secondary	Primary	
Availability of 3rd Party Software	Primary			
PIM		Primary		
Internet Capabilities		Primary		
E-mail Capabilities		Primary		
Cost	Primary			

FIGURE 5.5

Device Requirements Table

opinions about which devices they are willing to use. For example, executives may not have the patience or inclination to work with a multi-featured device. Repair technicians that will simultaneously operate other equipment may favor a device that can be used with a single hand. Salespeople may avoid stylus-oriented devices for fear of losing the stylus while traveling. In some cases, such as consumer applications or sales applications, your solution may have to support the devices already owned by your target audience. Try to develop a profile of your target audience so that a device can be selected that meets the group's rather than individual's needs.

The major criteria for defining device requirements are shown in Figure 5.6, the Device Selection Cheat Sheet. Refer to Chapter 10 for more detailed descriptions of each criteria.

When selecting requirements, remember that some device features are more important than others. For example, a color display may be a nice feature but add nothing to an order status application, and even be detrimental to battery life as compared to a monochrome unit. An application that relies primarily on voice processing may call for a device optimized for voice, such as a smart phone or communicator. In general, a laptop device will offer the richest functionality of all the mobile devices. When it comes to display size, data input mechanisms, processing power, memory, extensibility, wireless communications, and availability of third-party software, a laptop can't be beat. Laptops, however, may fail to satisfy user preferences. The large size and weight of a laptop, the long time to power on, and the complicated software and hardware environments lead many users to select a more portable, faster, and simpler device.

1. **Voice**   Will the device need to support voice communication instead of, or in addition to, a data application? As described in Chapter 10, wireless device features are starting to merge, with PDAs supporting voice processing and telephones supporting data displays and Internet access. Additional capabilities are required if the solution calls for other voice processing such as recording or speech recognition.

2. **Size/Weight/Portability**   These requirements—size, weight, and portability—are largely influenced by how the user expects to carry or wear the device. While some types of devices, like laptops, may be disfavored because of their size, other devices like PDAs may fit a wider variety of needs. In general, if a device is burdensome in terms of size, weight, or ability to be stowed, or limits the performance of other activities (e.g., requires two-handed use or cables), then a user will tend not to carry or wear the device.

3. **Display**   Our natural inclination is to select the largest and most visually appealing display possible. Selecting the right device display, however, is likely

Device Selection Cheat Sheet

Voice

| None | Communications (telephone) | Voice processing (recording, speech recognition, etc.) |

Size/Weight/Portability

| In a hand | In a pocket or on a belt | In a briefcase | In a vehicle | Not an issue |

Display

Display Size

| 1 | *Single line of text* | 2 | 3 | 4 | 5 | *Full screen display* |

Resolution/Color

| 1 | *Text* | 2 | 3 | 4 | 5 | *Color multi-media* |

Performance in variable lighting

| 1 | *Office lighting* | 2 | 3 | 4 | 5 | *Variable outdoor lighting* |

Data Input

Manual
    Large volume - keyboard
    Small volume *(circle options)* Keyboard Keypad Touch screen Handwriting recognition
Automated data capture

Processing Power

| Low | Medium | High |

Memory

| Low | Medium | High |

Durability

| 1 | *Office use only* | 2 | 3 | 4 | 5 | *Rough outdoors use* |

Battery Life - *Importance of long life*

| 1 | *Not important* | 2 | 3 | 4 | 5 | *Essential* |

Built-in Capabilities

Wireless communications (check all that apply)
    WLAN
    Bluetooth
    Infrared
    Wide Area Networks
Special features
    Bar code scanning
    Credit card processing
    Voice recording
    Other *(list)*_____

Extensibility

| 1 | *Not important* | 2 | 3 | 4 | 5 | *Essential* |

FIGURE 5.6   *Continued*
The Device Selection Cheat Sheet

to involve a compromise between display capabilities, portability, power requirements, and cost considerations. For example, a color display may be irrelevant to an order status application, and even be detrimental to battery life as compared to a monochrome unit. The best approach is to define the minimum requirements for the planned solution and to upgrade from those requirements when feasible during device selection. Three major criteria help to define

Device Selection Cheat Sheet   **Page: 2**

Bundled Software		
Not important	Nice-to-have	Specific Needs *(list)*

Availability of Third-Party Software		
Not important	Important as proof point	Specific Needs *(list)*

PIM		
Not important	Nice-to-have	Essential

Internet Access		
Not important	Limited access useful	Essential

E-mail Access		
Not important	Limited access useful	Essential

Cost *(Note approximate limits)*		
Not an issue	Important	Major concern

Approximate number of units required_____

Maximum acceptable cost per unit_____

Cost parameters (list)_____

FIGURE 5.6 — *Continued*
The Device Selection Cheat Sheet

device requirements: display size, display quality, and performance in variable lighting conditions. The volume of information that must be displayed at one time determines display size. At one extreme, a pager may display one line of text. At the other extreme, a laptop provides a full screen display. The quality of a display includes its resolution and color capabilities. Simple text can be displayed monochromatically in low resolution, while high resolution is necessary for graphics or video, or to display larger volumes of information on a small display. Variable lighting conditions require more flexible displays, such as a backlit display for dimly lit locations.

4. **Data Input**   How will the user interact with the device? How will information be entered into the device? How much data will be collected? Data collection may be automated through the use of scanners, readers, or sensors. In most cases, however, the user will interact with the device through a keyboard, touchscreen, or other data entry mechanism. The choice of mechanism depends on the volume of data being entered and the working style and personal preferences of the user. A full keyboard may be required to enter a large volume of text, but an inspector checking items on a form may work best with a stylus. A small, thumb-typing keyboard may be perfect for short e-mails, but difficult to use for someone wearing protective gloves. A user sitting at a desk can use a foldout keyboard, but a trader on the stock market floor needs a portable mechanism that can be used while standing.

5. **Processing Power**   Processing power is determined by the needs of the application(s) that will operate on the device. If the device merely displays data from other sources, processing requirements are low. Conversely, if the device has to support multiple concurrent applications or heavy calculations, it will need greater processing power. Limitations in processing power can sometimes be addressed through careful partitioning of functionality between device and server, a topic discussed in Chapter 12.

6. **Memory**   Like processing power, memory is determined by the needs of the application(s) that will operate on the device. Running a single, simple application takes far less memory than supporting multiple concurrent applications. Multimedia applications, for example, will require more memory than a standard word processing application. The ability to add external memory to a device can help ease storage demands, but will not alleviate a persistent lack of memory.

7. **Durability**   Mobile devices are often subject to adverse conditions and abuse. A stationary device within a controlled office environment faces the fewest durability issues, while a device used outdoors on a construction site must be rugged to survive. A device used in dusty or damp conditions better have good sealing. If the device is apt to be dropped, and most will be, can it withstand the impact? Will it be dropped on carpet (in an office), or on concrete (outside or in a warehouse)?

8. **Battery Life**   Battery life is not an issue for devices installed in offices or on vehicles with easy access to power, but becomes an important issue for mobile devices used away from convenient power sources. Factors that affect battery life are expected level of use (occasional versus constant use), power draw by the device and application, availability of recharge power sources, length of

downtime needed to recharge, and the convenience of carrying and inserting spare batteries. Figuring out power needs before a device and application are used in real versus simulated conditions is guesswork. If loss of battery power could cause significant damage or problems, however, it pays to consider battery options and mitigation strategies upfront.

9. **Built-In Capabilities**   An important tradeoff when selecting a wireless device is whether to have as many of the necessary capabilities already integrated in the chosen device or opt for an extensible device amenable to add-ons. Fully integrated devices tend to be bulkier, heavier, and more power-hungry than extensible general-purpose devices, affecting portability to a degree, but eliminating the inconvenience of carrying a device with separate attachments. When carrying and accounting for multiple pieces of equipment is problematic, simpler is better. For example, a package delivery person collecting signatures at each drop-off location, or a lineman climbing a telephone pole, won't want to carry any more components than necessary. Conversely, an executive user may prefer a smaller device, and be willing to keep extra components in a briefcase until needed. Consider whether the additional capabilities will be used constantly, such as a network connection, or only occasionally such as a foldout keyboard for an executive's PDA. If a wireless application requires a highly specialized set of features, the best approach may be a custom-built device that integrates all of the features. In addition to network support, special features that could be incorporated in an integrated unit include bar code scanning, credit card processing, and voice recording. One disadvantage of a highly integrated device is the need to scrap and replace it when capability requirements change over time.

10. **Extensibility**   Extensibility is the flip side of built-in capabilities. It assesses the ability and desire to add and attach new features to the selected device. Extensibility is a means of getting around constraints, such as adding a foldout keyboard to enable higher volume data entry into a PDA. Add-on features include flash memory for back-ups and data transfer, plug-ins for voice and games, modems and cards for network communications, and other data entry mechanisms such as scanners and credit card readers. Special-purpose devices such as e-mail appliances tend to be less extensible, whereas PDAs and laptops are more extensible. Extensibility adds to the lifespan of the device and solution by allowing it to evolve to meet new requirements. On the other hand, users may resist carrying separate add-ons for commonly used capabilities.

11. **Bundled Software**   Devices, such as PDAs, are provided with a set of bundled software. This software includes the operating system and applications such as a spreadsheet and word processor. In single-purpose solutions, such as e-mail on an e-mail appliance, the particulars of the bundled software are unimportant to the solution. In other cases, the availability of a particular bundled application,

such as word processing, may be an attractive feature that can differentiate between two otherwise close options. For other applications, the characteristics of bundled software are critical. For instance, the choice of application or middleware may necessitate the use of a particular operating system. If you plan on developing custom applications, your developers may come up with their own specific requirements for the device.

12. **Availability of Third-Party Software** Depending on the business problem being addressed, availability of third-party software for a given device may be the number one factor driving selection or may be totally unimportant. Third-party software can provide base application functionality, add-on applications, and/or development and support tools tailored to the specific platform. Some ERP, CRM, and SFA software vendors offer mobile extensions to their solutions for certain wireless devices. The availability of third-party software for a given platform can be an important proof point of its level of adoption within the marketplace. If the business requirements are sufficiently unique or leading edge to require a custom programmed solution, availability of third-party software may be unimportant.

13. **Personal Information Management (PIM)** PIM tools such as contacts, calendars, task lists, and alarms were the original applications provided on PDAs. These tools are increasingly offered on a variety of wireless devices. They may be important and attractive to sales and executive solutions, but less important or unnecessary for many special-purpose applications. If PIM features are desirable, it is important to ensure that the wireless versions and office versions are compatible.

14. **Internet Access** The ability to access and surf the Internet through a mobile device is not a requirement for many applications. In other cases, such as for executives and salespeople, it can be a "nice to have" feature that is used occasionally. Given their smaller display size and slow data transfer rates, devices such as WAP telephones offer access only to specially enabled web sites. For the mobile professional, Internet access may still be useful for information such as weather reports and stock quotes. If full Internet access is a necessity, a laptop and access to a high-speed network become requirements.

15. **E-mail Access** E-mail is the killer application for many wireless solutions. Devices vary widely in the quality and ease-of-use of their e-mail support. If e-mail is an essential application, it is important to evaluate how well the device can send/receive, display, and manage e-mail messages and attachments. If limited access to short text messages is sufficient, a wider range of devices can be considered. While e-mail access may be essential to an executive solution, it may be entirely unnecessary for a device used on a manufacturing shop floor.

16. **Cost**   If you are providing a wireless service to end users who already own devices, device costs are irrelevant. Similarly, the cost of supplying an executive with a PDA is insignificant compared to the value of fast response to business issues. Price sensitivity may be an issue, however, for users, such as students, that must purchase their own devices. Also, although wireless devices can be relatively cheap compared to desktop workstations, total device costs will mount quickly if there are thousands of expected users.

5.3.2 Wireless Applications

The requirements for wireless applications are driven primarily by two questions: **What** functions will the applications have to provide, and **Who** will use that functionality. The process of turning business requirements into functional specifications is fundamentally the same for wireless applications as it is for other business applications with a few nuances to cover differences in the wireless environment. Given the wide range of possible applications and the wealth of options for obtaining or creating those applications, a simple checklist approach is inadequate for specifying application requirements. Furthermore, choices affecting application functionality or design may drive or be driven by choices made in the other wireless architecture categories. Consequently, many important decisions affecting application selection, such as where to distribute functionality between the device, servers, and host systems, must be deferred until later stages of solution design. These issues and considerations are discussed in Chapters 9 and 12. Nevertheless, it is possible to offer some selection criteria to guide your considerations. As shown in Figure 5.7, these criteria are: functionality, approach, source, device/network support strategy, user experience, interface, host platform, and security.

1. **Functionality**   The business issues you intend to address define the functional requirements for your solution. The particular requirements for a telemetry application monitoring an ice-making machine, for example, are wildly different from those for a trading application for stockbrokers. Despite these differences, all wireless applications contain one or more of the following generic functions: communicate, access information, collect data, update, transact, monitor, and locate/track. Use these functions as a starting point for describing the high-level functional requirements of your solution. For example, the ice machine application must *monitor* conditions such as temperature within an ice freezer and *communicate* problems to the host organization. With this high-level definition in hand, you should be able to begin researching possible solution options. For instance, do any software vendors offer applications with those functions? As you progress further into your implementation, you will need to flesh out

Wireless Application Cheat Sheet

Functionality (check all that apply)

- ❏ Communicate
- ❏ Transact
- ❏ Monitor
- ❏ Locate/Track
- ❏ Access Information
- ❏ Collect Data
- ❏ Update

Approach

- ❏ Build
- ❏ Extend

Source

- ❏ Established (List)_____
- ❏ Buy Package
- ❏ Need to Research
- ❏ Custom Build

Device/Network Support Strategy

- ❏ Agnostic
- ❏ Native

User Experience

Navigation (circle appropriate level)

1 Simple   2   3   4   5 Power user

Level of Interaction (circle appropriate level)

1 "Hold my hand"   2   3 "I'll tell you"   4   5 Automatic

Importance of Visual Presentation (circle appropriate level)

1 Not important   2   3   4   5 Critical

Interface

- ❏ Languages Supported? (List)_____
- ❏ Voice Recognition?
- Support User Preferences
  - ❏ Not necessary   ❏ Important   ❏ Critical

Host Platform (list characteristics)

- Hardware platforms_____
- Operating systems_____
- System Software (security, network, etc.)_____
- Databases_____

Security

Authentication
- ❏ Not necessary   ❏ Important   ❏ Critical

Authorization
- ❏ Not necessary   ❏ Important   ❏ Critical

FIGURE 5.7

The Wireless Application Cheat Sheet

this high-level definition with more detail. To create those specifications, you can follow the IT methodology of your choice.

2. **Approach**   Will you need a new application to provide the desired functionality or can you extend an existing one? If your goal is to provide office functionality to field workers, adding a wireless channel to an existing application may be an option. Where feasible, extending an existing application can often be the

quickest and least costly method of gaining desired functionality. If a new application is required, you will have to build or buy it as described below.

3. **Source** How and where are you going to obtain your desired application functionality? If your functional requirements are new or unique, building a custom application is the only option. To do so, you will need to invest in tools and expertise to develop and support the application, or turn to an experienced third party to do the work. Over time, as wireless adoption increases and more robust and varied packaged solutions appear, the need to create custom applications will wane. If your functional requirements are more basic or generic—providing wireless e-mail access, for example—it is likely that a package or other "off-the-shelf" software will meet your needs. You may be able to use this software "as is" or may need to do some tweaking to get it to perform to your exact requirements. Finally, some of the enterprise applications that your company has already implemented will have their own wireless interfaces, and gaining wireless functionality is as simple as enabling some pre-existing code.

4. **Device/Network Support Strategy** Depending on your strategy, you may choose to tailor your solution to work with a single device or specific network architecture (the "native" approach) or opt to create a generic solution capable of working with multiple device and network types (the "agnostic" approach). By focusing on a single device or network, the native approach can take full advantage of the unique capabilities offered by those components. In contrast, an agnostic approach takes a compromise path, not exploiting the unique capabilities of a single device or network type, but seeking to work on as wide a range of components as possible. An agnostic approach is the only choice if your application must support users, such as consumers, who may already be using a range of devices, and it offers the advantage of easily supporting changes in devices or networks. A native approach makes more sense when your solution is focused, such as providing salespeople with e-mail access via RIM BlackBerry devices, and you are able to standardize on a particular device and/or network.

5. **User Experience** Different types of users have different needs for interacting with the wireless application. Application design factors such as navigation, level of interaction, and visual presentation have to match the needs of the intended users to encourage adoption of the solution. Navigation, the steps required to find desired information, requires a tradeoff between ease/speed of use, and power and flexibility. A harried emergency room doctor will not use a solution that requires many layers of slow navigation to find the desired information, while a power user seated in an airline lounge is willing to trade simplicity for a full range of features. Similarly, an occasional user of the application, such as a consumer making a wireless purchase, will not recall

application features and functions between sessions, and will need considerable handholding to complete the transaction. A stocktrader who constantly uses the application will want more control over his interactions with as little interference and overhead as possible. A person performing highly repetitive tasks will prefer a high level of automation. For example, a package delivery person will not want to enter a series of commands to upload delivery data every time he returns to his truck. Visual presentation requirements also vary. To convey one-line alerts on changes in commodity prices, plain text will suffice. Fuller visual presentation is needed to support more complex information demands, such as using color to highlight differences in order status or giving a demo to a customer. Presentation becomes critically important if a wireless application will deliver highly visual information such as x-rays or schematics.

6. **Interface**   Most wireless applications will depend on a visual interface for interaction. In some cases, it may be advantageous to supplement that interface with speech recognition, to permit, for example, a user to interact with the application while driving. Multiple language support will be important if the application is used in more than one country or targets multi-lingual populations. If a diverse set of users will use the application, the ability to tailor the application to individual working styles through user preferences becomes important.

7. **Host Platform**   The characteristics of your company's host platform are important selection criteria for wireless applications and middleware that must integrate with that platform. This criterion is not important for standalone wireless solutions, such as a wireless interface to an existing enterprise package. If you choose to purchase a package, it must be compatible with your existing platform or you will have to change the platform to support the application. Similarly, if you are developing a custom application, platform characteristics may affect your choice of development tools.

8. **Security**   Security is an important consideration in all four wireless architecture components—applications, information architecture, devices, and networks. From an application perspective, security is enhanced through authentication (Are you who you say you are?) and authorization (Are you allowed access to a given function or piece of data?). Given the high rates of theft and loss among devices, application-level protection is vital if the device contains confidential or valuable information and/or can gain access to sensitive applications or data behind the corporate firewall. Application-level security is less of an issue for many telemetry and consumer applications, but is critical for applications that perform financial transactions or other sensitive functions.

5.3.3 Information Infrastructure

Information exchange is the underlying purpose of virtually every wireless solution. The data source, volume, and confidentiality requirements of a wireless application affect many aspects of the solution architecture, and implicate back-end application integration, data security, and network choices. The solution's information infrastructure supports the wireless application's information needs. For certain types of wireless applications, the information infrastructure is pre-supposed. Wireless Internet access or WLAN access takes advantage of pre-existing information infrastructures. Conversely, special-purpose wireless applications will likely need a separately defined architecture to handle information needs. For example, providing a field service worker with access to current customer status may require capturing and integrating information contained within various back-office customer and support databases.

The main drivers for information infrastructure requirements are the answers to the **What** and **When** categories of questions. In many ways, application functionality sets data requirements, especially from a source and content perspective. When a user needs access to data and how long that data remains useful define the mechanics of data transmission. **Who** is using the data and **where** it will be accessed affect considerations such as data formatting and display and security and back-up requirements. Like wireless applications, information infrastructure requirements are not easily captured through a simple checklist; however, the criteria in this section will help you develop a high-level understanding of your needs. As shown in Figure 5.8, the criteria in this section are: data sources, type of data access, volume, format, latency, immediacy, data integrity, synching requirements, and security. Refer to Chapter 12 for additional information on these topics.

1. **Data Sources**   This criteria applies to wireless applications that exchange data with corporate information systems and other repositories of information. A source may be a host database, file, a web site, or a wireless interface into a host application. Information from a single source may be delivered directly to the application, for example, a query sent to the wireless interface of a CRM package returns a response specifically formatted for the application, or it may have to be drawn from several sources and merged or processed before sending. In the latter case, the information infrastructure will have to contain the additional hardware and software needed to process the data. Processing may involve calculation, extraction, summarization, or reconciliation of disparate data to make it useful for the wireless application. As a starting point, list each potential data source needed for your application.

2. **Type of Data Access**   Data may flow in either direction between host and device. Data may be intended for viewing only, or for update (added or modified). A more

Information Infrastructure Cheat Sheet

Data Source

Single Source _____

Multiple Sources - Merge _____

Multiple Source - Process _____

Type of Data Access *(select for each data source)*

View Only - Host Data

Update - Host Data

View Only - Device Data

Update - Device Data

Volume *(select for each data source)*

Very Low *(Examples: Short Messages, Alerts)*

Low *(Examples: text e-mails, Internet clippings, simple forms)*

Moderate *(Examples: Transactions, complicated forms, e-mail attachments)*

High *(Examples: Large files, program uploads and downloads)*

Very High *(Examples: Video files, Videoconferencing)*

Format *(select for each data source)*

As is   Style Sheet   Extract   Needs Processing

Latency

Shelf Life of Data *(Estimate time)* _____

Immediacy

Device to host

Data must be sent immediately

Data can be sent periodically *(approximate timing)* _____

Host to device

Data must be sent immediately

Data can wait for user request

Data Integrity

Not Important   Important   Critical

Synching Requirements

One way - Device to Host   One way - Host to Device   Two way

Security

Importance *(check highest applicable need)*

Not Important

Low - *data has little value to outsiders*

Moderate - *sensitive data*

High - *highly confidential or valuable data*

Exposure *(check all appropriate areas of risk)*

Authorization

Transmission

Caching/Storage on Device

FIGURE 5.8

The Information Infrastructure Cheat Sheet

complex architecture is needed to support data modification than simple data viewing. For each data source selected above, identify the appropriate data access requirements. For instance, an inventory tracking application may capture bar code information on the device for uploading into the corporate inventory database. A wireless banking application will allow customers to transfer funds between accounts. These types of applications require the ability to update host data.

3. **Volume**   The volume of information that the application will exchange has implications for data display, storage, and transmission. Given bandwidth constraints, data volume is an important consideration for network selection, but also affects application design. Application users may not tolerate the time and effort required to process and display large volumes of data, requiring creative design techniques to extract and present only the data most relevant for their needs. Volume of data ranges from very low, such as alerts and short messages, to very high for videoconferencing.

4. **Format**   Data format affects the level of processing needed to get the information into a form useful for the wireless device. Simple text data, web pages formatted for mobile display or files compatible with a given wireless application (e.g., a word processing document) may be exchanged "as is" without further processing. Other information exchange formats, such as XML and its variants, use stylesheets to allow the same data to be presented in different ways on different devices. This approach is used to "re-purpose" web-based data for simultaneous use on a wireless device. Given the limited data capabilities of many wireless devices, relevant data may need to be extracted from larger sources. At the high end, a complex query, such as providing a summary of sales activity across five territories, may require considerable processing to return the data to the wireless device in an acceptable format.

5. **Latency**   Data often has a useful "shelf life" or latency before it loses its value. Latency affects how and where data is stored and how frequently it must be refreshed to maintain its value. Some data, such as a list of inflation figures for 1990 through 2000, will never change and has infinite latency. Perishable information, such as the price of an individual stock during trading hours, changes on a moment's notice and loses value quickly. In contrast, an e-mail message may retain its value while it sits in an inbox for several hours. Estimate the latency of the major information exchanged by your wireless solution.

6. **Immediacy**   Immediacy addresses how quickly a given unit of data must be exchanged between the device and server. Stock price movements may require instant dissemination while inspection data may require only periodic updating. Serving the stockbroker with price information requires real-time processing and an "always on" network connection. The inspection application requires only hourly synching between device and server. Immediacy requirements affect both sides of the wireless exchange. A server-resident application that tracks packages may expect immediate notification of delivery problems from the field, but only periodic uploads regarding successful deliveries. Some server-resident information, such as stock price fluctuations, may require immediate dissemination while other information, such as last week's closing price, can wait until the broker specifically requests it.

7. **Data Integrity**   Data integrity assesses the risk to the organization if a piece of data is corrupted or dropped while in use by the wireless application. When data problems occur, if the symptoms are obvious rather than subtle, and if the data can be quickly and easily recreated, then data integrity is not really an issue. For example, if a download of the latest news from a web site is interrupted, the transmission can be restarted without concern for integrity. Conversely, executing a wireless fund transfer requires careful measures to ensure that the data is correct and reliable, and that all aspects of the transaction complete successfully.

8. **Synching Requirements**   Some data, such as a contacts file, may be maintained in multiple locations. Ideally, data should be identical at all times, across all locations, but this goal is not always feasible or practical. Synchronization is the method used to match data sources on a periodic basis. The period between synchronizations is determined by the latency of the data being shared and practical considerations for exchanging data. Synchronization may proceed in three ways. Data resident on a wireless device can be periodically delivered to a server. For example, a package delivery person sends collected delivery data to headquarters every few hours as he passes by a wireless access point in his vehicle. Server-resident data can be pushed out to a wireless device, replacing the version of the data kept on the device. For example, an engineering firm could send updated schematics to its field service workers' devices once a week. Or device-resident and server-resident data can be compared, matched, and merged to create updated versions at both ends, a process requiring a more complex synchronization scheme to deal with data conflicts or inconsistencies.

9. **Security**   The value and confidentiality of a piece of data determines the level of protection needed. Issues include who has access to that data, how it is transmitted and how and where it is stored. Authorization—access to a given function or piece of data—can be controlled at the information architecture level. For example, a user may be able to access personnel records from his department, but not from other departments. Highly confidential data may need additional security layers such as a secure network and/or encryption when it is transmitted. Given the high rate of theft and loss among wireless devices, a solution may require the encryption of confidential data stored on a device, or it may prohibit storing data on a device altogether.

5.3.4 Wireless Network

Answers to the **Where** category of questions are the predominate driver of network selection, however, the other categories help set other important network requirements. There are seven major criteria for defining network requirements as shown in

Network Selection Cheat Sheet

Type *(check all applicable)*
- ❑ Voice
- ❑ Data

Coverage/Range Requirements *(check all applicable)*
- ❑ Close Proximity (under 3 feet)
- ❑ Within an Office
- ❑ Within a Building
- ❑ Campus or Office Complex
- ❑ Regional
- ❑ National - Urban
- ❑ National - Full Coverage
- ❑ International

Bandwidth Requirements *(check highest applicable need)*
- ❑ Very Low *(Examples: Short Messages, Alerts)*
- ❑ Low *(Examples: text e-mails, Internet clippings, simple forms)*
- ❑ Moderate *(Examples: Transactions, complicated forms, e-mail attachments)*
- ❑ High *(Examples: Large files, program uploads and downloads)*
- ❑ Very High *(Examples: Video files, Videoconferencing)*

Latency *(check applicable in each category)*

User to Host
- ❑ Access when possible
- ❑ Access when convenient
- ❑ Access on demand
- ❑ Always connected
- ❑ No fail, always connected

Host to User
- ❑ Reach when possible
- ❑ Reach on demand
- ❑ Always connected
- ❑ No fail, always connected

Reliability *(check highest applicable need)*
- ❑ Minor importance - Can start over
- ❑ Moderate - Must recover from point of interruption
- ❑ Cannot fail

Security *(check highest applicable need)*
- ❑ Not important
- ❑ Low - data has little value to outsiders
- ❑ Moderate - sensitive data
- ❑ High - highly confidential or valuable data

Interoperability *(check all applicable)*
- ❑ Not important - single platform; highly homogenous
- ❑ Important - several platforms; some divergence
- ❑ Highly Important - highly dispersed users; multiple platforms; highly heterogeneous

Cost *(Note approximate limits)*
Cost of installation tolerance_____
Cost of usage tolerance_____

FIGURE 5.9
The Network Selection Cheat Sheet

Figure 5.9: coverage, bandwidth, latency, reliability, security, interoperability, and cost. Refer to Chapter 11 for more detailed descriptions of each criteria.

1. **Coverage**   The level of coverage required to support your desired solution is the major determinant of your network options. Since your final wireless solution may involve more than one network type to gain the desired coverage, be

sure to indicate all applicable requirements on the checklist. Where people will use the application in the field and how far they may roam are the determining factors for coverage. For example, if the users are waiters at a restaurant, the coverage area is obviously circumscribed. Sometimes there will be multiple locations of intended use. For salespeople using a wireless application designed to answer customer queries, coverage will be needed wherever customers are located. If salespeople also want to use their devices for e-mail, coverage may be needed at their homes or remote offices as well. If users are repair technicians servicing gas lines or power stations, coverage will be needed wherever these physical assets are located. If a "user" is a piece of equipment, coverage will be needed wherever the equipment may be located or moved. One approach to determining coverage is to create a map of the locations where the application is expected to be used, and compare this map to a potential carrier's coverage map. The more overlap, the better the coverage. The physical environment where the solution will be used also affects coverage. Environmental conditions influence the working as opposed to theoretical range of the wireless network and pose possible reception problems. For example, the range of an infrared network goes to zero if there are obstructions between communicating devices. The signals transmitted over a short-range network become progressively weaker as the distance between transmitting and receiving devices grows. In-building reception may also be poor with some WANs. Information about the environment can also help uncover interference concerns, and characteristics of the physical environment may also help narrow device choices.

2. **Bandwidth**   While short- and medium-range wireless networks can approach wired throughput, bandwidth is currently a limiting factor for wireless WANs. Although bandwidth limitations can be overcome to a degree by application design, transfers of large quantities of data over a wide coverage area remains slow and costly, and downloading of some applications, such as video, are not presently feasible over wireless WANs. Consider bandwidth requirements by the primary types of data transfers your solution will need and the user's toler-ance for waiting. Users may be willing to tolerate a longer transfer time for occasional activities, such as downloading a system upgrade, if their routine activities occur at a reasonable speed.

3. **Latency**   Latency, the timeliness of information exchange, is not really an issue for local, in-building wireless networks, like WLANs or Bluetooth. Their perfor-mance depends in large part on the performance of the wired network to which they connect. Latency becomes an issue when wireless WANs are used, and organizations want the ability to push information to users wherever they are, and want mobile users to be able to transmit information back to headquarters from any locale. Data that must be accurate to the second, such as stock price

quotes, requires an always on, "zero" latency approach. Less time-sensitive data, such as filing service reports, can be transmitted by storing and forwarding the information when convenient, perhaps through synching. Even when real-time information is desired, remember that coverage and data volume/network bandwidth can impede or prohibit immediate access.

4. **Reliability**   Reliability assesses the tolerance that a wireless application has for network-related problems. Applications aimed at ensuring personal safety are useless if the networks they run on are available only sporadically. Likewise, wireless applications that depend upon the integrity of transmitted data, such as processing a credit card transaction at a point-of-sale terminal, cannot afford a network that regularly drops data.

5. **Security**   Although security is ultimately an "end-to-end" issue that includes devices, servers, applications, and other solution components as well as networks, wireless networks differ in their inherent levels of security. WLANs have experienced some well-publicized security flaws, while WANs are generally highly secure.

6. **Interoperability**   This factor is more of a design "flag" than a network selection criterion. The preponderance of wireless components and platforms, from devices to modems to network equipment to cards, makes interoperability essential. As interoperability declines, costs and support overhead mount, and redundant, overlapping, and conflicting solutions emerge. Interoperability within a class of wireless network solutions, such as Wi-Fi WLANs, may be high, but interoperability between different network types is iffy at best. Interoperability is not an issue for standalone solutions that rely on a single network and limited device types, but can be a significant issue when designing a solution for rollout to a diverse and dispersed audience.

7. **Cost**   Networks and network services vary by cost of implementation and cost of usage. There may be a range of suitable wireless network options, with a range of features, available at different cost points. Knowing the correct one to choose may hinge on the amount of money budgeted for the solution.

5.4   Moving from Requirements to Design

With solution requirements in hand, you are ready to start your wireless architecture design. In keeping with the book's "Just Enough" concept, the steps described in this section will help you select the components for an initial high-level design. This design should be sufficient to roughly scope the project and provide a starting point for further research and more detailed discussions with your technical architects and consulting partners. The final, more detailed architecture and design can be a significant project and must be performed by the team that will be charged with its implementation.

The remaining chapters and appendices within this book provide a wealth of information to assist in solution selection. Figure 5.10 shows the relationship of the book's chapters to the layers and components of the Wireless Decision Process. Additional, up-to-date information can be found on the book's web site, *www.just-enoughwireless.com*, and the other web sites listed throughout.

Use the following steps to guide your selections for each of the four wireless architecture component categories. Once choices have been made in these categories, the same process can be used to support an initial design for the Implementation and Support Infrastructure component.

1. **Review the Technical Requirements**   Make sure you have a good understanding of the technical requirements for the component category that you are selecting. If you have used the forms provided in Appendix A, these requirements will be organized by the same parameters used in the descriptive chapter.

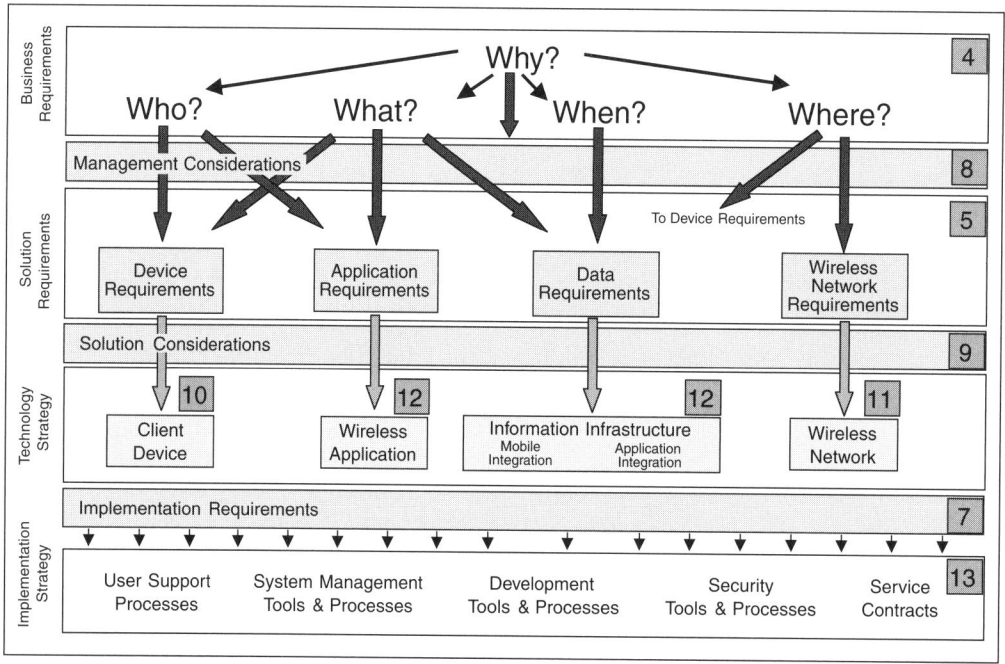

FIGURE 5.10
The Wireless Decision Process with Chapter References

2. **Gain an Overview of Component Features and Selection Considerations**   Each component chapter contains a wealth of information; however, many details may be outside your areas of interest. To focus your efforts, concentrate on the chapter introduction, considerations, and features and function sections.

3. **Identify the Options Best Suited for Your Situation**   Using the constraints imposed by your technical requirements, you should be able to identify quickly a narrower range of options. For example, if your desired level of wireless coverage is confined to an office environment, you can focus on short- or medium-range, "Do It Yourself" networks.

4. **Read the More Detailed Descriptions on Your Candidate Options**   Refer to chapter comparison tables and read the detailed descriptions on the remaining options. This information should further refine your options and may identify the specific one that best suits your needs. Note any issues or constraints that may conflict with choices made in other architectural categories.

5. **Perform High-Level Research**   When a preferred option or set of options has been identified, perform some additional research to ensure you have the latest available information to guide your decisions. Wireless technology is evolving so quickly that getting current information is essential. Identify the vendors offering the desired solution. Visit industry and vendor web sites, read analyst reports, and contact peer organizations. Check for newly announced standards, evolving features and functions, current pricing, and new, competing options. Again, note any issues or constraints that may cause conflicts.

6. **Finalize High-Level Selection**   The information from the previous steps should be sufficient to select the candidates for your initial, high-level design and provide cost estimates for cost justification efforts.

7. **Conduct Detailed Research**   For many wireless solutions more research will be required to support the detailed design and final selection of the components to be implemented. This research can include product and vendor evaluations, technical feasibility pilot projects, and visits to corporate users of the desired technology.

5.5 Working Around Wireless Constraints ──────

As you translate your business requirements into technical specifications, you will undoubtedly discover that the wireless environment is rife with constraints, incompatibilities, and compromises. You may discover that your original solution cannot be deployed as envisioned as limitations in areas such as wireless network bandwidth and device displays reintroduce constraints that have long been eliminated in the world of wired workstations. Don't let these issues push you into abandoning or unnecessarily delaying your foray into wireless technology. Although wireless technology has yet to reach the levels of maturity, standardization, and support found in other information technologies, corporations such as FedEx and UPS have used wireless solutions productively for years. Current wireless capabilities are sufficient to handle many practical applications, and wireless constraints can often be overcome through creative design. There are two principle categories of wireless constraints.

- *Evolving Constraints*   Some wireless constraints are temporary. Over time, the rapid evolution of wireless capabilities will reduce constraints in areas such as standards, bandwidth, coverage, security, and support tools. For example, the throughput of short and wide area wireless networks continues to improve, and the arrival of 2.5G and 3G networks will provide high-speed data exchange across a wide geographic area. Similarly, coverage is improving as wireless network operators continue to add infrastructure.

- *Inherent Constraints*   Other constraints such as device size, display limitations, and data entry capabilities, result from the demand for small, portable devices and are a permanent part of the wireless landscape. A typical mobile device will never support the robust applications found on a desktop. Display size and resolution determine the volume and type of information that can be presented effectively. Keyboard size or handwriting recognition schemes direct the types of interactions possible. Memory and processing power affect the architecture of the application. Size, weight, and useful battery life affect the portability and convenience of the device. These inherent constraints must be overcome by designing around them. For example, applications can use numbered menus or forms to minimize data entry, and intelligent partitioning of functionality between the device and an application server can circumvent processing and memory constraints.

There are many possible paths to take when designing a wireless solution. If the original solution is blocked by limitations, another approach or combination of approaches may give you the capabilities you need.

5.5.1   Rethink Constraints

Is the perceived constraint truly a limitation or is it imposed arbitrarily? A given solution may require sending forms between the user and a corporate system. If the entire form is sent on each transmission, the bandwidth of a wireless wide area network may be insufficient. If the application is designed to transmit only the changed portions, however, current bandwidth may be more than adequate. Does a traveling salesperson really need instant access to account file updates or can files be updated once a day? System architects and designers used to working on unconstrained wired office systems may need to rethink their approach to solution design to work around, or even take advantage of, the differences in wireless capabilities.

5.5.2 Switch Paths

Ask yourself if you can accomplish the same goals with a different set of technology options. For example, if your main goal is to relay sales orders to your corporate office, you may enter the information through a wireless web page, use a custom designed form to transmit data across a digital cellular network, or use e-mail templates through an e-mail device. While the form of the data and the design of the application may be very different, the end result is the same—the sales order is relayed.

5.5.3 Reprioritize Features

While the originally intended, fully featured content-rich solution would have met every need of your target audience, you may be able to gain most of the solution's benefits using a less ambitious design. As described in the Chapter 3 case study, Atlantic Envelope Company found that a small set of capabilities covered a very large percentage of their sales force automation needs. Focusing on the top ten customer queries, allowed the company to develop, deploy, and gain benefits from their solution now, using currently available and easy to implement technology.

5.5.4 Find Creative Workarounds

Look for creative workarounds if particular aspects or functions of your solution run into roadblocks. As an example, in an originally envisioned medical solution, emergency room doctors would have received an updated version of the emergency room's patient whiteboard on demand. Experience quickly showed, however, that the busy and highly mobile physicians could not tolerate having to initiate and wait for the updated whiteboard to load on their handheld devices. To overcome this issue, the project's design team switched to a "push" approach, automatically relaying updates once a minute to the doctors' devices. By restricting transmissions to only those items that had changed, transmissions were kept short and unobtrusive, enabling the doctors to simply look at their devices periodically to see up-to-date whiteboard contents.

5.5.5 Break the Problem into Smaller Parts

Sometimes a combination of approaches provides a more effective solution than a single approach. Penske Logistics uses wireless technology on its trucks and at its cross docks to keep customers informed about deliveries and manage driver routes and performance. An onboard computer manages and transmits data about arrival and departure times, delays, traffic conditions, and changes in schedules in near real time over a satellite network. More routine data about cargo pickups and drop-offs is captured via handheld devices used by delivery personnel. When a truck returns to a Penske terminal, the routine data on the handheld device is synchronized with the host system via a wired docking cradle. This hybrid approach provides immediate exchange of high-value information, with coverage across Penske's entire delivery range, with periodic bulk transfer of less time-dependent data at much lower network costs.

5.5.6 Phase Your Solution

If your solution is constrained by limitations in evolving areas, such as coverage or bandwidth, you can use a phased approach to implement a subset of features now and add new features as limitations disappear. For example, you can provide field service workers with a mobile solution for dispatch and work order invoicing with current technology and add capabilities, such as access to on-line repair manuals or videos of repairs in action, as network capacities increase.

5.6 Example Solutions

The best way to illustrate the principles of this chapter is to examine a few actual examples of wireless applications. This section describes four diverse types of applications: e-mail service (Figure 5.11), WLAN (Figure 5.12), data capture (Figure 5.13), and national tracking and location (Figure 5.14), and the considerations that influenced their designs.

5.6.1 Example Architecture: E-mail Service

- *Application*   Atlantic Envelope Company (AECO) wanted to give its salespeople wireless e-mail access and a custom application that would allow them to answer customer queries and submit changes to orders. AECO chose e-mail as the platform for person-to-person messaging, and as a method to exchange transactional data between client devices and a back-end server. To overcome expected device presentation and network bandwidth limitations, AECO designed its application to focus on customers' top ten queries and to use forms and templates to ease data entry and minimize data transfer.

- *Information*   AECO isolated the data needed to support the application to order, customer account, and inventory information. Without a central source of this information, AECO created a program to draw the data from several back-end systems, store it on an application server, and refresh it twice daily, a timeframe sufficient to provide reasonably current information to the field.

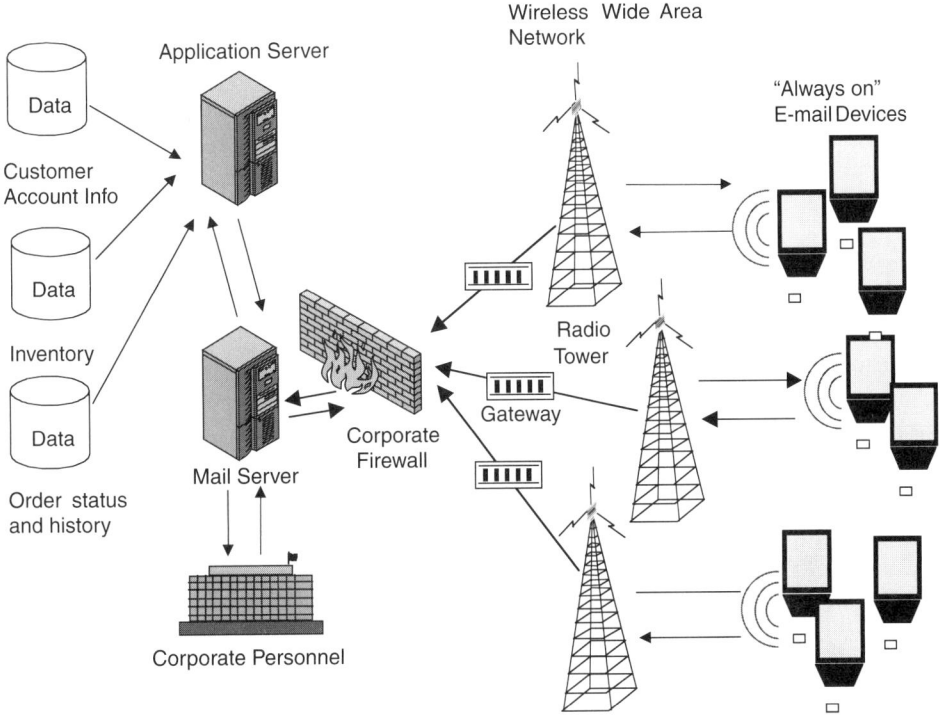

FIGURE 5.11
E-mail Service Architecture

- *Device*   AECO had two important device requirements: the device had to be optimized for e-mail and give sales reps the ability to type freeform e-mail messages using a keyboard rather than a cryptic handwriting scheme. These two requirements led AECO to choose a RIM BlackBerry device equipped with a keyboard.
- *Network*   Given the locations of its customers and the areas in which sales-people would roam, AECO had one viable network choice: a wireless WAN. Drawing a map of its customer locations, and superimposing this map on various carriers' coverage maps, led to the selection of Cingular Wireless as the carrier with the best overlap. Although maintaining the integrity of transmissions wasn't a key concern for AECO, the RIM devices and network allow for storing and forwarding of messages when a sales rep is out of the coverage area.

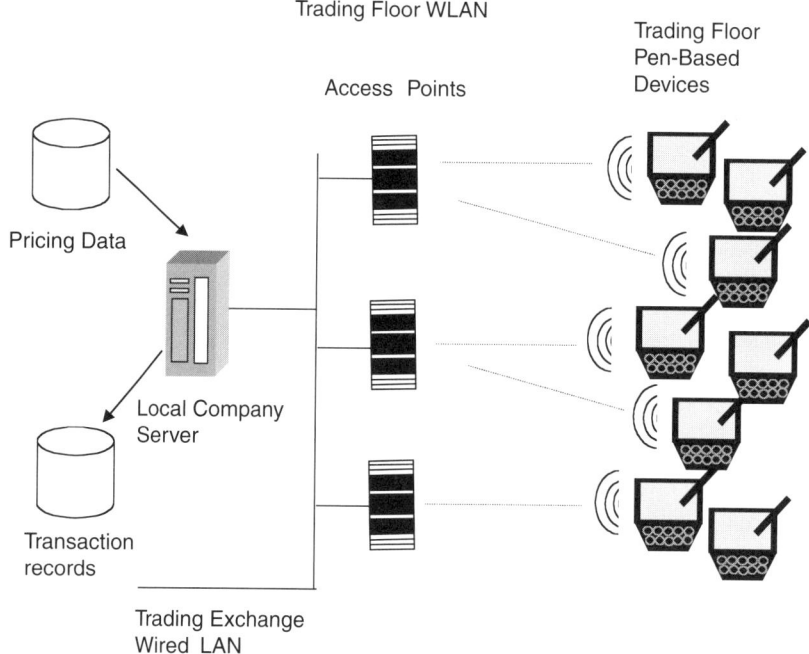

FIGURE 5.12
Wireless LAN Architecture

5.6.2 Example Architecture: Wireless LAN

- *Application*   Hull Trading wanted to give its derivatives traders access to current pricing information and to capture a record of trades as they occurred. As
  mobile workers on the floor of an exchange, these traders previously had to
  rely on hand signals and runners to obtain information from individuals using
  desktop systems.

- *Information*   The pricing information that traders needed included real-time
  bids, offers, and other proprietary information. This information was already
  available on site through a wired network connection, making it simply a matter
  of transmitting the data to the floor. As transactions completed on the floor, that
  data would be transmitted to a local server to be forwarded to a central database.

- *Device*   Traders needed to look up information and execute trades quickly.
  With speed as a primary requirement, data entry had to be fast, precluding
  keyboard-based input. Since traders would be physically standing on the floor,
  they needed a device they could easily hold in their hands, that wasn't too
  weighty, that would remain charged during the course of trading hours, and

that allowed them to access information using one hand. A pen-based hand-held PC from Fujitsu Personal Systems fulfilled these requirements.

- *Network* Traders were located within a main, indoor area, but were mobile within that space. Network choices were narrowed down to short- and medium-range solutions given the local coverage needed. With a lot of movement on the floor, an infrared network was obviously inappropriate—traders could not stay in close, unobstructed proximity to network transceivers. The range limitations of Bluetooth would not work within the exchange. A WLAN solution was selected for its ability to support mobile traders throughout the entire trading area despite physical obstacles.

5.6.3 Example Architecture: Data Capture

- *Application* UPS wanted to standardize its systems for tracking packages in its distribution centers. The wireless application had to be able to capture data for each package traveling through a center, to enable the company and its customers to track a shipment in real time. Package sorters would use the application to collect data encapsulated in the bar code affixed to each package.

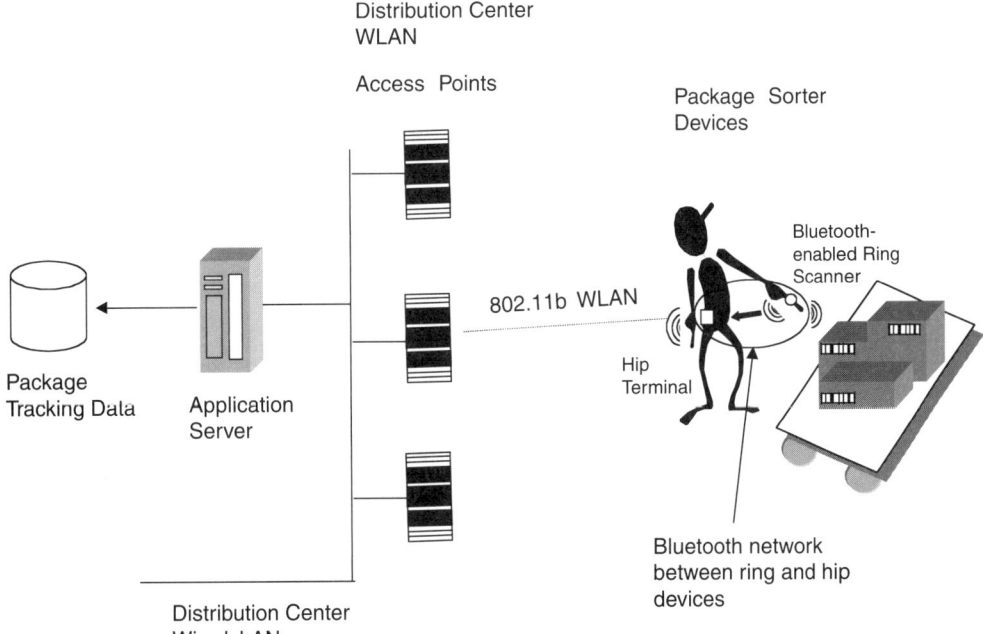

FIGURE 5.13
Data Capture Architecture

- *Information*   The goal of this application was to improve the way that data was transmitted from the point of collection on the sorting line to the back-end databases.   Since this data was already being collected and tracked via other means, the existing information architecture would suffice for the new wireless implementation.

- *Device*   The users for the application were package sorters handling the packages moving on belts through the distribution center. The chosen device had to allow the workers to be mobile at their station, permit hands-free usage so that workers could continue to handle packages, and be free of parts or cables that might snag in the moving equipment. The device also had to support bar code scanning, and contain appropriate integrated wireless communications capabilities.

To meet these requirements, two devices were chosen. The final selection was inextricably tied to the network options chosen (see below). The first device would perform the bar code scanning function. To free the workers' hands and avoid cabling, UPS chose a cordless, optical, bar code scanner mounted in a ring and worn on the finger. No commercial device was available that met these specifications, so Symbol Technologies was tapped to develop a custom device. To move the data scanned by the ring device to back-end systems, some type of wireless capability was needed. The size of the ring device precluded using a modem or WLAN adapter, but a small Bluetooth chip could be integrated without any problem. The disadvantage of using Bluetooth was that its range was insufficient to transmit the data straight to an access point located a distance away from the worker. Another device was needed to effectively "stage" the data and forward it via a more powerful network technology such as a WLAN.

The second device had the same requirements as the first device—it could not impede the workers' movements and could not have cables—but would necessarily have to be larger than the ring device to accommodate integrated WLAN capabilities. The final choice—a belt-worn terminal that would receive data sent from the ring device and forward it via a WLAN. Again, no commercial device existed with integrated Bluetooth and WLAN capabilities, so Motorola was chosen to develop a custom device. The cost savings anticipated from the wireless solution were deemed sufficient to justify the investment in these custom devices.

- *Network*   Given that the users of the solution—the package sorters—were located in a relatively contained physical space—a distribution center—short- and medium-range wireless networks were potential solutions. The distance over which data would be transmitted was within the range of a WLAN solution. But as noted above, device considerations (the small size of the ring scanners) precluded the use of a unified WLAN solution. The remaining possibilities, infrared and Bluetooth networks, could handle the size constraints, but faced other issues.

Infrared technology was eliminated because the workers' hands were constantly in motion and obstacles, such as packages and equipment, meant that line-of-sight requirements could not be met. Bluetooth, with its small chips, would work in the ring devices but was not robust enough to transmit data over the distances required in the distribution centers. A dual-network solution was needed to pick up the Bluetooth transmitted data, and then pipe it to the transceivers strategically located through the center. The best candidate for that task was a WLAN.

Choosing a dual-network solution posed device challenges since an integrated Bluetooth/WLAN device did not exist. Developing a device that could interact with both types of networks took some ingenuity. Since Bluetooth and WLAN networks operate in the same 2.4 GHz radio frequencies, signal conflict was a foreseeable problem. Symbol Technologies came up with a solution that allowed the device to pick up Bluetooth signals, and when the spectrum was quiet on the Bluetooth front, transmit the data over the same frequency to a WLAN access point.

5.6.4 Example Architecture: National Tracking and Location

- *Application*   Penske Logistics wanted to offer its customers more timely access to performance and logistics data, allowing them to monitor service levels among their supply chain partners. Penske designed three different applications to gather the necessary data: an application to improve communications between dispatcher and driver and transmit time-sensitive information; an application to track incoming and outgoing shipments at its cross docks; and an application to capture detailed data about goods picked up and delivered. The wireless applications needed the following abilities:

    - Locate trucks and communicate with drivers to provide route information, pickup and delivery information, and route or schedule changes

    - Track events such as arrivals, departures, and downtime, and communicate these events in real time to the internal logistics management system

    - Track, locate, and manage inbound and outbound freight at cross docks

    - Collect detailed delivery and pickup information for later uploading to the logistics management system

    - Capture customer signatures and print receipts, if desired

- *Information*   Several types of data were implicated in the solution, including:

    - Bar code and other detailed freight and package information collected in the field and uploaded to internal systems intermittently

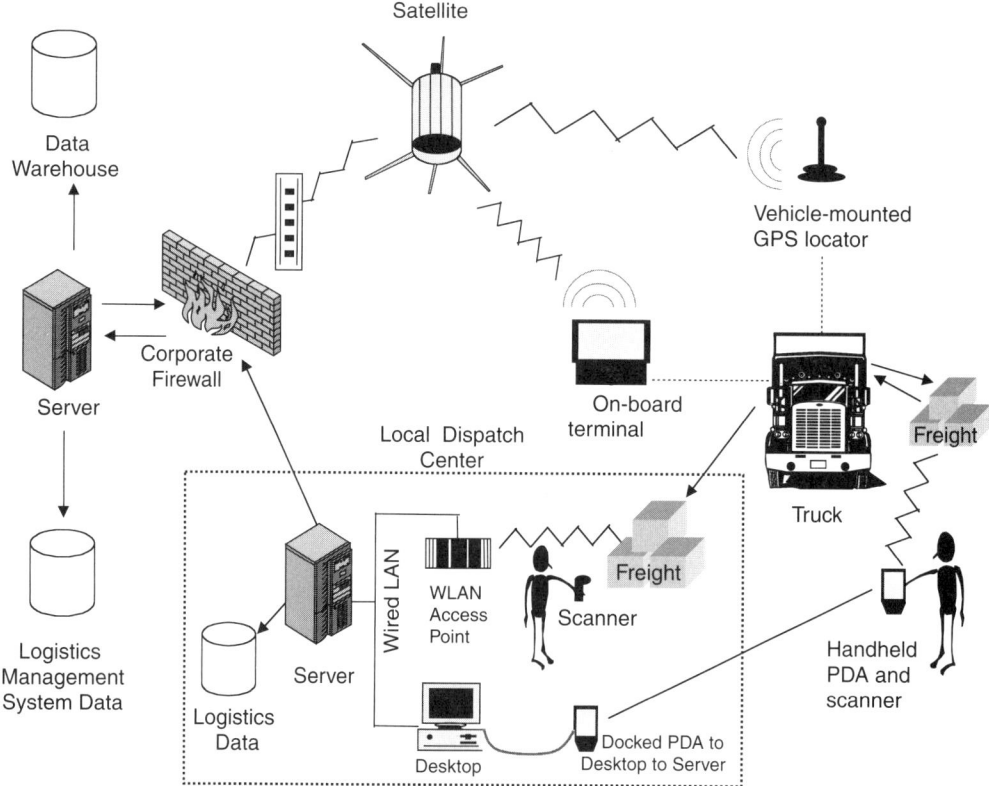

FIGURE 5.14
National Tracking and Location Architecture

- Arrival, departure, and other time-dependent information, originating in the field and transmitted to headquarters in real time
- Alerts, schedule changes, and route information, originating at head-quarters and sent to trucks in real time
- Inbound and outbound freight information tracked, collected, and managed at cross docks

- *Device*   Application requirements indicated that more than one device type was needed. For the truck fleet, two devices were chosen. First, GPS locators were installed on each truck to enable communications between dispatch centers and drivers no matter where the trucks were located. Second, to communicate route information, arrival and departure data, and freeform messages between driver and headquarters, another device was needed, one that would have data input capabilities for the driver. A larger-sized device with a keyboard was acceptable because the device would be mounted in the truck, drawing battery power from the engine. In addition, the device had to support real-time communications via a satellite network. QUALCOMM's OmniTRACS unit was chosen, a rugged device with integrated satellite communication capabilities. To collect detailed data on pickups and deliveries, a third device was chosen. These devices had to be portable so drivers could easily carry them, and had to be capable of capturing data on labels as well as customer signatures. A handheld PDA from Symbol Technologies was chosen for the task. A fourth device was needed at the cross docks to locate, track, and manage incoming and outgoing freight. This device had to be portable and rugged, and needed WLAN capabilities to transmit the tracking information to an internal system, leading to the selection of an Intermec unit.

- *Network*   With trucks traveling virtually anywhere in the U.S., relaying real-time information would require a WAN. Because trucks would often travel·in remote or rural places where digital cellular coverage was spotty or unreliable, a satellite network with ubiquitous coverage was chosen. To minimize the number (and expense) of satellite transmissions, only certain time-sensitive information (arrival times, departure times, downtime) is sent via satellite. At the cross docks, where incoming and outgoing packages and freight are scanned, tracked and managed, a medium-range WLAN solution was selected since it allowed dock workers and equipment to be mobile without the constraints of cables. Finally, the more detailed pickup and delivery information (number of parts, pieces, etc.) was not needed in real time, and could remain stored in the handheld unit until a driver finished his route. This information would then be uploaded into internal systems, using a cradle and wired network connection.

Justifying the Solution

Intuitive benefits aside, few companies are willing to make significant investments in any proposed wireless solution without some form of justification. At the very least, executives will want to know how much the project will cost and what types of benefits the company will receive. This information is essential for setting the project budget and for comparing it against competing investment opportunities.

The rigor and formality of your project's justification will depend on the needs, style, and policies of your organization. If your company is very entrepreneurial, it may be willing to trust its intuition and accept a less than rigorous justification. However, many organizations have formal processes for guiding their investment decisions. Potential projects must pass through a series of hurdles that consider factors such as ROI and payback period. If your organization falls into this category, unless the cost of your solution is very small, you will need to prepare a formal cost/benefit analysis. Even if a formal justification is not required, it is worth going through the exercise to understand the ramifications of your proposed solution fully. This knowledge will help you ensure that you have the proper level of funding, backing, and priority for your project. Moreover, it establishes parameters for "go or no go" decisions if the project encounters difficulties during implementation.

The difficulty of performing a project justification depends on the type and complexity of the proposed wireless solution. If the decision is whether to install a wireless or wired LAN in a new office building, justification is relatively straightforward. The cost of each option is easily determined. The "hard" benefit—avoiding the cost of future re-wiring—can be calculated from past history. A simple cost/benefit analysis comparing short- and long-term cost of ownership for each option is likely to be sufficient and sidesteps the need to quantify "soft" benefits such as user convenience. Conversely, deploying a wireless CRM solution to a large, national sales force presents the other end of the justification spectrum. Calculating its cost is laborious and time consuming as the solution has many complex hardware, software, and network components and requires integration with back-end IT systems. Training costs alone are likely to be a significant investment. The benefits, despite being intuitively substantial, are difficult to quantify. How do you calculate the dollar value of being better prepared for a sales presentation or being more responsive to your customers? The "hard" benefit—increased revenues—doesn't instantly follow the deployment of the new solution and is influenced by many unrelated factors.

As shown in Figure 6.1, the justification process presented in this chapter consists of four basic steps. The first step, quantify benefits, helps you determine whether your solution makes sense and establishes the economic boundaries for your cost calculations. If your solution has the potential to provide several million dollars of "hard" benefits, an investment of a few hundred thousand dollars is easy to justify. Conversely, if the benefits are relatively modest, they set the maximum "price" your organization should be willing to pay for the solution. The second step, calculate costs, computes the short- and long-term expenses associated with the solution. These costs include direct expenses for software, hardware, and network services,

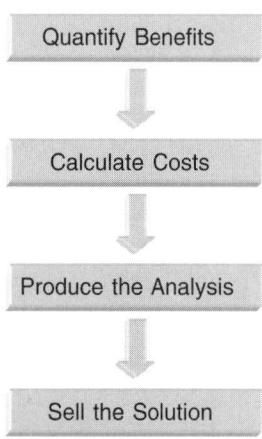

FIGURE 6.1
The Justification Process

deployment expenses, such as training and installation, and ongoing expenses for support and services. If you are part of an evaluation team, you can shorten the elapsed time of the justification effort by performing the benefit and cost steps concurrently. The third step, produce the analysis, combines benefit and cost data to calculate ROI, payback period, cash flow, and other analyses to provide an objective business justification for the project. The final step, sell the solution, completes the process by building support for the proposed solution and demonstrating that its benefits are achievable.

6.1   Quantifying the Benefits

Benefits are the heart of every project justification. Gaining them is the raison d'être for the project. The type and value of benefits will differ widely from solution to solution. Chapter 2 describes many categories of possible wireless benefits and is a good starting point for identifying important, but less obvious benefits. The project's goals define the primary benefits that the solution seeks, but secondary and tertiary benefits are possible and increase the value of the solution. The objective of this section is to provide some guidelines for identifying, selecting, and where possible, quantifying the direct and indirect benefits provided by your solution.

With a little creativity, it is possible to come up with an endless stream of plausible and perhaps not so plausible benefits for most projects. These benefits range from the highly tangible, such as cutting the cost of service delivery by 50%, to the highly intangible, such as creating a more attractive work environment. When quantifying benefits, there is a tradeoff between aggressive approaches that attempt to maximize the solution's value by including every possible benefit and conservative approaches that rely on strictly objective and quantifiable benefits. By increasing the potential value of the project, the aggressive approach helps justify larger efforts, but runs the risk of straining credibility. It sets high expectations for the project, which can lead to disappointment if the effort fails to achieve the anticipated benefits fully. In contrast, by using only irrefutable and non-controversial benefits, the conservative approach sets a solid, but potentially much lower bar for performance. If a project can be justified using a conservative approach, its feasibility is unquestionable and it has a good chance of outperforming expectations. The downside of a strictly conservative approach is that it may eliminate highly valuable solutions that lack easily quantifiable benefits. By taking a narrow view of benefits, it arbitrarily favors small-scale solutions that target straightforward objectives and is likely to pass over riskier breakthrough opportunities. The approach you choose may already be set by your company's project justification policies, but if not, you may want to consider an amalgam of both approaches. Be aggressive in identifying the full range of solution benefits, but be as conservative as possible in assessing the level of those benefits.

6.1.1   Identifying Benefits

Identifying the primary benefits of your solution should be easy, at least at a high level. As mentioned above, your solution goals state the main benefits desired for the project. For example, you want to improve the effectiveness of your sales force, reduce shipping losses, or increase responsiveness to customers. A successful solution will provide benefits that support these goals. The challenge in a cost justification is to translate these lofty objectives into a series of concrete benefits that support value estimations. Ask yourself what benefits constitute the larger objective. Improving the effectiveness of the sales force may mean reducing time spent on paperwork, being better informed during sales calls, and having the ability to respond to customer queries while on the road. Each of these benefits has implications that lead to secondary and tertiary benefits. Less time spent on paperwork means more time in front of customers. More time in front of customers means more sales. Other benefits may arise as intentional or unintentional side effects of the original benefits. Eliminating paperwork saves a salesperson time, but also provides better information for sales management. Figure 6.2 shows some of the benefits that could be obtained from a wireless data capture solution.

At this point in the exercise, try to identify as many potential benefits as possible. Include both "hard" and "soft" benefits. Understanding the full range of benefits provides a better picture of the project's true value and provides greater flexibility during the justification analysis. The list of benefits will be honed later as you prepare the cost justification. Review the solution's business requirements and interview the individuals affected by the project in order to identify positive and negative impacts. As the beneficiaries of the project, the intended users are in the best position to identify potential benefits and determine if a desired benefit is feasible.

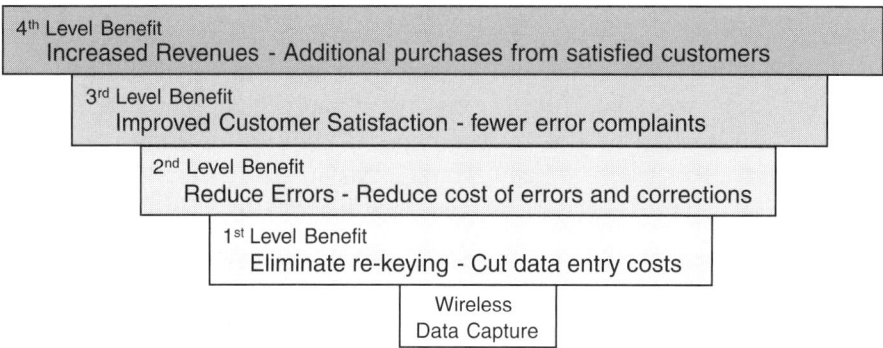

FIGURE 6.2
Pyramid of Benefits

6.1.2   Types of Benefits

In an ideal world, every benefit has obvious and indisputable value that is easily quantified within a cost justification. In the real world, many important benefits are impossible to quantify and many others fall in the gray area between these two extremes. Every project will have a mixture of these types of benefits. For the purposes of a cost justification, benefits fall into three main categories.

- *Tangible Benefits*   This category includes quantifiable "hard" dollar benefits. The value of these benefits are indisputable, the only question is their size. Examples include straight comparisons of costs between two alternatives, reductions in labor or material costs, new sources of revenue, reduced shipping losses, and tax savings. In each case, the benefit has direct economic value that is easily estimated and measured.

- *Semi-Tangible Benefits*   Semi-tangible benefits are the gray area between hard and soft benefits. They have intuitive economic benefits, but lack the historical data or measurement techniques to support solid "hard" dollar estimations. Examples of these benefits include improved safety, fewer errors, better use of people or equipment, more efficient operations, and faster access to information such as medical test results. While a benefit such as faster access to medical test results should translate into improved decision-making, better patient care, and ultimately lower insurance costs, the financial impact of this benefit is difficult to estimate. If quantification is available, such as knowing that a specific type of shipping error cost the company $100,000 last year and that the wireless solution will eliminate that type of error, the benefit becomes tangible. Cost avoidance—the reduction or elimination of a new cost that would otherwise occur—such as avoiding the need to buy $1,500 laptops by providing salespeople with $500 PDAs, is a semi-tangible benefit unless the original cost would definitely be incurred.

- *Intangible Benefits*   "Soft" benefits such as better employee morale, gaining technology experience, improved responsiveness, higher customer satisfaction, and enhanced corporate image have definite value, but are very difficult, if not impossible, to quantify economically. Despite the inability to quantify intangible benefits, they are important and are worth including in a project justification at least as back-up for more tangible benefits. In some cases, the perceived value of intangible benefits may be sufficient to justify a project. While there are techniques for grossly estimating their value, such as using industry statistics, many executives will discount the validity of those estimates. Further, attempting to quantify intangible benefits may cause executives to question the credibility of your other estimates.

While some benefits will always fall cleanly into one of these categories, others, such as productivity, depend on the situation. For instance, take an evaluator whose sole job is to fill out forms at a defined value of $20 apiece. This evaluator takes twenty minutes to fill out a form, producing three per hour for an economic value of $60 per hour. If a wireless data capture application reduces the time expended to five minutes per form, the evaluator's output is quadrupled to $240 per hour for a tangible net benefit of $180/hour. In a different scenario, if the evaluator fills out an average of four forms per day interspersed with other work, the wireless solution frees one hour per day of the evaluator's time for other activities. If the evaluator's time is billable or if he can use it to fill out more forms, a tangible value can be placed on the improvement. If the extra hour is spent on activities with unspecified value, the benefit is semi-tangible and is likely to show up over time as an improvement in another activity (perhaps as a reduction in work backlog). Finally, if the evaluator fills out forms only occasionally, the small volume of time freed, while beneficial in some manner, is unlikely to be spent on any quantifiable activity, resulting in an intangible benefit.

This categorization becomes important when selecting benefits for inclusion within a cost justification. Organizations that accept only "hard" dollar justifications will use only tangible benefits in their calculations, while less conservative organizations may also accept ballpark estimates for semi-tangible benefits. Entrepreneurial organizations are likely to consider all three categories.

6.1.3 Benefit Characteristics

The more tangible a benefit's financial value, the more useful it is in a cost justification. Several other factors, however, affect usefulness including the potential size of the benefit, timing, beneficiary, ability to be quantified, and ability to be attributed. Use these characteristics to select the most appropriate benefits from the overall list discussed in Section 6.1.1.

- *Size*   The importance of this characteristic is obvious. In a perfect world, your main benefit is quantifiable, directly attributable to your solution and large enough to justify the cost of the project.

- *Timing*   When is the benefit received? Is it a one-time benefit or is it recurring? How long does the benefit last? Answers to these timing questions are important when computing project cash flow and calculating payback periods.

- *Beneficiary*   Who reaps the benefits of the solution? A wireless solution that transfers customer billing to the field service organization saves resources in the billing department and improves cash flow, but may increase costs within the field service organization. While the net value to the company is high, the field service organization may have little incentive to fund the project.

- *Ability to Be Quantified*   By definition, tangible benefits can be quantified. It may be possible to quantify intangible and semi-tangible benefits in other ways. For instance, although it is difficult to directly tie customer satisfaction to financial benefits, it can be measured through customer surveys, making it conceivable to assess whether a given solution positively impacts customer satisfaction.

- *Ability to Be Attributed*   Some benefits, such as the availability of a wireless data capture application to reduce back-office data entry time by 75%, can be directly attributed to the solution. At the other extreme, higher level benefits, such as increasing revenue per customer, are affected by many factors and may be impossible to tie to a given solution. The more closely a benefit can be attributed to your solution, the greater its value in the cost justification.

Using the wireless data capture solution portrayed in Figure 6.2, Figure 6.3 illustrates these characteristics in action. This solution offers four levels of potential benefits. Its primary goal is to cut data entry costs by eliminating the need to re-key data manually from paper forms. This benefit is direct and easily quantified. It may also be large enough to justify the project by itself. Eliminating re-keying has the secondary effect of cutting re-keying errors, saving both the cost of the error and its correction. In situations where the cost of an error is high, such as a stock trade or a drug prescription, the value of this benefit is potentially quite large, but is more difficult to quantify. Fewer errors mean less negative impact on the customer, providing the third level of benefit—improved customer satisfaction. Although this benefit can be measured, it is difficult to attribute changes in customer satisfaction to the wireless data capture solution. Also, customer satisfaction is a "soft" benefit that is difficult to translate directly into economic value. That economic value is derived from the fourth level of benefit—increased revenues from additional purchases by satisfied customers. Like customer satisfaction, purchasing patterns are affected by many factors making it virtually impossible to isolate and attribute a change to the wireless data capture solution. Yet, in a business where trust is essential, such as prescription filling at a pharmacy, the indirect revenue benefit of eliminating manual data entry may dwarf the more easily quantified benefits.

Benefit	Type	Size	Timing	Beneficiary	Quantifiable	Attributable
Cut data entry costs	Tangible	Significant	Immediate	Field Service	Yes	Yes
Reduce errors	Semi-Tangible	Potentially large	Immediate	Multiple Depts.	Partially	Partially
Improve customer satisfaction	Intangible	Unknown	Long-term	Company	Possibly	No
Increase revenue	Tangible	Potentially large	Long-term	Company	Yes	No

FIGURE 6.3
Comparing Benefits

6.1.4 Measuring Benefits

After using the characteristics described above to select a likely set of benefits for your cost justification, it is time to finalize the process by estimating the impact of those benefits. Theoretical benefits look great in a cost justification, but accurately estimating benefits using real-world data and a reasonable and defensible set of assumptions provides credibility and gives executives confidence that the project can achieve its objectives. Before selecting a benefit as a final candidate for your cost justification ask yourself:

- *Is the Benefit Realistic?* If a given solution frees one hour of time per day, is that freed time truly going to be used productively? If the hour is freed in twelve, five-minute intervals randomly scattered throughout the day, it is unlikely to result in more work being accomplished. Conversely, if the solution allows a worker to perform one extra service call per day, a tangible and measurable benefit has been gained.

- *Is the Benefit Truly Achievable?* Too many cost justifications are based on assumptions that are unlikely to materialize completely in practice. If a solution cuts the effort to perform a given service by half, it theoretically either doubles your capacity to deliver that service or allows you to perform the same level of service with half of the current staff. Before you include the full level of either benefit in a cost justification, consider whether staffing levels would really be reduced by half to gain a cost benefit or whether customers will really buy twice as much service to support the increased capacity.

  When calculating the benefits for its field force initiative described in Chapter 3, Honeywell's ACS division took a conservative approach, using only indisputable hard benefits. The project team, with members primarily drawn from the operational areas that would use the solution, had to reach unanimous agreement on each line-item benefit before it could be included in the cost justification.

- *Can Attainment of the Benefit Be Verified?* The true test of a good benefit is the ability to measure its attainment objectively after the project has been deployed. Too frequently, benefit measurement is forgotten after a project is approved. If a given solution is supposed to increase billable time, can you put the measurements in place to review billable hours periodically to assess the solution's performance? A regular review of project benefits provides data to tweak the solution to maximize its performance and information for future cost justifications. Semi-tangible benefits become tangible over time if measurements are put into place. Plus, the ability to demonstrate a track record of meeting project benefit commitments will increase your credibility for future projects.

Many methods for measuring current performance and estimating the value of selected benefits exist. To quantify its benefits, Honeywell's ACS division deployed three teams to collect data by using questionnaires, conducting interviews, and observing field performance. These efforts produced conservative estimates based on actual performance. While estimates based on a combination of historical performance and validated performance improvements are ideal, any of the following techniques can provide data for your justification efforts.

- *Questionnaires*   Questionnaires are a quick and simple means of collecting data to set current performance baselines and estimate potential improvements. On the downside, questionnaire responses are often opinion-based rather than factually assessed and can be highly variable. Still, they provide a good starting point.

- *Interviews*   Interviews with the future users of a solution can provide considerable insight into the type and size of benefits that can realistically be attained. Since interviews are interactive, additional points of interest or potential discrepancies can be explored in detail as they arise.

- *Observation/Experimentation*   Live measurement of an activity can provide excellent and quantifiable insight into expected performance. This approach is especially useful for highly repetitive activities. For example, a series of warehouse workers could be timed performing their routine inventory tasks to obtain an average level of effort per task. The measurement can be repeated using simulations or a pilot solution to obtain expected future performance per task. Multiplying these respective factors against the number of times the task is performed provides an accurate sizing of expected benefits.

- *Historical Data*   Where available, historical data on past performance in areas such as billable hours, production output, budgets, inventory levels, and number of service calls, is likely to provide a more accurate baseline than is possible through other approaches. If your organization is "measurement-minded" or process-oriented, it may have a wealth of other performance statistics that can be used as a basis for estimates.

- *Company "Rules of Thumb"*   Your company's accounting and finance area is likely to have "rules of thumb" figures for use in estimates and budgeting. These figures include items such as fully burdened costs for employees, expected work hours per year, opportunity cost for capital, cost of carrying inventory, tax rates, and average loss/theft rates. Since your company already accepts these figures, using them will save time and allow your cost justification to be compared correctly against others.

• *Industry Statistics*   Industry analysts, trade publications, industry associations, and government agencies produce a stream of useful statistics that can be used to support a cost justification if the appropriate internal data is unavailable. These statistics cover the gamut of categories from average costs to future market growth predictions. While these statistics may be easy to obtain, their basis is often suspect at best. Proceed with caution unless you are sure of the quality of the analysis underlying the statistic.

6.2   Calculating the Costs

Costs are the flip side of the economic analysis, representing the investment that your company has to make to get the benefits identified in the previous section. Unlike many benefits, most costs are tangible, identifiable, and directly attributable to your solution. The two biggest issues in cost calculations are ensuring that all major costs have been identified and that cost estimates are accurate. Costs will differ significantly by the scope and type of solution. While some cost generalizations are possible, for instance, an infrared network is less expensive than a satellite network, wireless technology prices fluctuate rapidly and continually as the technology and its market evolves. For these reasons, this book does not attempt to offer specific costing information, but focuses instead on cost issues and cost areas to consider. Note that the cost categories in this section may not be complete or address all possible solutions, however, they should provide a starting point for your cost analysis.

6.2.1   Costing Issues and Considerations

Costing a wireless solution has both simple and subtle components. For instance, computing the cost of wireless devices is a straightforward matter of multiplying the number of devices that are needed by the unit price. Conversely, estimating the impact of learning curves on short-term productivity is an educated guess at best. As is true for other projects, hidden and unexpected costs exist. The following points cover some of the more common issues and considerations that you may face when trying to estimate the costs of your wireless solution.

• *Ballpark vs. Detailed Estimates*   Costing can be produced at many levels of detail. Your company's policies and practices will determine how far you will have to go to produce an acceptable estimate. A "ballpark" estimate that computes a high-level sizing of costs may be sufficient for "go or no go" decisions. A more detailed estimate may be necessary to produce the project's budget (Figure 6.4).

Cost Checklist

Hardware

Client Devices
Laptops
PDAs
Tags
Transceivers

Device Accessories
Spare batteries
Memory cards
Network cards
Keyboard
Printer
Scanners
Synching Cradles
Carrying Cases
Mounting brackets

Information Infrastructure
Application Servers
Mobile Integration Servers

Network
Routers
Hubs
Access points
Transmitters
Transceivers

Software

Device Software
(if not included
  in device cost)

Application Software
ASP
Packages
Custom

Infrastructure Software
Network utilities
Operating systems
Security
Databases
Middleware

Support Tools
Backup and Recovery
Device Management
Network Monitors
Issue Tracking
Problem Resolution Databases

Development Tools
Development Environments
Testing tools
Project Management Tools

Project Management

Program Manager
Project Managers

Development and Integration

Business Process Redesign
Process Analysis
Process Design

Application Development
Analysis & Design
Coding
Testing
Consulting Fees

Application Integration
Analysis & Design
Coding
Testing
Consulting Fees

Documentation
Application and Device
Policies
New Process

Deployment Costs

Installation Charges
Hardware
Software

Device Preparation
Loading Applications
Testing

Support Infrastructure
Setting Up Help Desk
Training Help Desk Staff

Communications
Communications Specialist
Materials

Training
Training Development
Trainers
Attendee Time

Travel
Installation
Training
Support

Recurring Costs

Maintenance Fees
Hardware
Software

Leasing Fees

Service Charges
ASP fees
Airtime
E-mail usage

Support Costs
Telephone and E-mail Support
Backup and Recovery
Device Management
Software and Hardware
  Upgrades and Replacement

Application Maintenance
Production Support
Enhancements

Intangible Costs

Lost productivity
Learning curves
Distraction

FIGURE 6.4
Cost Checklist

- *Who Pays the Costs?*   In the simplest case, all project costs occur in, and are borne by, a single organization. In reality, many solutions are cross-organizational, requiring you to identify where costs appear and who is responsible for paying them. For example, shifting a back-office function to your field service force reduces costs in the back-office while creating extra costs for the field service organization. Or deploying a new system in the sales department increases user support effort and costs in the IT organization. In some cases, others will pay for components of the solution. A college may require its students to purchase their own PDAs to access the college's wireless class scheduling information. Although the cost to students does not directly affect the college's project budget, it is a consideration for the students whose acceptance is crucial to the ultimate success of the effort.

- *What Are the Real Costs?*   One of the more difficult aspects of estimating project costs is identifying the true cost of a given item. A good example is accurately computing people costs. How much does a person's time cost? Among the many options are straight salary, departmental chargeback rates, fully burdened salary, and average charge rate for billable employees. Should small increments in time be charged or will they be absorbed within normal operations? Total Cost of Ownership studies by analyst organizations such as the Gartner Group, for example, show that the true organizational cost of a tool such as a laptop PC is far higher than the purchase cost of the equipment and software. Subtle costs abound, such as the productivity losses that occur when more experienced users are asked to provide handholding to novice users. The best approach for addressing these issues is to follow your company's accepted cost estimation practices wherever possible. Be sure to think about how a cost will be used as you determine "true" costs. For example, the straight (unburdened) salary rate of an internal employee is not an accurate comparison against the billing rate of an external consultant.

- *Cost Precision and Variance*   When producing a cost justification, it is important to note differences in precision between different types of costs. More precise expenses, such as the cost of hardware purchases, are unlikely to vary significantly from their estimates. In contrast, application development costs may have a 20% to 50% or higher variance from original estimates. Depending on your organization's policies, you may choose to use a high estimate, a cost range, or provide best case, most likely case, and worst case scenarios.

- *Depreciation and Expense*   The accounting treatment for different components of your solution will vary based on their type, cost, and company policies. Large capital purchases of hardware and software are usually depreciated across their useful life, while other costs may be expensed when incurred. These treatments affect project cash flow and income statement calculations, as well as corporate financial statements. Be sure to seek guidance from your company's financial organization to determine the proper treatment for your project's components.

- *One-Time and Recurring Costs*   Every project has a mixture of one-time costs, such as hardware purchases and deployment expenses, and recurring costs, such as software maintenance and network fees. Consider whether any of your one-time costs may recur. For example, if your solution has a five-year antici-pated life, are you likely to upgrade or replace PDAs within that time frame? Be sure to spread the cost of fixed, recurring fees across the life of the project and include provisions for potential rate increases where appropriate. Some recurring fees, such as network airtime charges, may be variable and difficult to estimate accurately over the life of the project, requiring the use of one of the precision techniques described above.

- **Source of Cost Information**   Consider the sources of your pricing information. Be careful when using vendor-supplied figures other than straight product prices. Make sure that vendors include all possible costs and that prices are not "low-balled" to get in the door only to be raised once the project has reached the point of no return. The same caveat applies to industry cost statistics as previ-ously described.

6.2.2 Hardware Costs

Hardware constitutes the largest immediate cost for most wireless solutions. Costs mount quickly when deploying large numbers of devices even if the cost per unit is relatively low. Fortunately, estimating hardware costs is straightforward once you have determined your solution architecture and overall scale of the project. Since a given solution may use many different hardware components deployed in many places, the biggest risks are accidentally missing one or more components or acces-sories or forgetting to equip an important, but peripheral area. Hardware costs fall into the following major categories.

- *Client Devices*   Wireless device expenses include the basic cost of each device along with upgrades, accessories, and supporting peripherals. End-user devices such as laptops and PDAs may have numerous add-on items such as spare batteries, cards for additional memory and network interfaces, carrying cases, scanners, printers, keyboards, carrying cases, and synching cradles. In the case of telemetry and telematic solutions, costs include tags, transceivers, and perhaps mounting brackets. To calculate the overall cost, you will need the cost per component and the number of each type of component required. If your solution uses large numbers of components, check to see if quantity discounts are available. When calculating the number of devices required, include the needs of the development and support organizations as well as the solution's users. Many companies also choose to keep an inventory of spare devices and accessories to replace missing or malfunctioning units.

- **Information Infrastructure**   If your company is hosting its own applications, you will need one or more servers to support application software and provide mobile integration. Consider whether extra servers are needed for back-up or to equip your development and support organizations. IT organizations in larger companies are usually very experienced in calculating server pricing and deployment costs.

- **Network**   Depending on the type of network(s) selected, your solution may need a variety of network equipment such as routers, hubs, access points, transmitters, and transceivers. Consider the number of devices you will need to cover all supported locations adequately.

6.2.3 Software Costs

Software costs can range from relatively small for single-purpose solutions that rely on generic packages to very high for complex, custom solutions that are tightly integrated with corporate back-office systems. Software requirements fall into the following areas.

- **Device Software**   Some device software such as the device operating system, browser, and PIM software, is included in the price of the device. If additional software is required, it will usually be licensed and priced per device.

- **Application Software**   The cost of application software varies widely based on its type and source. Custom-built applications are usually the most expensive. The factors affecting their costs are described in Section 6.3.5. Packaged software is generally purchased or leased, and may consist of separate license and maintenance fees. If an application service provider (ASP) provides the application, its costs are usually rolled into the monthly service fee. Software vendors can be very creative with their pricing schemes and discount policies, so check with your vendor rather than using list prices for your estimates. Packaged software may also require customization and/or integration with corporate applications. The vendor may provide this service for a fee or your development organization may perform it. Account for these costs separately in the cost justification.

- **Infrastructure Software**   A wireless solution may need a variety of network and server software and utilities to support its network operations and information architecture. This software includes network utilities, operating systems, security software, databases, and middleware. Although pricing models vary, this software is usually licensed by hardware platform or number of users, and consists of license and maintenance fees.

- *Support Tools*   This category includes the tools and utilities used to manage and support the wireless solution. Examples include backup and recovery software, device management tools, network monitors, and help desk support tools, such as issue tracking tools and problem resolution databases. These tools are licensed in a similar manner as infrastructure software.

- *Development Tools*   If the solution relies on a custom-developed application, customized package and/or integration with corporate applications, software development tools will be required. This category includes tools to support analysis, coding, and testing as well as configuration management and project management. These tools are usually priced per seat, and may be one-time purchases or a combination of license charges and maintenance fees.

6.2.4 Project/Program Management

Every project needs management to track progress, monitor budgets, control risks, resolve issues, and respond to changes. A small project may have a single project manager, while a large project may have a fully staffed program management organization. Be sure to include specific costs for the required level of project management. Use historical information from the company's IT department to determine the appropriate level of staffing and time commitment for a project of your size.

6.2.5 Development and Integration Costs

Unless the wireless solution is very straightforward, such as a WLAN or a simple e-mail application, the company will incur development and integration effort and expenses. These expenses are not restricted to IT effort. Changes to business processes require analysis, design, and documentation effort. Internal staff, consultants, or a mixture of both can perform the activities in this category. The best and most accurate method for gauging these expenses is to request estimates from the individuals who will lead the respective tasks. If your company plans to use outside consultants, they will usually provide cost estimates as part of their proposal. More information on consulting appears in Chapter 7. If the project will use internal resources, follow your company's cost justification policies for determining the proper cost per resource.

- *Business Process Redesign*   This category includes the effort to redesign company processes to support and take advantage of the efficiencies offered by the wireless solution. Activities include analysis of existing processes, identification of potential process changes, and the design of new processes. Expenses associated with the rollout of the new processes are covered under deployment.

- *Application Development*   Application development costs include analyst and developer effort for application analysis, design, coding, testing, and production deployment. If developers are not familiar with the solution's technologies, additional time and budget must be allocated for developer training.

- *Application Integration*   Application integration costs cover similar areas as application development; however, since changes affect production applications, separate estimates may be required from the development teams supporting each affected application.

- *Documentation*   Documentation will be required to support training and the rollout of the solution. Documentation needs include materials to support application and device use, policy changes, and new business processes. Documentation may be provided in printed form, as a help feature within the application, and/or over a corporate intranet. Document costs include writing, review time, and production expenses.

6.2.6 Deployment Costs

Deployment is a major expense bucket for many wireless solutions, and is particularly large for solutions that involve major process changes or whose users are dispersed over a wide geographic area. Deployment costs fall into a variety of disparate areas and are best estimated by the individuals responsible for their delivery.

- *Installation Charges*   This category includes charges for physically setting up networks, installing servers and other hardware, mounting devices where required, and loading and setting up infrastructure and support software.

- *Device Preparation and Setup*   Many of the companies described in Chapter 3 pre-loaded and tested all client devices before distributing them to users. If large numbers of devices will be deployed, this effort can be significant.

- *Support Infrastructure*   The user support infrastructure for the solution has to be in place before deployment. This effort may be as simple as training existing help desk or support personnel or as extensive as setting up, equipping, and training an entirely new organization.

- *Communications/Awareness*   Large projects benefit from the use of a communications person or group to build awareness, excitement, and support as the solution approaches rollout. Costs include staff time and materials.

- *Training*   User training is an important aspect of any solution deployment and is a critical success factor for solution acceptance. For this reason, be sure to request sufficient funds for a quality education effort. Costs in this category include training development, production of training materials, trainer time and expenses, and attendee time and expenses.

- *Travel*   Travel expenses can be significant for solutions that will be deployed over a wide geographic area. Travel expenses may be incurred for evaluation and pilot efforts, as well as for installation and training activities.

6.2.7 Recurring Costs

The previous cost categories focus primarily on one-time charges. A number of additional cost categories will recur regularly through the life of the project and must be factored into the cost justification.

- *Maintenance Fees*   As described above, some hardware and software components may have annual maintenance fees of up to 12–18% of their license cost to cover support and upgrades.

- *Leasing Fees* If the project's hardware and/or software is leased or rented rather than purchased, periodic costs will be charged, typically monthly or quarterly.

- *Service Charges*   Service charges are usually usage based and cover items such as network airtime, e-mail usage, access to ASP applications, etc.

- *Support Costs*   User support is one of the largest long-term costs for a wireless solution. This category includes help desk operations, regular back-ups and recovery of application data, device management, software and device upgrades, and replacements for lost or damaged devices. Seek help from your company's IT organization in estimating appropriate cost allocations for each area.

- *Application Maintenance and Enhancement*   If your solution includes custom-developed applications, customized packaged applications, or integration with corporate applications, change is inevitable, requiring a separate budget for long-term maintenance and enhancement. The size of this budget will depend on the complexity of the application(s) and the predicted level of change. Have the developers of the application include an estimate of future maintenance effort as part of their development estimate. If external consultants develop the application, they may be willing to provide ongoing support on a fee basis.

6.2.8 Intangible Costs

A wireless implementation project may incur a number of intangible and difficult to quantify costs, ranging from short-term productivity losses as users learn to use the new solution, distraction from getting a new toy, to disruption in routines from changed processes. Whether you decide to identify and include these potential impacts in your cost justification will depend on your organization's policies and practices.

6.3 Producing the Justification _____

Once you have your solution's costs and benefits in hand, you can begin the project justification analysis. Most companies require some type of formal analysis before evaluating and approving a project, although you are unlikely to proceed with this analysis if your cost and benefit data already indicates the project has dubious merit. This analysis has three major steps. The first step organizes the information in a way that allows executives to assess the impact and timing of project expenses and benefits quickly. The second step uses the data to compute the project's financial value to the company in terms of ROI and payback period. The final step creates a project justification report to back-up the analysis. This report includes information such as the assumptions made when performing the analysis, a discussion of intangible benefits, and an assessment of project risks. Many companies have standard, pro forma approaches for producing project justifications and you should use your company's preferred method. The method discussed in this section is generic and draws upon standard techniques used for other common financial analyses.

6.3.1 Analyzing Costs and Benefits

One effective approach for analyzing project costs and benefits is to prepare income and cash flow statements such as those commonly used by the investment community to assess corporate performance. Depending on your company's justification approach, an income statement alone may be sufficient. The purpose of the income statement is to analyze whether the project stands to make a profit. It is prepared by listing the value of the benefits earned during the analysis period, listing the expenses incurred to earn those benefits, and subtracting expenses from benefits to determine profit or loss. Benefits include inflows of revenue and savings produced by the project. Income statements differ from cash flow statements by the way they treat benefits and expenses. In an income statement, expenses and benefits are recorded when they accrue rather than when funds are actually paid or received. For example, if a hardware purchase is capitalized, its value is depreciated over its useful life. Rather than incurring a cost at the point of purchase, the income statement includes these costs as depreciation expenses as they accrue throughout the project. This method smoothes the project's financial performance by showing smaller losses at project initiation and smaller gains over the project's life.

In contrast, a cash flow statement analyzes the project from the perspective of cash inflow and outflow. Expenses are shown as cash outflows when they are paid and benefits are shown when received by their impact on the company's cash position (revenues are cash inflows, savings are cash not spent). Since most projects incur expenses at their start and receive benefits over their life, they have a strong negative cash flow at the onset and strong positive cash flow later. This cash flow impact is important when the project is large. The size of a long-term benefit is irrelevant if the company runs out of cash before achieving it.

	2003	2004	2005	2006	2007	Total
Benefits						
Additional Service Revenue	94,500	157,500	220,500	252,000	375,750	1,100,250
Data Entry Labor Savings	114,048	114,048	114,048	114,048	114,048	570,240
Error Correction Labor Savings	86,400	86,400	86,400	86,400	86,400	432,000
Total Benefits	294,948	357,948	420,948	452,448	576,198	2,102,490
Expenses						
System Depreciation	58,000	58,000	58,000	58,000	58,000	290,000
Maintenance Contracts	18,000	18,000	18,000	18,000	18,000	90,000
Service Charges	11,985	11,985	11,985	11,985	11,985	59,925
User Support	75,000	75,000	75,000	75,000	75,000	375,000
Application Support	50,000	50,000	50,000	50,000	50,000	250,000
Miscellaneous	10,000	10,000	10,000	10,000	10,000	50,000
Total Expenses	222,985	222,985	222,985	222,985	222,985	1,114,925
Income						
Pre-Tax	71,963	134,963	197,963	229,463	353,213	987,565
Income Tax (34%)	24,467	45,887	67,307	78,017	120,092	335,772
Net Income	47,496	89,076	130,656	151,446	233,121	651,793

FIGURE 6.5

Example Income Statement

6.3.1.1 Preparing an Income Statement   Preparing an income state-
ment is relatively straightforward once your cost and benefit data is organized. Any good accounting textbook will contain detailed tips and techniques to assist your preparations. Computing depreciation expenses and tax impact are the two most challenging aspects of the income statement. Seek assistance from your company's finance organization to answer questions and ensure that you are following company standards and guidelines. In particular, company rules for depreciation vary widely, with each company setting its own policies for which costs should be capitalized or expensed. Variables include type of cost, dollar thresholds, and depreciation time periods.

Figure 6.5 shows an example income statement. Use this income statement as a starting point, but feel free to change the individual line items to match the needs of your solution.

6.3.1.2 Preparing a Cash Flow Statement   The cash flow statement is
prepared in a similar manner as the income statement, except expenses are listed when incurred rather than accrued. Figure 6.6 shows an example cash flow statement.

	Initial Investment	2003	2004	2005	2006	2007	Total
Benefits							
Additional Service Revenue		94,500	157,500	220,500	252,000	375,750	1,100,250
Data Entry Labor Savings		114,048	114,048	114,048	114,048	114,048	570,240
Error Correction Labor Savings		86,400	86,400	86,400	86,400	86,400	432,000
Total Benefits		294,948	357,948	420,948	452,448	576,198	2,102,490
Expenses							
Hardware Purchase	(35,000)						(35,000)
Software Purchase	(85,000)						(85,000)
Development Consulting	(120,000)						(120,000)
Deployment	(50,000)						(50,000)
Maintenance Contracts		(18,000)	(18,000)	(18,000)	(18,000)	(18,000)	(90,000)
Service Charges		(11,985)	(11,985)	(11,985)	(11,985)	(11,985)	(59,925)
User Support		(75,000)	(75,000)	(75,000)	(75,000)	(75,000)	(375,000)
Application Support		(50,000)	(50,000)	(50,000)	(50,000)	(50,000)	(250,000)
Miscellaneous		(10,000)	(10,000)	(10,000)	(10,000)	(10,000)	(50,000)
Income Tax (34%)		(24,467)	(45,887)	(67,307)	(78,017)	(120,092)	(335,772)
Total Expenses	(290,000)	(189,452)	(210,872)	(232,292)	(243,002)	(285,077)	(1,450,697)
Net Cash Flow	(290,000)	105,496	147,076	188,656	209,446	291,121	651,793

FIGURE 6.6
Example Cash Flow Statement

6.3.2 Assessing ROI

When deciding whether to fund a project, companies use a variety of calculations to assess the level of return that they can expect from their investments. Three common tools include calculating the project's payback period, computing ROI, and calculating the overall value of the project in today's dollars (net present value). Companies generally set internal hurdle rates, such as having a payback period of three years or less and/or providing an ROI of at least 25%, in order to receive approval. Check with your finance department to determine the company's preferred calculations and standard hurdle rates.

6.3.2.1 Calculating the Payback Period   The payback period is the length of time required for your company to recoup its investment in the project. This calculation is derived from the project's net cash flow and is expressed in terms of years (Figure 6.7).

	Initial Investment	2003	2004	2005	2006	2007
Initial Cash Outflow	290,000	0	0	0	0	0
Net Cash Savings		105,496	147,076	188,656	209,446	291,121
Payback Period	2.2 Years					

FIGURE 6.7
Example Payback Period Calculation

Annual Cash Benefit (5 year Average)	$	188,359
Capital Investment	$	290,000
Average Return on Investment		**65%**

FIGURE 6.8
Example ROI Calculation

6.3.2.2 Computing Straight ROI   Where payback period focuses on cash outlays, ROI provides a means of assessing the financial benefit returned by the project's capitalized investments. Expressed as a percentage, it provides a measure for comparing an investment in the project against other possible uses for the same funds. For example, if a project returns 27% on its investment while the corporate bank account pays a 5% interest rate, the company's money is better invested in the project.

ROI is calculated by dividing the five year average of the net income (after tax savings) shown in the Income Statement by the project's capital investment. (Figure 6.8)

6.3.2.3 Calculating Net Present Value   Net Present Value (NPV) is a method of calculating the total future financial value of your project in today's dollars. It is based on net project cash flow and provides executives with a means of comparing the project's returns against other investment options. Given inflation, a dollar in the future is worth less than that dollar today. If a project returns an average of $200,000/year in benefits, its total five-year return is $1,000,000. To calculate the value of that total return in today's dollars, we need to know the timing of each benefit. We then apply a **discount rate** that typically contains an adjustment for inflation and a risk-adjusted return on the use of the money. This rate is compounded over the time period in the calculation (Figure 6.9). If we use a discount rate of 5%, the $1,000,000 return is worth $865,895 in today's dollars. Ask your finance department to provide the discount rate used by your company.

	Initial Investment	2003	2004	2005	2006	2007	Total
Net Cash Flow	(290,000)	105,496	147,076	188,656	209,446	291,121	651,793
Net Present Value of Cash Flow	(290,000)	100,472	133,402	162,968	172,311	228,101	507,254

Discount Rate   5.0%

FIGURE 6.9
Example Net Present Value Calculation

6.3.3 Writing the Project Justification Report

The objective of project justification reports are to present your solution in the best possible light to the executives who will decide on its approval. To win approval, the report must clearly spell out the project's tangible and intangible benefits, provide confidence as to its feasibility, size the cost and effort of its implementation, and establish the project's value relative to other investment opportunities. Remember that this report is a selling document; to capture the interest of its readers, it must be accurate, well-written, and radiate enthusiasm about the project.

Many companies have pro forma approaches to project justification and your finance department is likely to have good examples of successful justification reports to serve as a model. While the format and contents of the report will depend on your company's preferences, project justifications commonly contain the following major sections.

- *Background*   The background section explains the rationale behind the project. It describes the circumstances that triggered the investigation of the solution and identifies the issues that the solution will address. It discusses the business and technology drivers that shaped the solution design and creates urgency by describing why the solution should be considered now.

- *Solution Description*   This section briefly describes the solution and its use. It explains the solution's technology, how that technology will be deployed, and how the solution will enhance company performance and operations.

- *Solution Benefits*   This section discusses the solution's expected benefits. It introduces the tangible benefits that will be the basis of the cost/benefit analysis in the Financial Analysis section of the report and describes the additional intangible benefits that strengthen the value of the project.

- *Financial Analysis*   This analysis provides the economic justification for the project. Using tables and spreadsheets, it compares benefits and costs to determine project payback period, ROI, and other measures for assessing investment value. This section also includes a description of the assumptions used when calculating benefits and costs.

- *Project Implementation*   This section summarizes the solution's high-level project plan. It describes how the solution will be implemented, along with a suggested approach, timeline, and anticipated resource requirements. It discusses the technological, organizational, and procedural impacts of the project and how those impacts will be addressed during implementation. It identifies project risks and possible mitigation approaches.

- *Back-Up Materials*   Analyst reports, trade press articles, and other materials that support the choice of solution and strengthen its benefit arguments can be included as appendices.

After writing the project justification report, review it carefully with members of your company's finance, HR, and IT organizations, as well as with the solution's intended users. This review verifies the accuracy of the report content and ensures that the project's constituents agree with its conclusions.

6.4 Selling the Solution

Even the best solutions do not sell themselves. A thoughtful, carefully researched, and well-written project justification is an important starting point, but gaining approval for a full implementation and rollout requires building consensus and support across a disparate set of constituents. Each affected group has its own needs and priorities that don't necessarily match the goals of the project, nor each other. Your sales force may be excited and impatient to receive a new PDA-based solution, while your IT organization is backlogged with other pressing priorities, and your finance organization would rather direct project funds elsewhere. Part of the challenge of implementing a new wireless solution is getting each group to understand why the solution is important and how it will benefit them as well as the company. The larger the project, the greater the number of groups that will be affected and the larger the selling effort. The intended users of the solution may need the most selling. A solution that excites management as a means of cutting costs and increasing efficiency looks like more work and fewer jobs to the line workers. Even when the intended users intellectually understand the benefits of the solution, they might still resist the changes needed to make the solution work. Most of these issues can be dispelled through education and involvement. A combination of presentations, working meetings, and proof-of-concept projects help the constituent groups better understand the project and provide an opportunity to enlist their support in planning and executing the implementation.

6.4.1 Presenting the Solution

In all likelihood, you will have to present your proposed solution to many different groups. This presentation should cover the project's background, describe the solution, and explain its benefits. Use a demonstration if at all possible. Wireless devices have high appeal for many users and demo well. While the project justification report can be the basis for all of these presentations, it is important to know each audience and tailor the presentation accordingly. Ask yourself which benefits will have the greatest resonance for the intended audience and which aspects of the project cause greatest concern. Understanding each audience's needs and biases helps you to explain the solution in a way that is compelling for them. For example, sales executives may find the management reporting features of a wireless SFA solution its most compelling benefit, while sales representatives are most attracted to the ability to receive e-mails and check order status at a customer site. The sales representatives may even fear the management reporting features and consider them a drawback. Using the same presentation for sales representatives and sales executives may spawn unnecessary objections and resistance. Allow plenty of time at the end of each presentation to answer questions, alleviate concerns, and accept recommendations.

6.4.2 Demonstrating Feasibility

We can't blame executives for being skeptical about costs and benefits of a wireless solution. Historically, technology projects have a less than stellar track record for arriving on time and within budget and for delivering promised benefits. Use of new technologies exacerbates this trend. Unexpected issues arise and solution components rarely work together as first advertised. Use of proof-of-concept exercises and/or a pilot project can demonstrate the feasibility of the project while providing actual data to back-up cost and benefit claims. These initiatives strengthen the project justification, surface potential issues, and provide an excellent demonstration for solution presentations.

- *Proof-of-Concept*   A proof-of-concept is a quick, small-scale exercise that tests the feasibility of one or more aspects of a solution. It does not require a complete implementation to be useful. For example, the usability aspects of a wireless SFA solution may be tested by giving a group of sales representatives PDAs loaded with a prototype of the client application. During this proof-of-concept, the back-end operations needed to support the application are simulated manually. This exercise helps to build sales representative support while demonstrating the feasibility and value of the solution.

- *Pilot Project*   A pilot project is a larger version of a proof-of-concept. It consists of a controlled rollout of a portion of the solution (or possibly the full solution) to a small group of participants. For example, a company may test the feasibility and usability of wireless e-mail access by providing a test group of executives with RIM BlackBerry devices. This solution is fully functional and allows the company to assess the value of the solution and identify potential issues before approving a wider rollout.

6.4.3 Building (and Keeping) Internal Support

The real work of building and keeping support begins after the project is approved. It is critical to have the full support of the project's constituents throughout the implementation to overcome issues and guarantee a fast and successful deployment. Actively seek the support of individuals from all of the organizational areas that are affected by the project. Within each major constituent area, try to enlist the support of the key influencers. Highly regarded by their co-workers, these influencers tend to be early adopters of new solutions and their acceptance of a solution (or lack thereof) has significant impact on its adoption. Rely on these individuals to help refine the project's goals and objectives and to assist in identifying and solving project issues. Giving these constituents a feeling of ownership in the project and its outcome builds commitment and encourages active involvement. Likewise, regular communication of project progress and successes is essential for building and maintaining momentum. The communications process should begin as early in the project cycle as possible. Larger projects should consider using a full- or part-time communication specialist to keep constituents informed, involved, and excited.

Implementing the
Solution

7

Implementation is the process of turning your wireless solution into reality. Depending on the scope and type of solution, implementation can be as straightforward as negotiating a contract and providing users with training and devices, or it can be a highly complex, multi-step, multi-year effort involving hundreds of people. The purpose of this chapter is to describe the activities and issues of a wireless implementation from a business perspective. Its goal is to provide readers with just enough information to understand relevant implementation issues and avoid major pitfalls. Additional details about specific technology issues are found in later chapters and in the many widely available technical texts on wireless technology.

Wireless technology implementations share many issues and characteristics with other technology implementations. Challenges such as technology immaturity and user resistance to change are common to many types of projects. The newness and evolving nature of wireless technology exacerbates these issues. Standards are still evolving and the best methods for areas such as user interface design have yet to be discovered. Device limitations and a lack of supporting tools require application developers to return to austere programming techniques used in earlier days. Add to these issues the challenges of dealing with a mobile user base potentially dispersed over a wide area and you will understand why wireless solutions require some additional planning and care during implementation. Fortunately, these challenges are surmountable and many companies have successfully implemented solid and valuable wireless applications.

This chapter covers how to plan the implementation, manage the project, and redesign the underlying business processes. It offers implementation and deployment tips and techniques and describes how to get assistance from the right wireless service provider.

7.1   Planning the Implementation _____

When it comes to implementing a wireless solution, planning is everything. Even a relatively small implementation may involve selecting and integrating solution components, devising a support structure, installing hardware and software, setting up devices, training users, and deploying the solution in multiple locations. Each of these activities has resource requirements, dependencies, and scheduling considerations. Add business process redesign, involvement of several internal organizations, solution components from multiple vendors, and a team of consultants, and the need for thorough and careful planning becomes obvious. Planning helps to ensure that resources are available when needed, schedules are reasonable and achievable, risks are identified, and contingencies and mitigation strategies are in place. Planning provides the roadmap by which your wireless solution is implemented.

7.1.1   Planning Considerations

The goal of the planning effort is to identify the tasks to complete, the resources needed for those tasks, the level of effort required from those resources, and dependencies between tasks and resources. Its end product is a project plan that includes a budget, resource requirements, and an actionable schedule complete with milestones and interim deliverables. The attributes of your project will guide the development of this plan. Factors, such as the number and geographic location of implementation sites, affect timing and resource requirements. For example, if the wireless solution is deployed at ten locations in the continental U.S., will you use one team flying from location to location to perform the installation or dispatch ten teams to work simultaneously? The following list covers some of the more important planning considerations.

- *Scale*   The level of effort devoted to planning will depend on a number of factors, such as the size and scale of your project. Deploying an e-mail solution based on a RIM BlackBerry device to a group of employees based out of a single facility is obviously more straightforward than a solution that supports hundreds of end users, dozens of device types, and a wide geographic area. Project scale is affected by the following items.

- *Scope* Scope is determined by factors such as the number of users, devices, and sites supported. The larger the scope, the more resources required for preparation, deployment, and support.

- *Complexity* The complexity of the solution affects everything from project risk to development effort, skill levels, and travel costs. Three attributes of complexity include:

  - *Technology* A single "off-the-shelf" solution is far less complex than a multi-technology solution, such as the one deployed by UPS for its package sorting operation.

  - *Integration* How many back-end data sources have to be tied together?

  - *Rollout* How many sites will be deployed with the solution and where are they located? How many activities are required for each location?

- *Level of Change*   The greater the level of change imposed by the wireless solution, the more effort that must be devoted to activities such as communications and awareness, training, and documentation. Providing a new application to current PDA users is a much lower level of change than a total redesign of a major business process.

- *Timing* Timing considerations abound. How quickly must you deliver the project? What is the lead-time to obtain a desired piece of equipment, negotiate a network contract, or find and train resources? Five traveling training teams, for instance, will cover twenty locations faster than a single team, but a six-week lead time for a network hook-up has to be factored into the project schedule rather than addressed through more resources.

- *Dependencies* Perhaps the greatest challenge in building a plan for a complex project is dealing with the dependencies between individual tasks. For example, before application development can begin, developers must be trained and ready, the development tools installed, and the development infrastructure setup. This challenge grows exponentially as the number of tasks and resources grow.

- *Capabilities* Does your organization have the capabilities to handle the range of tasks needed to implement the solution? Does it have the right skills? Does it have enough free people? Does the opportunity cost justify diverting those people from another effort? The availability of skills affects cost, lead times, project risk, and the decision of whether to hire consultants or outsource portions of the project.

7.1.2 Implementation Approach

The next major decision point in the planning process is determining *how* to implement the solution and *who* will perform the implementation. Implementation options include source of functionality, source of expertise, and responsibility for operations.

- *Functionality*   You can develop application functionality from scratch, purchase and use it "as is," or purchase and customize it. While packaged wireless applications are still limited, vendors are developing new applications at a rapid pace for many functional needs. Even when full applications are unavailable, functional components may exist that can address part of the need. Similar options are available for devices and other hardware. While most solutions will use "off-the-shelf" equipment, it is possible to have custom-designed hardware to handle special-purpose needs.

- *Expertise*   You can have the project delivered on site, off site, or even offshore. On-site projects are staffed by internal employees, consultants, or a mixture of the two. Off-site and offshore projects usually involve the development and delivery of a full solution.

- *Operation*   An external company can provide all or part of the wireless solution. For example, airtime on wide area wireless networks is typically purchased as a service from a regional or national communications provider. Wireless Application Service Providers (WASPs) offer solutions that include applications and full back-end operational support. Other companies will take responsibility for device management and deployment. Alternatively, your company can install and operate its own solution.

Choosing the right implementation approach is a tradeoff between business needs and the cost and difficulty of implementation. If the solution is generic, such as wireless e-mail using a RIM BlackBerry device, the implementation is quite simple. Using a fully packaged service eliminates application development, many aspects of infrastructure deployment, and reduces future support costs. Conversely, a highly customized solution may require building many components of the solution from scratch, call for highly trained specialists, and have major implications for long-term support. Given the difference in cost and risk, it is tempting to select solutions that are quick and easy to implement. Unfortunately, generic, packaged solutions are not available for all needs, and, even if available, provide no competitive advantage over other solutions. Before selecting your preferred implementation approach, consider the following questions.

- *How Unique Are Your Requirements?*   If your business requirements are new or unique, you will have little choice but to develop a custom solution. Other business requirements are common across companies and industries and are often supported by packaged functionality. For example, you would find it difficult to justify building your own e-mail or contact management applications. In some cases, a packaged application may cover most, but not all, of your business needs. In this situation, you have to decide whether the missing functionality is worth the cost and risk of custom development.

- *Is the Solution Strategically Important?*   A solution that provides your company with major strategic benefits, such as first mover advantage in offering a new service or creating a barrier to competition, will justify building a custom version to gain additional functionality or enable earlier deployment.

- *Does Your Company Have the Skills to Create the Desired Solution?*   The newness of wireless technology means that many companies lack the internal skills to build complex wireless applications. Companies face the choice of training their employees in the desired skills, recruiting new employees, or hiring consultants. If wireless technology will be an important part of your company's business, hiring or training your own resources is probably advantageous. If wireless technology is merely a means to a business end, it does not have to be a core competence, making it more advantageous to use consultants or purchase services. Even if it is advantageous to build your own capabilities, using consultants early in the process can shorten delivery and provide mentoring to internal staff.

- *What Type of Technology Is Required?*   If your solution involves a single location, 802.11b WLAN, you may want to own, install, and maintain that network. If the solution involves national network coverage, developing your own network is not only too costly, but infeasible as well.

- *How Closely Is the Solution Integrated with Internal Systems?*   Freestanding solutions are easy to outsource to consultants or to application service providers (ASPs). Solutions that require tight integration to back-end corporate applications will need internal assistance and are likely to require a higher level of customization.

7.1.3 Deployment Approach

Determining how to deploy the solution is another important aspect in designing the project plan. The deployment approach affects project staffing, schedules, risk levels, and costs. To a large degree, the choice of deployment option is driven by the size and complexity factors described in Section 7.1.1, but other business objectives may affect the desirability of one approach over another. While many variations exist, deployment falls into three basic categories.

- *"Big Bang"*   This approach completes the entire project and deploys the solution to all users *at one time*. Small wireless projects, such as deploying an intra-office WLAN, are usually deployed in this manner. This approach has the shortest timeframe and least overhead, and releases team members right after deployment. It requires a high level of confidence in the quality of the solution and its ability to be easily assimilated. It is less feasible for large, complex projects due to the logistics of deploying the solution to many users over many locations. While it shortens the project timeframe and lowers overhead in earlier phases of the project, it requires a large deployment team in its final stages.

- *Pilot* A pilot approach softens the "big bang" approach by testing the completed solution on a smaller scale before a large-scale rollout. The pilot allows the project team to test all components of the solution, including training and support procedures before progressing to a larger audience. It builds confidence in the solution and smoothes the larger phase of the rollout. On the downside, it lengthens the project schedule.

- *Phased Rollout* A phased rollout breaks deployment into smaller pieces. It may involve the rollout of a complete solution area-by-area or a rollout of functionality layer-by-layer. Phasing the rollout by area or location allows for smaller deployment teams and lowers the stress on user and system support areas. Issues discovered in the first locations can be ironed out before progressing to subsequent areas. A drawback of this approach is the length of time needed for overall deployment; however, it may be the only feasible option for large, widely dispersed solutions. Rolling out functionality by layer spreads development effort over time, permits the use of smaller development teams, allows users to start gaining experience using the solution, and enables the company to start reaping partial business value sooner.

7.1.4 Implementation Roadmap

Developing a project plan for any complicated project is intimidating. If the project is viewed as a whole, the number of details appears overwhelming and it is difficult to know how and where to start. Happily, even the largest project can be broken into "bite-sized" pieces by project phase and activity category. Given their size and scope, defining and estimating the activities within each of these pieces becomes quite manageable. Once defined, these pieces are rolled up into a complete project plan. Figure 7.1 shows a simplified deliverable of this process. It illustrates a high-level roadmap for a larger-scale wireless solution that involves business process redesign, contracted network services, and custom-developed applications integrated into central, back-end data sources.

The example uses six phases: planning, preparation, implementation, testing, deployment, and operation. While the basic activities included in these phases are similar for all projects, many other breakdowns and naming conventions are possible.

	Project Management	BPR	Infrastructure				Support	
			Hardware	Networks	Information	Applications	User	System
Planning	Select Program Manager Create Project Plan Setup PMO Recruit PMO Team Finalize Proj. Plan		Finalize Reqs	Finalize Reqs	Finalize Reqs	Finalize Reqs	Finalize Reqs	Finalize Reqs
Preparation	Setup Delivery Infrastructure Finalize Implem. Plan	Recruit BPR Team Provide BPR Training	Order Development Hardware	Evaluate Network Suppliers Negotiate Service Contract	Purchase Middleware	Recruit Development Team Train Development Team Purchase Components Purchase Tools	Develop User Support Polices	Develop System Support Policies Purchase Support Tools
Implementation	Manage Implementation Setup Testing Infrastructure Finalize Testing Plan	Analyze Current Processes Redesign Processes Document Processes	Install Development Hardware Order Support Hardware Install Support Hardware	Install Development Networks	Setup Information Infrastructure Integrate Data Sources	Setup Devel. Environment Develop Applications Client Back end Unit Test	Recruit & Train Staff	Recruit SupportStaff Develop Sys Supp. Environment Security Backup/Recovery Device Management
Testing	Manage Testing Setup Deployment Infrastructure Finalize Deployment Plan	Simulate Process Revise Processes Revise Documentation	Order Production Hardware	Install Test Networks		Integrate Apps & Info Integration Test System Test	Setup Help Desk Prepare App. Training Prepare Process Training	Test Support Environment
Deployment	Manage Deployment Setup Production Infrastructure		Install Servers	Install Production Networks		Setup Devices	Provide User Training Provide Intensive Support	Train administrators
Operation		Measure Performance Continuously Improve		Monitor Network Performance	Application Maintenance & Enhancement		Provide User Support	Provide System Support

FIGURE 7.1
The Implementation Roadmap

Follow the method preferred by your company or, if you are using a formal project methodology, stick to it. Similarly, the category columns vary by project and who is responsible for implementation. While the activities within each column are likely to have interdependencies with activities in other columns, they focus on a specialized area of the implementation, and are usually assigned to an independent team. For example, the BPR team is composed of business analysts and users and focuses strictly on business processes, while the application development team is composed of programmers and focuses on coding the client applications. The number of discrete teams will depend on the size and scope of the project. In a small project, a single team may be responsible for applications, information infrastructure, and client devices, and the support team may be responsible for both user and system support. Where possible, use representatives from each team to devise their own portions of the project plan. The resulting plans are likely to be more accurate, and team members have a greater investment in meeting schedules that they helped to develop.

7.2 Managing the Project

Although project management is by no means unique to wireless technology, it is essential to the success of your implementation effort. The best tools, technicians, and technologies are useless without clear objectives, adequate support, and solid management. Given their unique nature, wireless technology projects bring their own coordination and alignment issues in addition to the standard challenges faced by a technology implementation project. These challenges can include the learning curve and risks of new technology, combining solution components from many sources, and managing the solution's deployment to a highly dispersed set of users. Choosing and implementing the right project management structure and environment helps to surmount these and the myriad of other challenges the project will inevitably face.

7.2.1 Why Projects Fail

As any book on project management will attest, technology projects fail at an alarming rate. These failures do not have to be inevitable. Understanding the most common factors that cause projects to fail allows you to mitigate those factors, greatly enhancing the project's likelihood of success. Studies by the Standish Group, the major analyst firms and experts such as Ed Yourdon and Capers Jones, highlight that failing projects share one or more of the following traits.

- *Inadequate Support* To succeed, a project requires a committed executive sponsor and full organizational support. Without this support, the project cannot get the budget, resources, and cross-functional support it needs to meet its objectives. Addressing a real and recognized business issue gains executive interest. Executive commitment is cemented through a strong business case backed by a solid project justification.

- *Unclear Objectives* A successful project needs a clear vision and well-defined goals and objectives. While some degree of change is normal, constantly shifting requirements and inadequately specified project deliverables are a recipe for failure. Technical specifications may evolve as the project progresses and knowledge about wireless technology grows, but the business goals and functional objectives should be defined and agreed upon at the onset of the project, providing a single, constant target throughout its duration.

- *Unrealistic Expectations* Naïve estimates, the pressure to minimize costs and resources, or the desire to gain the solution's benefits as quickly as possible lead to overly optimistic delivery schedules and unrealistic performance expectations. Disaster occurs when the project team inevitably fails to meet its commitments.

- *Lack of Planning and Management*   Despite the best intentions of team members, projects are not self-managing. Adequate planning is required to set realistic and achievable schedules, and basic project management practices are necessary to provide the oversight and discipline to remain on time, within budget, and tracking to specifications.

- *Inexperience in the Technology*   Lack of experience in a new technology leads to poor code, unworkable architectures, and missed schedules. This risk is especially high for organizations implementing their first wireless solutions or deploying new or more complex versions of the technology. To avoid this risk, engage experienced external resources to deliver the project or to train and mentor your company's staff. Implementing small proof-of-concept projects is a lower risk method of gaining expertise before attempting a larger-scale implementation.

- *Lack of Resources*   Failing projects are often understaffed from the start; their only hope of success lies in heroic efforts by team members. A strong business case combined with solid planning and an accurate estimate gives executives the information they need to provide an adequate budget and assemble a properly staffed and trained team.

Implementing the right project management structures and practices at the earliest stages of the project can help avoid these pitfalls. An experienced project manager has the expertise to identify and mitigate project risks. Factors that enhance project success include solid executive support, user involvement throughout the project, use of a formal project methodology, regularly scheduled deliverables, change control processes, and regular project reviews.

7.2.2 Project and Program Management Structures

Wireless implementation efforts range widely in size and needs. While the basic project management functions remain the same, the size and scope of a management structure for implementing a small intra-office WLAN is very different from that of a multi-year, cross-functional project that affects a national fleet of trucks and dozens of dispersed distribution centers. The former implementation is an example of a project. It is a separately defined and managed undertaking delivered by its own team following a plan. Most likely, the company's IT organization implements and manages it. The second implementation is an example of a program. It consists of a series of multiple, concurrent, complex projects that accomplish an enterprise-level business objective.

Most companies have adequate project management expertise to handle small- to medium-scale wireless implementations. An experienced project manager has the skills to identify and develop mitigation strategies for project risks, knows how to plan and estimate, and can manage a project within budgetary and schedule con-straints. Choose a project manager who is comfortable with the uncertainties associ-ated with new technologies. The ideal wireless project manager is someone with a "can do" attitude who believes that every problem can be solved and is willing to experiment to find the right solution. Classic, operational-oriented managers can support the solution after deployment, but are likely to become stressed by the num-ber of factors that are out of their control during implementation. Actual wireless experience is obviously beneficial, but is likely to be in short supply for the foresee-able future. If these skills are not available within your organization, consider retain-ing an experienced manager from a consulting firm or outsourcing the implementa-tion effort altogether.

Divide larger wireless implementations into a series of smaller projects imple-mented in series or parallel by separate teams. For example, a wireless solution that involves significant business process redesign along with a custom-developed wire-less application would use separate teams for each task. Coordinate the individual projects through a Program Management Office (PMO) structure. PMOs are well suited for large, cross-disciplinary implementations and contain the full range of project management functions needed to oversee complex efforts. Many companies used PMOs to manage their Year 2000 compliance efforts. Figure 7.2 shows the func-tional structure of a typical PMO. The size and scope of the wireless project will determine the number of people staffing each function. A small PMO may consist of two to three individuals responsible for several functions apiece, while a large PMO supporting a major implementation may have several people dedicated to each function.

A PMO has the capabilities needed to overcome the risks described in the previous section. The Executive Steering Committee ensures strong sponsorship, provides oversight, and maintains alignment between the project teams and executive objec-tives. Financial and Performance management controls project budgets and tracks overall performance to objectives. Quality Assurance oversees the quality of program deliverables. The Resource and Infrastructure Management function is charged with procuring the right resources and equipment for the effort. It is especially important in complex, multi-vendor wireless implementations. Delivery Management oversees and coordinates the individual implementation teams. Process Management and Improvements is responsible for finding tools, training, and techniques to enhance the functioning and productivity of the program's project teams. The Enterprise Plan-ning and Risk Management function ensures that the program remains firmly aligned with the solution's business objectives, handling shifting priorities and identifying and mitigating project risks. Communications and Awareness coordinates informa-tion flow across dispersed project teams and helps build organizational excitement about the solution.

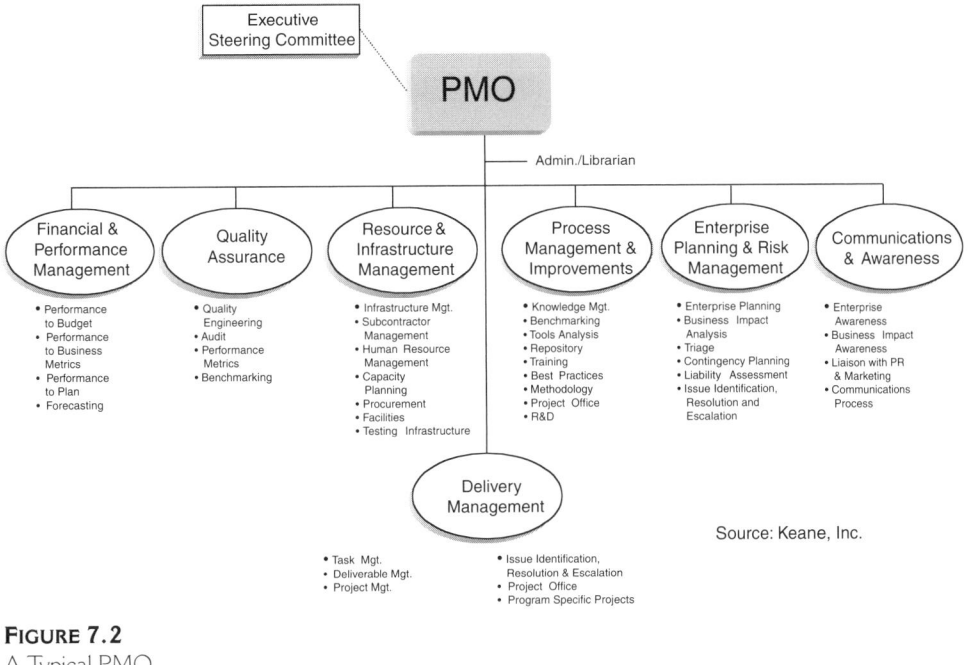

FIGURE 7.2
A Typical PMO

7.3 Business Process Redesign _____

With the exception of the simplest wireless network implementations, the real leverage in wireless technology is its ability to improve the efficiency of business processes. Whether the wireless solution improves responsiveness to customers, increases the efficiency of delivery routes, or shortens billing cycles, it is the change in the business process—how people do their work—that provides the benefit, not the implementation of the new technology. Success depends on using wireless technology to support a desired process change not vice versa. To gain benefit, a company must redesign the process to support its desired goal, train users in the new process, and ensure that it is accepted and adopted once implemented. A failure to do so guarantees that your wireless solution will not deliver its promised benefits. As business process redesign is an essential component of a wireless solution implementation, this section will provide a brief overview of its concepts and basic steps. Since our focus is on wireless solutions, we will forgo descriptions of the initial mission setting and planning aspects of a business process redesign effort. Presumably, these activities were covered in the early stages of your wireless solution design effort. Further information on business process redesign is available in scores of articles and books.

7.3.1 How to Look at Processes

What is a business process? It is a series of logically related tasks, activities, or functions designed to bring about a specific business outcome. Business processes have identifiable start and end points and defined objectives. For example, performing a service call is a process in a field service organization. Processes have interfaces into other processes and can cross organizational boundaries. They may be formal and well documented, or a set of practices passed from employee to employee. If your organization adheres to any of the process-oriented standards such as ISO9000, Six Sigma, or the Software Engineering Institute's Capability Maturity Model, you are probably very familiar with processes and process documentation. If not, a few simple concepts can help you get started.

A process diagram shows the sequential layout of activities within a process. There are many ways to draw these diagrams, and at varying levels of detail. Figures 7.4 and 7.5 are examples of the high-level field service process diagrams described in Chapter 2. Each box in the diagram is an activity and these activities are the building blocks of a business process. An activity is a specific set of steps or actions that create one or more work products. As shown in the examples, a field service organization's activities may include: diagnose a problem, perform repair, and complete work order paperwork. Figure 7.3 shows the components that make up an activity.

Activities are connected to each other through workflows consisting of work products, information, or the results of a decision point. Each activity receives one or more work products as input. A work product could be a product specification, an assembly line component, or a telephone call relaying a change in interest rates. The activity uses its input(s) to create its output work product(s). The output work product of one activity is the input work product for the next activity in the

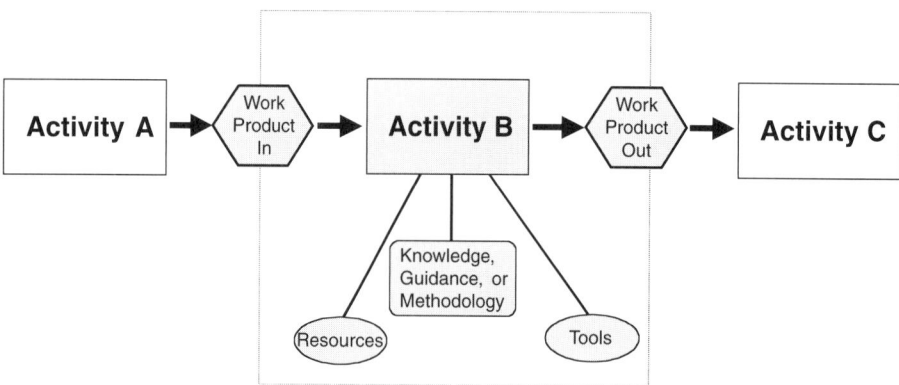

FIGURE 7.3
Process Building Blocks

sequence. Each activity has associated people or equipment resources that perform its actions, and the same resources may be used across multiple activities. Actions are conducted by a set of rules. In an informal process, these rules are based on the undocumented knowledge of the people associated with the activity. In a more formal process, written guidance or even steps from a methodology will specify how to perform all aspects of the activity. Finally, an activity may be supported by a set of tools.

A wireless solution can affect a business process in many ways. It could eliminate the need for one or more activities. It could enable a more effective activity to replace an existing activity or set of activities. Within an activity, it might supply the guidance or documentation, be used as a supporting tool, or even serve as the repository of the work product. For example, a building inspector may use a wireless tablet to capture information in an electronic inspection form. The wireless tablet is a tool for that activity, the completed electronic inspection form is its work product output, and the use of an electronic form may eliminate several subsequent data entry and paper-handling activities.

7.3.2 The Current Situation

If your company has not done so already, the first step in business process redesign is to document and measure the process or processes that you intend to improve. This exercise provides several important benefits. First, it ensures that you understand all aspects of the process, including its activities, flows, work products, interfaces, and performance. This knowledge is essential for ensuring that the redesigned process is complete and workable. Second, it helps to identify areas of opportunity for improvements. Finally, it provides a baseline for estimating the potential benefits when designing the project and for measuring whether those benefits were obtained after project implementation.

Documenting the current process involves interviewing the individuals who use that process. In the interview, begin by understanding the activities and the work products that flow between the activities. You may need to interview many people before you have a complete, end-to-end view of the process. As you delve into the details of the various activities, expect to find considerable variation. In one large organization, the author found over twenty versions of one relatively simple support function. When conflicts exist, try to select the most common version, or failing that, the most effective version. Document your findings in a process diagram. You can use any of the common methods for charting processes. While you can draw process charts by hand, many good process documentation tools can help with the chore as well as capture many additional types of information. Some tools even support process simulations to test the effect of changes made during the process redesign.

Figure 7.4 shows a simple example of a process chart for a field organization. This version shows only the major activities, workflows, and decision points. As part of your redesign effort, you will also want to include the work products that pass between activities. Review the diagram with the process users to ensure accuracy and completeness.

A variety of different measurements can be applied to the process. These measurements include:

- *Time*—elapsed time per activity, total cycle time, etc.
- *Effort*—resource hours per work product, resource hours per activity, total resources needed to support the process, etc.
- *Cost*—cost per work product, cost to support the process, etc.
- *Volume of Activity*—number of work products, number of customers supported, etc.

If possible, pick metrics where historical data is available. Otherwise, estimate the measurements with the assistance of the process users.

7.3.3 Redesigning Processes

Many practitioners advocate designing new processes starting with a blank slate. This approach will probably result in the most efficient processes that make the best use of the wireless solution, but will also be the most costly and disruptive to implement. Most companies do not have the luxury of taking this approach, but must instead evolve from their existing processes in a phased manner. In this case, using the existing process as a base, look for opportunities to perform the work in a different manner to increase efficiency, remove unnecessary steps, and/or reduce cycle time. One effective method for accomplishing this task is the use of facilitated workshops with the current process users. Their experience with the process is invaluable, and involving them helps ensure their acceptance and support for the redesigned process.

Figure 7.5 shows the improved version of the field service process shown in Figure 7.4. The description of this process and its improvements can be found in Chapter 2.

7.3.4 Testing Your Design

Once the new process design is complete and accepted by its future users, test it for workability. Do not view the new process as an end point, but rather as a prototype for subsequent improvement. At a minimum, compare the original and redesigned processes in a paper walkthrough. If the processes have been documented in a process tool that enables simulations, and data from the original process is available,

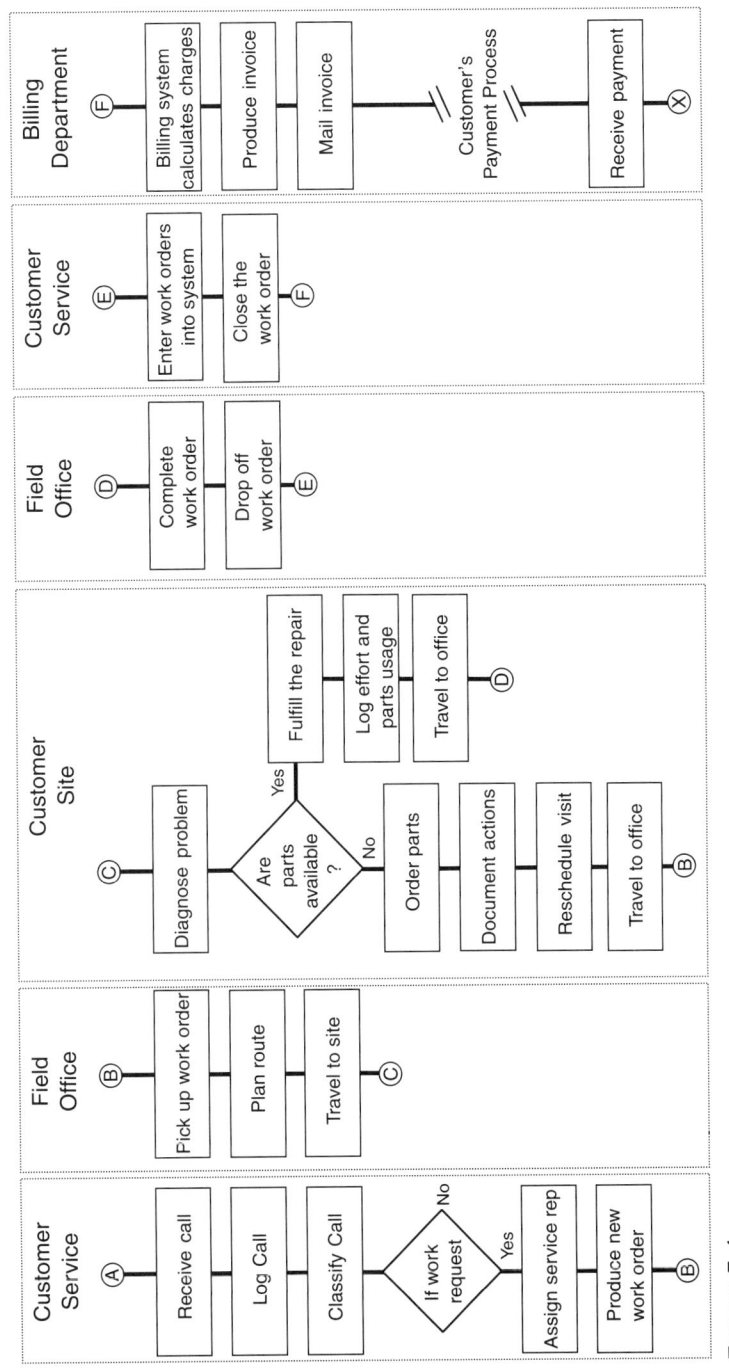

FIGURE 7.4
An Existing Field Service Process

203

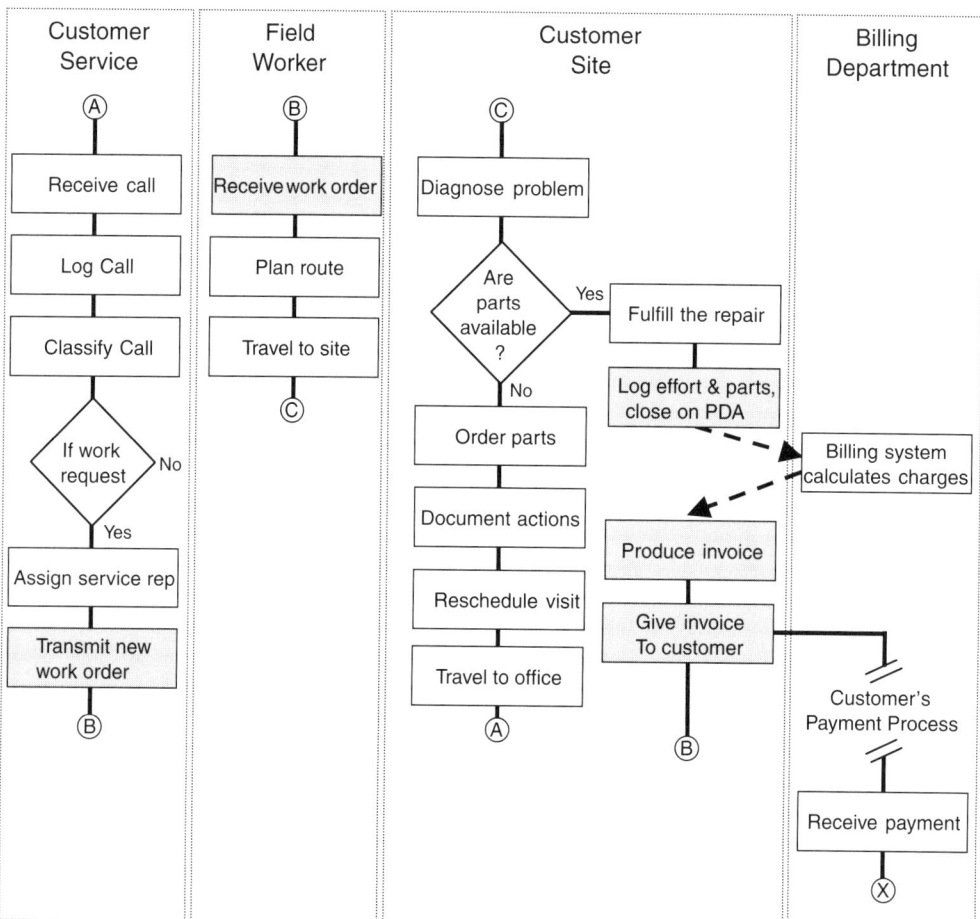

FIGURE 7.5
Improved Field Service Process

iterative simulations can provide substantial improvements while building faith in the process design. Another approach is to use role-playing exercises where teams of users try the process in practice, manually simulating any technology components or system interfaces that are not available. The most robust approach is to exercise the new process as part of a pilot project that tests the full solution on a subset of users.

Process tests are key for gaining buy-in from future users. Since they modify employees' job responsibilities, process changes are threatening and frequently resisted by the affected individuals. To prevent resistance, market the positive aspects of the new process before deploying it. Pilots and role-playing exercises also help to make new processes more tangible and less threatening.

7.4   Implementing the Project

The activities of project implementation depend on the characteristics of your solution and the decisions made in the planning process. If your solution is primarily a freestanding application obtained through a WASP, your implementation effort will be minimal. Conversely, a groundbreaking solution that relies on new, customized applications and the latest in device and network technology will involve a major implementation effort and require a wide range of skills. The details of implementing a particular technology can fill a book and are not covered here. Before embarking on your implementation effort, conduct a search for available books, web sites, and vendor-provided resources that can offer guidance on your chosen technology and implementation approach. The sections below provide some high-level tips on activities shared by most wireless implementation efforts.

7.4.1   Choosing a Team

The quality of your implementation effort is only as good as the individuals that deliver it. Given the breadth of knowledge needed to implement a wireless solution successfully, the implementation team will need to draw upon resources from many areas of your company. Some of these individuals will participate full-time through the life of the project, while others can serve part-time or in an advisory capacity. If specific expertise is not available within your company, you must either hire or train staff or seek consulting assistance. Even if skills are available within your company, incorporating consultants on your team can bring valuable experience and an outside perspective.

- *Program/Project Management*   The quality of the management team leading the solution effort is one of the most important factors in project success. Every project needs an enlightened and involved executive sponsor to provide guidance, help obtain resources and funding, represent the project to other executives, and resolve conflicts. Ideally, this sponsor comes from the corporate organization that will receive the greatest benefit from the wireless solution. If the project is mostly focused on technology, a project manager from the IT organization can lead day-to-day efforts. At the other extreme, a large-cross-disciplinary effort requires a PMO led by a program manager with experience and internal connections to lead projects composed of technical and business people. The program manager needs a team of project managers to direct the individual efforts that compose the program. These project leaders are drawn from the areas supporting each effort.

- *Sponsoring Business Area(s)*   The business area or areas sponsoring the project need to be heavily involved in its design, implementation, and testing. Individuals from these areas provide business expertise and represent their peers in functionality and user experience design issues. They must directly participate in the business process redesign efforts as well as user testing and pilot projects.

- *IT Operations*   In addition to hardware, operating software, and network installation and execution, the operations area of an IT organization usually has extensive experience in hardware and software evaluation and acquisition, and in negotiating service contracts. Also, unless the solution's operation is outsourced to a WASP or other service provider, this area is responsible for the operation and system support of the solution after deployment. Use representatives from IT operations for infrastructure acquisition, setup and testing, and to support the creation of development and support environments. If IT operations is responsible for system support, it should write support policies and setup the appropriate processes and tools.

- *IT Applications*   Some aspects of the solution's information infrastructure will fall under the purview of IT operations, however, access to corporate data will likely require the involvement of members of the IT organization's application development and support groups. These individuals are responsible for the preparation and integration of application data used by the wireless solution. While not trained in wireless technology per se, application team members often have deep knowledge of the processes and practices in the business areas they support, making them ideal candidates for development and business process redesign teams.

- *Wireless Development*   If your organization has an advanced technology group, it may already have some wireless development expertise. If not, obtain this expertise externally through hiring or the use of consultants. If your organization decides to build its own skills, consider training a mixture of technical specialists and application specialists from the team(s) already supporting the involved business areas.

- *Help Desk*   If your company has a separate help desk or user support group, involve this organization in the early phases of the wireless implementation. Since wireless technology brings many unique support challenges, this organization will need lead time to develop the tools, policies, and training to support the solution's users through deployment and long-term operation.

- *Legal*   Your company's legal team will be involved in contract preparation and negotiation, and depending on the solution, may help to define corporate policies for the use of wireless applications and devices. See Chapter 8 for more details on legal issues.

- *Human Resources*   Any project that requires significant changes to employee job requirements will involve your company's human resources department, particularly if those requirements affect pay scales or union rules. Other HR functions useful to a wireless implementation effort are recruiting and training.

7.4.2 Choosing a Methodology

IT organizations use various methodologies to guide their analysis, design, development, and testing efforts. Methodology types include top-down structured analysis, waterfall, iterative development, and agile and extreme programming. Each approach has its adherents who defend their choice with religious fervor. At the time of writing, no universal methodologies for wireless technology exist, although most commercially available methodologies will undoubtedly be modified to support wireless projects. Wireless consulting firms and systems integrators usually have their own methodologies, but rarely license them separately from their consulting engagements.

When considering the use of a methodology to support your implementation effort, research its source. Wireless technology projects have a number of unique attributes that are difficult to address simply by extending a standard methodology. For example, the iterative development methodologies used extensively for Internet projects have merit for wireless efforts, but within narrower constraints. They are helpful for evolving user interface designs and application functionality once the technology platform has been chosen, but are not as well suited for the earlier stages of the project. The cost of components makes it too expensive to select, master, prototype, and then throw out unsuitable candidates.

7.4.3 Implementation Preparations

Unlike standard IT projects, many aspects of the wireless solution's development environment will be entirely new to your organization. This development environment includes the hardware, software tools, and network support to build and test the whole wireless solution. Ordering, installing, and learning the appropriate hardware and software can require significant lead-time. Try to start this process immediately after receiving approval for the project.

Similarly, do not underestimate the effort required to build the appropriate system support environment. While your IT organization may be well versed in the operational details of data back-up and recovery and desktop management, wireless technology requires new policies, tools, and techniques. These tools and procedures have to be integrated with the methods used for existing applications and desktops.

Glitches and incompatibilities will invariably arise. If your solution uses standard devices and off-the-shelf software, setting up the system support environment may be the critical path activity in the implementation phase.

7.5 Preparing for the Rollout

The deployment or rollout of the complete wireless solution is where all the planning and hard work comes to fruition. A well executed deployment goes a long way to ensuring the overall success of the wireless solution. It ensures the solution's users are equipped, trained, and prepared to use their wireless applications to the fullest advantage. Too many otherwise good solutions are sabotaged by poor rollouts. Inadequate training, technical glitches, and logistical errors during rollout can lead the intended users to reject the solution before it has a chance to prove itself. Overcoming this initial rejection is tough, even after the problems are solved. All too often, the deployment phase of a project is treated as an afterthought. As the last major step in the project, it bears the brunt of previous schedule and budget overruns, and is delivered too quickly and with inadequate resources. Universally, the executives interviewed for this book stressed the importance of rollout preparations, and in hindsight, many would have devoted more resources to training and support. Consider the following topics when preparing for your solution's deployment.

7.5.1 Planning

All deployments, regardless of size, require planning. The rollout of any new technology is disruptive, causing lost productivity and anxiety among users until it is assimilated. Planning helps to minimize these negative effects and increase the speed of adoption. Many activities occur during the deployment phase including the installation of production equipment, the loading and distribution of wireless devices, training users on the application, making changes to devices and processes, and production rollout of user and system support procedures. Support requirements are especially high for the first weeks of deployment as users learn their application and devices and the inevitable problems get worked out. For many solutions, the overall effort devoted to deployment is as high or higher than the resources applied to the previous implementation phases. Large-scale deployments to hundreds of mobile users scattered over dozens of locations require the precision planning of a military campaign. While deployment is but one phase in the overall project plan, it is strongly recommended to create a separate, detailed plan to cover rollout activities. Use this plan to identify and schedule all deployment tasks and resources. Resource planning and scheduling is especially challenging for mobile employees and involves coordinating travel and time for training and installations. Rollout activities depend heavily on the activities in previous phases, putting them at risk for schedule slippage. For example, application training development cannot be completed until the application features have reached stability. User support teams cannot be fully trained until they have access to at least a beta version of the complete solution. Having a separate project plan makes it easier to identify and manage the impact of schedule slippage on other rollout activities.

7.5.2 Communications

Every project should devote resources to a communications/awareness function. Depending on the size and complexity of the effort, a single person may perform this function part-time or a small team may work on it full-time. The communications function operates throughout the life of the project, creating awareness and building excitement about the solution and its goals in the affected organizations. The communications function keeps every interested party informed about the project and its progress and successes. The communications person may distribute a newsletter, maintain a project web site, host question-and-answer periods and brown bag lunch sessions, and provide giveaways to attract attention. While this function may seem self-serving, it fulfills an essential role in preparing for deployment. By keeping the solution's intended users up to date, it lowers fear and resistance to pending changes, simplifies training, and creates excitement.

7.5.3 Training

Training is an important and potentially costly activity. Many types of training are required for a typical wireless solution. The solution's users need education on how to use the device and new applications, changes in their work processes, and wireless policies and support procedures. The help desk area needs similar training in advance to prepare to support the solution's users. Since wireless solutions typically involve mobile and dispersed workers, logistics pose the greatest difficulty in arranging training. Depending on the technical proficiency of the users, self-training through the web, tool-based training or videos are options. In many cases, however, face-to-face training is highly desirable or even essential. Face-to-face training or pairing trainers with users can result in potentially significant travel and productivity costs. Large implementations may require training sessions in many regional locations. The number of trainers required will depend on your deployment approach. A phased rollout uses fewer teams over a longer time period while a "big bang" rollout may require many simultaneous training teams. Where possible, combine training and deployment with another event to minimize cost. For example, Atlantic Envelope Company, featured in a case study in Chapter 3, combined the training and deployment of their wireless SFA application with a national sales meeting. To maximize their effectiveness, develop and test training programs well ahead of their rollout. Trying a new training program on a sample group of users helps ensure that the training is complete, understandable, and meets its objectives.

7.5.4 Installations

Most of the companies interviewed for this book pre-loaded and tested the applications on their client devices before distributing them to their users. This effort can be significant if a large number of devices are deployed, but ensures a smoother rollout. The production deployment of some wireless solutions will require additional installations of servers, network access points, tags, truck-mounted devices, or other hardware and software. If the installations are made in remote locations, they face the same types of travel and logistics issues as described above for training. If installation activities are a significant component of your wireless deployment, schedule them carefully to allow sufficient time and resources to install, test, and stabilize the hardware before training and user deployment.

7.6 Getting Assistance

Few companies will face their first wireless project alone. Nascent and complex technologies, evolving standards and protocols, a plethora of device options, and a lack of homegrown expertise are prompting companies to seek outside help for their projects. Wireless service providers offer a range of assistance from helping to navigate options, select implementations, deploy technologies, customize, host and support applications, negotiate carrier agreements, and integrate wireless and other systems. Unless your company already possesses significant wireless expertise, you are well advised to retain a wireless system integrator to help you assemble an end-to-end solution. Wireless system integrators combine a familiarity with the components of a wireless solution, including network options, packaged applications, hosting services, and device alternatives, with an ability to integrate these diverse components into a complete solution. Integrators mentor internal staff to help build skills and offer advice on how to select and use the services of other wireless providers. This section provides advice on how to select the right wireless services firms to assist your project.

7.6.1 Consulting Strategies

Companies frequently think of consulting services as an "on/off" choice—either you use consultants or you don't. In practice, the options are almost boundless. The categories below highlight a few of the available consulting options. You can mix and match these options to suit your project.

- **Strategy** Wireless integrators, strategy consulting firms, and high-end independent consultants can help your company identify the best targets for wireless technology and develop a solution, design and implementation strategy to address those needs.

- *Mentoring* If your company will use wireless technology long-term, and sees strategic value in supporting that technology itself, it makes sense to build internal skills. Hiring a few highly experienced consultants to bring in the desired expertise and mentor your staff is an excellent means of building skills while accomplishing a real project.

- *Project Management* System integrators are highly skilled in managing and delivering complex projects. Since they understand the nuances of wireless technology implementations, they can help your company create a workable project plan to meet your schedule and budget requirements and manage the project to that plan.

- *Skilled Resources* Various types of consulting firms from system integrators to independent consultants can offer skilled resources to extend project teams, fill in knowledge gaps, or handle highly specialized tasks. Consider supplemental staff to support short-term project needs, free internal staff for other assignments or handle tasks that do not fall within your long-term skill needs.

- *Outsourcing* If wireless technology is a means to an end, it makes little sense to develop internal skills. Systems integrators and other wireless service providers can handle any and all aspects of designing, delivering, and operating your wireless solution. You can outsource development, training, device management, user and system support, and/or full operations of your solution.

- *Specialists* Some consulting firms specialize in particular aspects of wireless technology. These firms can help meet highly unique and important requirements in areas such as security, location-based services, network design, and custom device creation.

- *Mixed Services* Other types of wireless integrators such as WASPs and Wireless Internet Service Providers (WISPs) are similar to their wired counterparts and provide a mixture of services, including application hosting, network connectivity, and integration. These firms are described in more detail below.

7.6.2 Choosing the Right Partner

Given the relative newness of wireless technology and the promise of wireless consulting dollars, the wireless services market is still evolving but attracting high numbers of new entrants. Quite a few excellent firms are operating within this shifting market, and with a little research you can find one or more firms to meet all of your consulting needs. Consider the following selection criteria.

- *Fit within Strategy*   Every wireless consulting firm has particular areas of expertise. These areas may focus on types of technology, such as applications for Palm devices, or vertical industries such as field service. Decide on your high-level strategy and technology directions first (use an independent strategy consultant to assist, if necessary), then select firms that specialize in the chosen strategies. A supplemental staffing firm is the wrong choice for designing and managing an extensive wireless deployment, and a management consulting firm is the wrong choice for WML coding expertise. If a given project is a stretch for a given firm, select someone else.

- *Project Management Capabilities*   Project management is the number one factor in project success. If your company is not going to assume full responsibility for project management, select a partner with demonstrated expertise in managing and delivering wireless projects.

- *Expertise*   Wireless technology projects require highly specialized expertise and trained resources for execution. Ensure that any selected consulting firm has sufficient quantities of both to support the types of consulting it offers. While a firm may have impressive marketing literature, its actual practices are more telling. Strong staff training programs, use of formal development methodologies, and solid relationships with tool vendors demonstrates the firm's experience and commitment to the wireless market.

- *Reputation*   The best method for judging a consulting firm is by its track record. While there are no infallible criteria for evaluating the potential success of a given project, the following criteria provide a good indication of a consulting firm's track record.

  - *Length of Time in Market*   While new, high quality consulting firms are founded on a regular basis, the longevity of a firm, and its ability to withstand economic upheaval, is a testament to the quality of its work.

  - *Customer Base and References*   Checking references is essential. Do not restrict reference checks to customer names provided by the prospective firm. Network at tradeshows and through professional organizations to find and chat with other customers of that firm.

- *Financial Performance*   Will the consulting firm be around to support the solution it delivers? If the firm can't manage its own business, should it be advising yours? Pick firms with solid financial performance.

7.6.3 Types of Consulting Firms

As described above, the wireless services market is evolving rapidly and the categories of providers as well as the skills and services of any given provider are subject to change. The solution provider list in Appendix B lists examples of firms in each of these categories.

- *Wireless System Integrators*   Wireless system integrators specialize in integrating the various wireless components, technologies, and providers to create a seamless, end-to-end solution. As mentioned above, these integrators typically are capable of taking complete responsibility for a wireless project. Many offer high-end strategy consulting, and most perform design and development work. Their implementation services include custom development, installation of packaged software, identification and recommendation of third-party vendors, and integration of wireless applications and data with other enterprise systems. Integrators often have experience implementing wireless projects in a number of vertical industries. For example, ArcStream Solutions has worked with Staples Inc. to provide e-mail access to senior executives; has helped Harvard Medical School roll out a mobile curricula-based application to faculty and students; and has developed a mobile sales force workbench for pharmaceutical and biotechnology companies.
- *Wireless Application Service Providers (WASPs)*   WASPs offer outsourcing services that generally consist of hosting, and providing remote wireless access to, web-based applications and services. Rather than build, support, and maintain wireless applications themselves, companies can turn to WASPs to obtain desired functionality while avoiding ongoing support burdens. Like ASPs, WASPs' services may include application hosting, system monitoring, diagnostics, problem detection and resolution, customer services, and support for a multitude of device types. Most WASPs own the software that they host, and will perform consulting work to customize the software and integrate it with other enterprise systems. WASPs offer a number of value-added services, eliminating headaches by ensuring seamless network coverage, and by managing devices, warranties, application upgrades, and third-party licensing agreements. The WASP category is currently struggling along with their ASP counterparts in the wired world. However, this category is likely to survive in some form, as few IT organizations have wireless technology expertise, and many do not have sufficient internal demand for wireless work to justify

obtaining dedicated wireless resources to build and operate peripheral wire-less applications. Also, unlike some early and unsuccessful ASPs, almost every WASP owns the software that it hosts. By avoiding the need to pay costly license fees to third parties, WASPs ensure that their financial returns are higher.

- *Wireless Internet Service Providers (WISPs)*   WISPs offer Internet and e-mail access to subscribers. Armed with a laptop, PDA, or handheld device, a snap-on modem, and a WISP subscription, a user can access a corporate intranet, send and receive e-mails, conduct commerce, view wireless-enabled web sites, manage documents, send faxes, and more. The business case for WISPs has been somewhat shaky. To date, WISPs have filled a void in the wireless data market, offering wireless Internet access for handheld devices. But demand for these services has not blossomed as predicted. OmniSky, one of the industry pioneers has already failed and other firms are struggling. GoAmerica, along with several other new entrants, is expanding its services by offering "hot spot" public WLAN access to its subscribers. When demand for these services picks up, expect the network carriers, who already offer wireless web services through phone handsets, and large portals and ISPs like AOL, Yahoo!, and EarthLink, to muscle their way into the market. EarthLink, in fact, is already pursuing the wireless market through its subsidiary, Boingo, that is engaged in building out public WLANs in areas such as hotels, airports, and other places where the public congregates.

- *Other Wireless Service Providers*   Many more wireless service providers exist to support a range of special-purpose wireless applications. Companies inter-ested in monitoring or tracking the whereabouts of equipment, goods, cargo, people, or other movable assets can work with a telemetry provider. For municipalities and utilities, telemetry applications include automated meter reading programs that transmit usage information wirelessly from a meter in a home or building to a nearby transceiver.

Management Considerations

8

What are the consequences of mobilizing your workforce, arming them with devices and data, and providing connectivity to corporate networks far from the management oversight and conventional controls present in an office setting? The very features that make wireless technology attractive as a business solution—devices that are cheap, small, and powerful, easy access to corporate information, and the ability to use these tools at almost any location—bring a host of management and legal concerns. How and where an employee uses a wireless application raises potential safety, privacy, data confidentiality, liability, and support issues. Companies can significantly lower their level of risk by proactively addressing these issues before any problems arise. This chapter provides a high-level summary of some of the more pressing business and legal issues that managers should consider and address before deploying a wireless solution.

The concept of a mobile workforce may or may not be new to your company. The sales force, field force, doctors, drivers, delivery personnel, and students have been mobile for years. What is new, however, is our ability to hook these workers into corporate systems and data via a variety of mobile devices, and to load handheld units with a plethora of corporate information. Whether these devices are personally owned or officially sanctioned and distributed, they bring a host of thorny problems. Who owns the data on a device? How will data and access be protected? What types of standards should be put in place? What kinds of legal ramifications may arise from mobile device usage?

This chapter urges companies to actively identify and address the business and legal concerns that may materialize from going wireless. By recognizing and dealing with these issues up front, companies can avoid being blindsided down the road. The first subsection looks at some of the topmost business considerations arising from wireless solutions and recommends adopting corporate policies and guidelines to deal with them. These policies should address issues of ownership, support, usage, standards, privacy, and security. The second subsection highlights related legal concerns including disclaiming liability, monitoring employees, and determining ownership of devices and data.

8.1 Business Considerations

Mobilizing your workforce and giving them access to corporate data and systems has ramifications beyond the technical. It affects how employees perform their jobs and their day-to-day workflow. Just like anything else that is used on a daily basis, employees will develop good and bad habits, tricks, conventions, and shortcuts in how they use their mobile devices and wireless applications. Every company has a responsibility to see that employees develop the *right* kinds of habits, and that they use their wireless applications in the *right* kinds of ways consistent with the company's overall business objectives, principles, and culture. An organized and structured approach to conveying this kind of information to employees is through corporate policies and guidelines.

Corporate policies and guidelines serve a useful purpose. They:

- Clarify the company's position on certain matters
- Explain the company's and employees' obligations, rights, and responsibilities
- Lay out the processes and procedures that employees must or should follow
- Specify how the policy will be enforced (optionally)
- Specify the consequences for non-compliance (optionally)

If you are wavering as to whether your company needs policies for its mobile and wireless deployments, consider these two factors. First, company employees will purchase their own devices, if they haven't done so already. As personal devices proliferate within a company, they create a hornet's nest of problems. Employees may start out using the devices for purely personal purposes, but over time, corporate data (contact lists, e-mails, work documents) will wind up migrating onto these devices. A laptop may even end up loaded with corporate software, files, marketing materials, and more. Defining corporate policies well in advance helps avoid crises caused by tacitly giving device-wielding employees unrestrained and unregulated access to corporate data and software.

The second reason to create mobile policies relates to risk mitigation. Once corporate data and software is dispersed and distributed in unknown places, and used in unknown ways by unknown individuals, a slew of risks arise. Lacking management control and oversight, companies must rely on their employees' judgment to use and safeguard information responsibly. Leaving these decisions up to the individual employee puts a company at tremendous risk should damage or harm materialize.

Finally, the exercise of defining policies forces a company to envision and play out various scenarios, think about the consequences, and develop measures to avoid, lessen, control, and monitor the risks involved. This process allows a company to be proactive, to determine in advance rather than in hindsight the best way to handle a given situation.

When creating mobile policies, companies should cover the entire class of mobile devices including laptops, PDAs, handheld computers, smart phones, pagers, e-mail appliances, and special-purpose devices like scanners. Ideally, these policies will complement existing ones. For example, companies commonly have e-mail policies that they can easily extend to apply to mobile devices. New policies are also needed to deal with uniquely mobile and wireless issues such as synchronization, back-ups, and support.

Developing these policies is a multi-departmental effort involving legal, HR, IT, and business community representatives. Legal and HR personnel should craft the actual language of the policies, as they normally do for other types of policies. The IT organization can provide invaluable input, but its role is mainly advisory.

Concentrating on the policies alone misses the mark. The real goal is to achieve high compliance with those policies. Compliance has three facets, as illustrated in Figure 8.1.

- Employees cannot comply with policies that they do not understand. Create policies that are clear, concise, and intelligible.
- Employees cannot comply with policies if they don't know about them. Every employee must receive copies of mobile and/or wireless policies. In some circumstances, a company will need a written receipt, or a written acknowledgement by the employee that he or she agrees with certain provisions.
- Compliance is attained and maintained only through enforcement.

After crafting and distributing policies, a company must enforce them. If policies are not enforced, employees will quickly learn to ignore them. Enforcement is a combination of monitoring compliance (through spot checks or other measurement

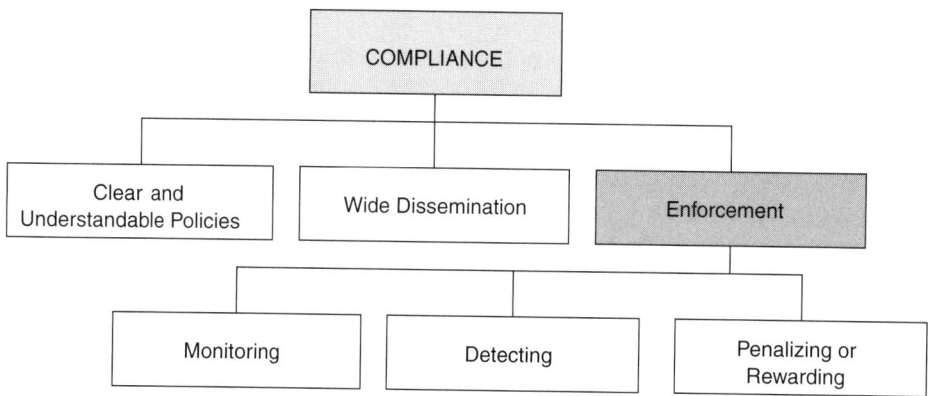

FIGURE 8.1
Achieving Compliance

programs), detecting non-compliance, and sometimes, meting out penalties or even rewards (Figure 8.1). Effective penalties are neither too severe nor too lenient. They should be in proportion to the offense (an accidental or intentional violation of a policy) and the consequences of the non-compliance. Penalties might include a reprimand, confiscation of a device, or even termination. Check with your HR and legal departments as to the advisability and type of penalty in each situation.

Lastly, the contents of every policy may need to change over time as work and business conditions change too. Periodically reevaluate your mobile and wireless policies, measure their efficacy, and adjust them to boost compliance and achieve their end goals.

The following subsections cover categories of policies that your company will want to consider in developing its own set of guidelines. They include ownership, support, standards, usage, privacy, and security policy categories.

8.1.1 Ownership

Will your company allow employees to use personally owned devices on the job? Or will they be required to use company-purchased and owned devices? As mentioned above, employees will purchase their own devices. Many workers already have cell phones, pagers, and PDAs that they use for personal and business purposes. Over time, corporate data will wind up on these devices, which is when trouble starts. Who owns the data on the device? What happens when the device is lost? Since the device wasn't company property, will employees even inform the company of the loss?

The best way to head off these issues is for the company to purchase mobile devices for employees. Although this tactic won't necessarily stop employees from purchasing their own devices, it does give them an easy way to segregate their business and personal use. By establishing clear-cut ownership of the physical device, the company is in a stronger position (legally) to dictate what can and can't be done with the device and how to use it. Ownership also gives the company better control over the data and programs loaded onto the device, and allows it to view, manage, retrieve, and delete content in accordance with its policies. The downside of company ownership is that it adds to deployment costs, which is a tradeoff that companies may be willing to make to gain increased control.

8.1.2 Support

Will the IT organization, call center, or operations group support mobile devices as they do other types of equipment, and to what extent? If devices are personally owned, the answer to this question is not clear-cut. The company may have no firm responsibility to support personally owned devices, but if it allows employees to use these devices for business use, then there is a good argument for supporting them.

If the company owns and distributes the devices, then it has a vested interest in supporting and maintaining them. But if the company also condones some small amount of personal use with these devices, the support issue again becomes blurred. Will the company support, back-up, and restore personal data on the device? Will it allow this data to be synchronized and commingle with business data on company desktop PCs and servers? Does the company have an obligation not to divulge the personal information on an employee's mobile device? How does this obligation coincide with existing employee privacy policies?

Even when a company will provide support, it must determine what does and doesn't fall within the scope of support. Will it limit support to repair or replacement of devices? If devices contain corporate data, then employees should not expose that data by sending devices out for repair to third parties. The IT organization must also decide the extent to which it will get involved with carriers or service providers when employees experience network or coverage problems. And the company must determine whether to allow individual employees to negotiate their own service contracts or whether to centralize these contracts to obtain preferential terms or volume discounts from providers.

8.1.3 Standards

Another method of controlling the portfolio of devices and software is to develop a list of standards. These standards cover the hardware and software that the company will support and dictate required activities that all device owners should follow.

- *Supported Devices and Accessories* A wireless application or solution will often determine the devices that employees receive. For example, to track deliveries, a driver is given a bar code scanning device. An emergency room doctor will use a Palm device to work with the custom whiteboard application developed expressly for that platform. In other cases, employees, students, or other constituents will have leeway in choosing a device. In that case, an organization should specify a list of devices that it will support, either by model or by operating system version, and a list of any prohibited devices. This list should contain only those devices that the company knows will work with its wireless solution and existing platforms, and that it knows how to support, within reason. The list may also contain required accessories, such as docking cradles, WLAN cards or modems, that will inter-operate with the supported devices. For example, a company may support only Windows CE-based devices because of its reliance and familiarity with Windows operating systems in general. Or, if WLANs are widely used, a company may prohibit the use of Bluetooth devices in its office buildings out of interference concerns. If the company has expertise in developing Palm OS-based applications, then it may restrict support to only those devices running the Palm OS.

- *Supported and Required Software*   Companies may also specify supported and required software. Supported software might include home-grown applications and off-the-shelf products. Required software might cover synchronization, security, and other task-based applications. The company may also prohibit loading or running any unsanctioned software on the device.

- *Recommended Wireless Service Providers*   Although an employee's location will influence the selection of wireless network services, a company may also maintain a list of preferred or recommended providers based on existing contractual relationships or past experience.

- *Synchronization*   If outdated data on devices is a concern for your company, consider specifying the minimal timing and frequency of synchronizations between mobile devices and servers or desktop PCs, or even automating synchronizations if possible. The actual synchronization products used should appear on the supported software list.

- *Anti-Virus Scans and Updates*   Depending on the device type used, and the inherent risk of virus attack or infection, consider specifying the minimal timing and frequency of virus scans. If the device is loaded with anti-virus software or files, indicate the frequency of updates.

- *Back-ups and Restores*   Similar to synchronizations, specify the recommended frequency of back-ups to a desktop PC or server. It may also be useful to explain how restores are performed, including the scope and types of data and software that are reloaded on a device upon a restore.

- *Reporting Lost or Stolen Devices*   To protect its networks and data, and to assess its exposure and potential risks, companies must emphasize the importance of promptly reporting the loss or theft of a device. Specify how an employee can report an occurrence, and the questions typically asked to scope the risk of such an event.

- *Training*   Depending on the wireless application, different types and levels of training may be required. Specify the minimal training needed for a new user, and the types and availability of training that may include on-line courses, performing simulations, attending classes, etc.

8.1.4 Usage

A company may not want to micromanage *how* an employee uses a wireless application or device day-to-day, but it may want to instill a few best practices for sound, secure usage. These best practices help lessen support burdens by ensuring more consistent use among employees, and help to avoid potential problems.

- *Personal Use*   A company must take a consistent stand on personal use of the mobile device and ownership of the device. If employees are permitted to use their own devices for business tasks, then a company cannot conceivably ban personal use. Conversely, if employees use company-owned devices, the company can circumscribe the amount and types of personal use permitted. When limiting personal use of a company-owned device, a reasonable approach is best. Prohibiting employees from using a cell phone for personal calls, for example, is just as difficult as banning personal use of office phones. If your company has dealt with this issue regarding mixed use of cell phones, laptops, or e-mail, see whether those policies are equally applicable, with or without tweaking, to mobile devices and wireless applications. If both business and personal use are permitted, consider whether to implement a chargeback/reimbursement policy based on the type of usage.

  The second facet of personal use relates to installing or saving non-business software or data on a mobile device. Again, it is difficult to have a blanket prohibition on personal data since mobile employees may need access to some personal information during the day. Conversely, it is entirely appropriate to prohibit loading and running unsupported software such as games or productivity tools on a device. If some personal use is inevitable, a sensible course of action is to acknowledge the use and disclaim responsibility for maintaining, managing, backing up, or restoring personal data.

- *Security*   Every company should specify "safe" usage methods intended to protect devices, applications, and networks from unauthorized access, viruses, and attacks. Make these practices mandatory, not optional, and enforce them rigorously. Require employees to adhere to security measures, such as password-protecting devices, and prohibit bypassing or disabling of security routines. Do not store passwords or keys on devices, and advise employees not to write them down. If a device is lost or stolen, have employees report the event upon discovery and assist the company in establishing the information contained on the device.

- *Local Laws and Policies*   Although your employees are legally bound to observe all local, domestic, and international laws relating to use of mobile devices, make it a corporate policy as well. In this way, your company is free to take suitable action when employees violate such laws. Some jurisdictions, notably New York state, ban the use of mobile devices while driving, unless they can be used hands-free. Airlines prohibit using electronic devices during portions of a flight. Hospitals may forbid using handheld devices in areas where interference is possible.

- *Etiquette*   Laws may not cover all aspects of mobile device usage, but custom does. There are courteous and discourteous ways of using a mobile device. Make sure your employees know the difference. For example, it is polite to turn off or discontinue use of mobile devices in certain venues or during certain events (classes, conferences, religious services, official hearings, etc.). Do not use a mobile device or application where others can eavesdrop or peak at trade secrets or confidential information about the company or its clients. Do not use offensive or vulgar language when using a device or an application like e-mail. Use appropriate ring tones, screen backgrounds, file names, etc.

- *Safety*   Encourage employees to use their mobile devices and applications in a safe manner. Unless they can be used hands-free, refrain from using mobile devices while driving. For devices that support voice applications, consider supplying earpieces as a required accessory. Do not use the device while operating machinery or in a situation where distraction can cause accidents. If GPS features are meant to protect employees while working in remote or rural areas, make sure they do not disable or circumvent the feature. Finally, do not use a device in a public place if it might inspire someone to steal it.

8.1.5 Privacy

Your company likely has privacy policies relating to disclosure of employee information. See if you can extend these policies to cover information contained on mobile devices, especially if personal data might be backed up to desktop PCs or servers. If company data will be stored on a device, ensure that employees understand the company's right to review the contents of the device. If trade secret or other confidential information is stored on a device, it may warrant a policy requiring data encryption.

If your wireless solution has GPS features to track the whereabouts of employees, you can expect resistance. Many employees balk at using tracking and location devices, fearing an invasion of privacy. While companies may have legitimate reasons (productivity, safety, etc.) to monitor the location of employees, it may make sense to spell out how the information will be used as way to allay concerns. If employees belong to a union or collective bargaining body, regulations or contracts may impact the ability to track and pinpoint the location of workers.

8.1.6 Security

Every company is well aware of the security risks posed by our inter-networked systems and applications, some of which are cited in Chapter 9. Companies are advised to perform a risk analysis and security assessment prior to rolling out any wireless solution or technology, and to develop policies or guidelines to maintain security. It is relatively easy for user areas to purchase a network-capable device and install a wireless short-range network, whether Bluetooth, infrared, or WLAN. Rogue deployments can make otherwise secure networks vulnerable to hackers, and disrupt existing security policies. Identify steps that employees can take to help maintain security, such as reporting device theft, encrypting content on devices, not sharing passwords, using virus protection software, and then turn these steps into proactive policies. As with any policy, make sure that these security measures are not so onerous that they discourage usage. Strike a balance between ease-of-use and amount of security required.

8.2 Legal Considerations

Companies and vendors are just starting to understand the legal ramifications of introducing wireless technologies into the workplace. Many of the business issues cited above—privacy, ownership, and misuse of information—raise corresponding legal concerns. To protect itself, a company can use a combination of tactics from implementing and enforcing policies to disclaiming liability and establishing incontrovertible ownership of wireless components.

8.2.1 Disclaiming Liability

While not always a foolproof method of avoiding claims and damages, establishing and disseminating written disclaimers of liability can measurably cut risks. What types of liability disclaimers are useful?

- *Health Hazards*   A company might disclaim liability for health hazards from using mobile devices. Numerous studies have found that mobile device usage does not pose heightened health risks; however, the question continues to crop up regularly.

- *Misuse*   A company might disclaim responsibility for an employee's knowing or reckless misuse of a device, application, or data. A disclaimer might also be appropriate for damages arising from device theft or from data interception.

- *Privacy Violations*   Companies are prohibited by law from disclosing certain classes of confidential information, such as medical records. They cannot use disclaimers to avoid liability or penalties for disclosing this type of information. With other information, privacy rights or expectations are not as clear cut, and companies can attempt to use disclaimers to limit their exposure.

- *Reliance on Outdated Information*   Many wireless solutions rely on data stored on a mobile device. While the company may have policies regarding the frequency with which this data must be synchronized with the host, the information may become outdated. Decisions made on the basis of outdated or inaccurate information (incorrect price quotes, inaccurate shipping times, etc.) may be the basis for lawsuits. A company can attempt to disclaim liability in these cases, but also has to weigh the negative PR and impact on goodwill if it tries to use these disclaimers to its advantage in dealings with its customers.

8.2.2 Monitoring Employees or Customers

As mentioned in Section 8.1.5, your company may have a legitimate reason to monitor and track employees and customers using wireless technologies, but that still does not mean that the oversight will be welcome. If monitoring is used to enforce policies or exact penalties, you can expect resistance or even legal challenges. How far can a company go in monitoring its employees? Precedent exists for monitoring e-mail use, as well as the contents of company-owned computing equipment. It is fairly simple to extend these policies to mobile devices and wireless e-mail use; just make sure that policies are updated and conveyed to employees.

If you intend to use wireless monitoring and tracking technologies to enforce certain behaviors and impose fines or penalties for non-compliance, check out the legality of your approach beforehand. Several cities are using wireless applications to aid in law enforcement by capturing drivers who speed through red lights with wireless video technology. Some states have gone so far as to require a picture of the driver's face in addition to the license plate, a move alarming privacy advocates. And, in a policy later deemed illegal, a New Hampshire car rental company used satellite technology to track its customers' rate of speed and assess fines for speeding. The New Hampshire Department of Consumer Protection ordered the fines refunded, finding that the company did not explain the policy adequately to customers, failed to provide due process, and levied fines far in excess of any damages.[1] An open issue is whether the car rental company also violated the privacy rights of its customers by secretly tracking their whereabouts during the rental term. But car rental companies are routinely using these types of technologies to monitor cars that venture outside of designated geographic boundaries or that reach a maintenance mileage threshold.

8.2.3 Device and Data Ownership

The ownership of devices and data, especially if ambiguous, can present interesting legal questions. If a company owns the mobile devices, then it has grounds to create and enforce policies regarding device usage, and can monitor and control the data loaded onto the device. As owner of the device, the company owns the content on the device. It can also take actions to enforce its ownership rights, such as confiscating and inspecting the contents of the device.

When the employee owns the device, the company clearly does not have any ownership rights to the device. However, if company data is stored on the device, or if the company credibly suspects that this is the case, it does have the right to investigate the device and repossess or destroy any data that it owns.

If an employee terminates his or her employment (or is terminated), a company has the right to demand the return of company-owned property including any mobile devices. If an employee owns his or her device, a company may still demand the right to investigate the device for possible corporate data, since the company is the owner of the data. Many employees do not fully understand this right, and are surprised when a company decides to exercise it.

If devices are used for personal and business use, disputes may arise over who is responsible for charges. If reimbursement policies or chargeback formulas are in place, make sure that they are well understood and enforceable.

---

1. "Officer Calls for Refund of 'Speeding' Fines," *Boston Globe*, February 6, 2002.

Solution
Considerations

You are about to take responsibility for a new wireless project. Although you have led many technical projects in the past, you have never dealt with wireless technologies, devices, and networks before. You have assembled a competent team of employees and consultants to perform the design and development work, and have a motivated group of users lined up to provide input to the project. You are on the verge of putting the solution together. What else would you like to know?

If you are like every other manager, you could use a little sage advice from someone who has been there before. You would rather know the top few issues in advance, rather than figure them out in hindsight. What can you do to make sure that the solution is proceeding on track? Are you building a solution that will scale and evolve? What kinds of security risks will your solution pose and what can you do about them?

Although you may welcome the chance to work with something as new and exciting as wireless technologies, it is still stressful to cope with a constantly changing environment. Components *should* work together, but will they? Those screens *should* let your users perform every task, but are they usable? Networks *should* provide coverage, but where do your users actually want to do their work?

Rapid change makes for a stressful development environment right now. It also makes it harder to know or predict what will happen in the future. Uncertainty aside, you still must lay the best foundation you can to ensure that your solution can adapt easily and quickly when the need arises. You may start out supporting just Palm devices, but have to master smart phones and tablet computers once your intrepid users begin buying them. You may think a 90% mobile rather than pure wireless solution is a safe bet given the state of WANs only to discover three months later that AT&T Wireless, Verizon Wireless, and Cingular Wireless have upgraded to 2.5G, higher-speed data networks in your area.

To compound the stress, there is always the issue of security. You can't wait until the solution is deployed to worry about security issues. Remote devices loaded with data pose a new type of security risk. Add hundreds of wireless access points, rogue wireless networks, and unprotected data transmissions, and the security picture quickly becomes complicated. While it is extremely difficult to secure your enterprise systems and data from every threat, you will want to take reasonable steps to minimize your level of risk.

The purpose of this chapter is to give managers new to wireless technologies a "heads up" in laying the groundwork for their solution. First, the chapter recommends performing one or more reality checks during the course of your project to help select the best components, determine the feasibility of your approach, make sure users approve the solution's design, and find out how well the solution works in practice. Next, the chapter briefly introduces the concept of agnostic development as a way to create scalable and extensible wireless solutions. Lastly, the chapter examines several security risks raised by wireless deployments, and offers methods to minimize their occurrence.

As a newcomer to wireless technologies, there are hundreds of issues and considerations deserving your attention. This chapter could not possibly hope to cover all of them. Chapters 10, 11, and 12 highlight some of the more salient issues affecting devices, networks, and applications, respectively. A handful of considerations, however, overlay the entire solution. These considerations include the following: testing the feasibility and operability of the solution at several key stages, designing extensibility into the solution from the outset, and securing the solution from end-to-end.

9.1   From Concept to Pilot _____

Assuming that you have followed the path described in the first part of this book, you will have identified your wireless opportunity, per Chapter 4, and have defined an initial solution, per Chapter 5. At this point, you have a general idea of the devices, networks, information architecture, and application(s) that will comprise your solution. You have performed a cost justification, per Chapter 6, and anticipate sufficient benefits to merit proceeding with your solution. After reading Chapter 7, you have figured out your implementation plan and approach. Now what? Do you go ahead and buy the devices, purchase network services, license some software, and cross your fingers, hoping that the solution will work out in the end? Of course you don't.

From the initial concept of the solution, to the final, finished product, it is wise to perform a series of "reality checks" to make sure the project is on track, that components are working as expected, that users can use the application, and that the anticipated benefits are being achieved. The outcome of each reality check will either justify moving forward or call for a re-assessment. Although these checks go by various names, the four touchstones illustrated in Figure 9.1 and discussed below are benchmarks, proof-of-concept, prototype, and pilot.

Iterative development models have become de facto standards in Internet and other enterprise projects, with many projects proceeding directly to the prototype phase to flesh out the requirements for the solution. While these techniques have merit, they do not apply wholesale to wireless development projects. Why? It is simply too expensive to randomly select, learn, prototype, discard, and then repeat the process with wildly different devices, networks, and other wireless technologies. Benchmarking and proof-of-concept steps help winnow options ahead of time and avoid more costly re-assessments and re-design at the prototype or pilot phases. In addition, since most companies will end up owning only small parts of their total wireless environment (being dependent on network carriers, software vendors, and sometimes even users who personally own their devices), it is wasteful and time-consuming to jettison solution components well into the development phase.

If You Need to Know...	Then Try This Checkpoint
Which Components?	Benchmark
Will It Work?	Proof-of-Concept
What Does It Look Like?	Prototype
How Well Does It Work?	Pilot

FIGURE 9.1
Perform Reality Checks

9.1.1   Benchmark Components

On the surface, several candidate devices, networks, and/or applications may appear to meet your solution's requirements. Benchmarking helps answer the question, "Which components should we use in our solution?" A benchmark assesses each of these candidates in more detail to allow your company to make final selections for its project. Vendors or independent groups often perform benchmarks under controlled conditions using pre-defined scripts or scenarios, although vendors always have a built-in conflict of interest. Your company can use benchmark studies from these groups or conduct its own. A benchmark consists of a series of carefully designed tests that measure certain attributes of the test group, and a final comparison showing how the contestants performed vis-à-vis each other. For example, with wireless devices, a benchmark may assess durability by conducting drop, temperature and dust tests, and performance by running the same Java application on each form factor. The benchmark should provide sufficient insight to let your company select the specific components that will comprise its solution.

9.1.2   Perform a Proof-of-Concept

The benchmark is helpful in choosing solution components. Now the question is, "Will the solution work?" In an unsettled wireless environment, where technologies are still in flux, it is not uncommon for products to fall short of advertised capabilities or to work in unknown or unpredictable ways when combined. A proof-of-concept lets you determine the feasibility of your solution. It enables an organization to assemble the discrete components sufficiently enough to see how well they work together. Will the selected devices, networks, and applications inter-operate? Will the network transmit messages at an acceptable speed? Will the packaged application run speedily on the designated device? What types and magnitude of problems are exposed? The proof-of-concept helps reinforce or undermine some of the initial assumptions in the cost justification, and can also provide baseline data to clarify estimates. At the end of the proof-of-concept, you will either have the confidence to move forward with your project or will need to reconsider your solution and component selections.

As described in the case study in Chapter 3, Honeywell decided to conduct a proof-of-concept to test the feasibility of its field service solution. Honeywell hooked up the devices, wireless application package, and networks just enough to enable it to perform some tests. Rather than spend the effort at this early stage to perform back-end system integration, Honeywell opted to simulate server-side functionality by having employees back at the data center send messages to the devices, and intercept and process incoming requests.

9.1.3 Prototype the Application

A prototype translates the solution's design specifications into something that users can see, touch, and feel. It answers the question, "What will the solution look like?" Prototypes perform a reality check between what designers and developers perceive users want, and what users desire in actuality. Prototypes screen some of the technical aspects of the solution—is it possible to display this information in a legible way on this PDA screen—and on the usability of the solution—are these menu selections meaningful to users and presented in the right order?

Prototyping is typically performed iteratively. Without expending a lot of effort, developers create screen mockups and perhaps the ability to perform a basic set of transactions, usually simulated. Users evaluate the prototype and provide feedback. Developers use this feedback to modify and fine-tune the prototype, which is again presented to users for another round of review and commentary. Theoretically, the "final" prototype should reflect what the finished solution is expected to look like.

9.1.4 Pilot the Application

The pilot phase is really a mini rollout, allowing a company to test its solution before unleashing it on a wide audience. The pilot helps answer the question, "How well does the solution work in practice?" The testers consist of a subset of representative users, and the solution may contain full or only partial functionality. A pilot allows a company to test not only the wireless application, but the logistics surrounding it—deployment, training, and support processes.

Pilots provide insurance against the risk of discovering overwhelming problems in production. On the technical side, they allow an organization to detect defects when it is cheaper to fix them—before they cause expensive problems in production. But, just as important, pilots allow companies to test their perceptions of usage patterns, user preferences, and working conditions. The wireless environment is so varied that it challenges many of our assumptions about how, and even where, users work. A pilot helps expose what the users' experience is *really* like. Do users stand or sit while they use the application? Do they use one, two, or even no hands? Is it bright or dim, hot or cold, noisy, quiet, dusty or clean? Do users have adequate network coverage wherever they roam? Which functions do users use the most, and in what order? Are the screens and instructions intuitive? These factors affect how the user will interact with the application, and the level of satisfaction derived from the experience. In many wireless projects, the full panoply of user experience issues does not surface until the pilot stage.

9.2 Building in Extensibility—"Agnostic" Design ___

The high rate of change among wireless device platforms and network technologies puts a premium on extensible and scalable design and development techniques. It is no secret that manufacturers are in a race to produce the latest, greatest devices. And, WAN providers are working frenziedly to upgrade their infrastructures to 3G technologies. A wireless application is barely launched before users want new functionality and new types of devices. As the saying goes, "Change is the only constant." It is best to design wireless solutions from the outset to accommodate change in devices, networks, and technologies.

Organizations can avoid the pitfall of having to rewrite code entirely as device classes and network technology change by using extensible design and development techniques. By properly designing, coding, and partitioning business rules, application logic, presentation logic, data access, and wireless communications, organizations can hope to re-use components as devices and networks shift or as new types are added to the portfolio. One way to accomplish this objective is to adopt **"agnostic"** coding techniques.

Agnostic coding techniques avoid device-specific or network-specific conventions. The result is that an application can run over the widest possible range of devices and networks. Introducing new devices or networks is quicker and easier because fewer coding changes are required. The disadvantage of this approach is that it cannot exploit worthwhile and unique device features. Applications are effectively coded to the "lowest common denominator" using only the subset of features common across all device types. If you would like your application to use the specific voice recording or multimedia features of a Compaq iPAQ device, for example, but still want it to run on a RIM BlackBerry, you cannot do both with a single, agnostic application.

Any solution that involves an application or agent running on a device is, per se, not device-agnostic. Code that operates on a smart phone, for example, won't run on a pager. To run on a device, an application must invoke application programming interfaces or APIs specific to that device, which is precisely why it cannot be ported "as is" to a new class of device. Many wireless or mobile middleware solutions claim to be device agnostic, but they are not. While it is true that they support all device types, it is because they generate (or have prepackaged) applications for each device type. As new devices appear, middleware providers must create additional device-specific modules.

An organization cannot arbitrarily decide which path to take—agnostic or specific—without considering what its wireless application and users need to accomplish, and the capabilities of the chosen device and network. The final decision will involve a trade off between the burden of supporting and maintaining multiple sets of code as new device types come into the fold, and the value of exploiting unique device features. Whichever path you choose to follow, strive to make your solution more extensible by designing components for maximum re-usability, partitioning components wisely, and separating business rules and application logic from presentation code. If device-specific or network-specific code is needed, try to use middleware products to perform the task instead of requiring your developers to learn the details of multiple platforms and technologies.

9.3   Securing the Solution

With a wireless solution, companies have three over-arching security goals. They want to secure the data stored on mobile devices, and to monitor, restrict, and revoke the access enabled by those devices. They want to keep their networks, wired and wireless, secure from unauthorized access. They want the data exchanged between clients and host to remain private and tamper-proof. Let's examine these three security risks and some ways to mitigate them.

9.3.1   Device Misuse and Theft

Any mobile, portable device is vulnerable to theft. Laptop thefts alone reached 387,000 in 2000, and the financial losses from misuse and attack were second only to virus attacks.[1] Because handheld devices are even smaller and more easily concealed, they are particularly susceptible to loss and theft, a risk that every company faces. Depending on the data stored on the device, which might include user IDs, passwords, encryption keys, and confidential information, the company may face significant exposure if the device falls into the wrong hands. The user IDs and passwords could give a snoop access to the company's wired network, including enterprise data and systems, especially if the theft is not discovered or reported immediately and the company does not have time to revoke access.

Many devices come with power-on passwords, implemented at the hardware level; such passwords are the first line of defense against unauthorized use. Companies are advised to activate these passwords prior to deploying devices. As an added precaution, adopt policies that prohibit storing passwords on devices and periodically monitor compliance with the policy. Consider encrypting and password-protecting data stored on the device, including on any Compact Flash cards or other accessories. In addition, some device management tools claim the ability to send messages to a device to lock it or destroy its contents, an attractive security feature.

---

1.   "Sensing a New Movement in Laptop Security," *Security Magazine,* August 2001.

9.3.2 Unauthorized Network Access

Device theft can lead to network intrusions, as mentioned above, if access codes or passwords are stored on the device. But anytime a wireless network connects to a wired one, the wireless network can, unwittingly, serve as a conduit for a hacker to gain entry into an otherwise secure wired network. This risk is especially high if the wireless network is not sufficiently secured in its own right.

Before the emergence of the Internet, hackers generally had to be physically present within the corporate complex to gain entrée to a wired network. The thousands of remote access points enabled by the Internet now allows hackers to perform their transgressions from a distance. This new type of threat has spawned a variety of different security techniques from firewalls to virtual private networks to complex cryptographic algorithms. Like the Internet, wireless networks offer hundreds of remote access points to corporate networks, creating similar vulnerabilities.

Several techniques are available, separately and in combination, to secure wireless networks from unauthorized access:

- *Authenticate Access Points*   To verify the client device or node attempting to connect to the network, maintain a list of valid access point names and regularly check to ensure that unknown names do not crop up. Several commercial products are available to help in this task.

- *Authenticate Users*   Rather than authenticating via access points as described above, consider using Layer 3 authentication, as specified by the IEEE, based on user IDs and passwords combined with 128-bit encryption. In this type of authentication, a user connects to a network via an access point that in turn connects to a RADIUS (Remote Authentication Dial-In User Service) server. The server challenges the user, who must supply the correct response to gain access.

  Other methods of authenticating users include using smart cards, tags, or portable tokens to generate one-time passcodes. Biometric techniques, which rely on unique physical traits to authenticate a user, are also gaining in popularity. Although biometrics solutions are not yet widely deployed, they include techniques for fingerprint scanning, voice recognition, and iris scanning. Biometrics can be used both to control access to the device itself and to the network.

- *Centralize Encryption Keys*   A straightforward piece of advice that companies often neglect to follow is to refrain from storing encryption keys on client devices or on WLAN adapter cards.

- *Enable Default Security*   Most commercial WLAN products ship with security features turned off. Depending on your organization's overall security strategy, it may make sense to enable these security features. For WLANs, the default security is called WEP (Wired Equivalent Protocol), which in its latest incarnation creates encryption keys for each packet of data transmitted. Standard WEP is based on 40-bit keys, although many vendors offer their own alternatives based on 128-bit keys. Note that the more bits contained in a key, the more secure it is; it theoretically takes longer for a hacker to break the key. Since reports have surfaced of hackers breaking 128-bit keys, it is safe to assume that key lengths will change in the future. The IEEE, a proponent of WEP, will likely come out with a successor to WEP offering even tighter security.

Another area of exposure with WLANs has centered around something called SSID (Service Set ID). WLAN manufacturers use the SSID as a way to identify the wireless networks to clients. Each manufacturer chooses its own SSID that is common across its WLAN products—Intel uses "Intel" and Cisco uses "tsunami." Since these default SSIDs are a matter of public knowledge, it is wise to either turn off SSID broadcasting or to change the default name.

- *Supplement with Third-Party Security Products*   Many third-party vendors offer products that address the different security risks inherent in wireless networks. If the default security features contained in your devices and networks are inadequate, consider purchasing these products to augment baseline security. These packages generally cost less to implement than a Virtual Private Network (see below) and combine authentication and encryption techniques.

- *Virtual Private Networks (VPNs)*   VPNs provide a secure "tunnel" between client and host through which all communications are sent, as illustrated in Figure 9.2. The tunnel works by wrapping communications within Internet protocol (IP) packets. A VPN relies on compatible software or hardware loaded at the client and host ends. If a client device does not use the same tunneling protocol as the host, it is denied access to the network. Various VPN standards exist, including PPTP from Microsoft and L2TP from Cisco.

9.3.3 Data Interception

A separate concern from unauthorized network access is data interception. Data interception brings with it loss of privacy, theft of information, and the ability to tamper or alter data to destroy its integrity, sometimes in subtle, undetectable ways.

Besides wanting to gain admittance to the corporate network, a hacker may simply want to eavesdrop on data transmissions. Just as the Internet gave hackers the cover of distance and anonymity, so too do wireless networks. Many wireless networks, especially the 802.11x WLAN networks, transmit data outside the physical confines of a company using radio waves. Depending on the strength of the signal and the amount that "leaks" outside the physical premises, hackers can potentially

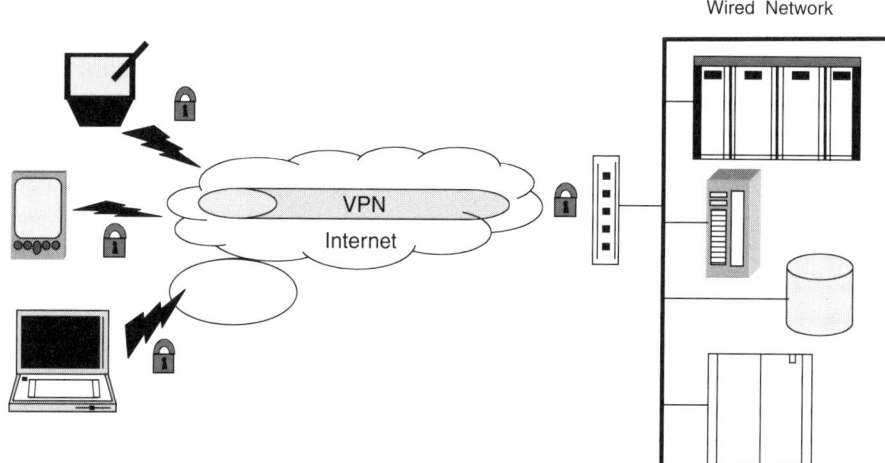

FIGURE 9.2
Virtual Private Network

intercept and decode transmissions, and even gain entrance into wired networks. The publicity surrounding these security flaws has led many companies to pull back and re-assess their WLAN plans. At the same time, the IEEE has been working diligently on developing and promoting more secure standards, and many WLAN vendors have software patches to strengthen the security of their network products.

Chapter 11 briefly mentions the relative security risks present in Do It Yourself wireless networks, including Infrared, Bluetooth, and WLANs. Wide area cellular networks are quite secure, although in the past, exposure has been reported at gateways, such as the WAP gateway, where data is decrypted momentarily, translated to a new format, and re-encrypted. The risks present in this brief interlude, known as the "WAP Gap" have since been overcome.

If data interception is a concern with your wireless solution, consider taking these security measures.

- *Beef Up Network Security*   As discussed above, most network products and third-party software have algorithms to encrypt data packets. Starting at the network layer, use built-in capabilities to protect transmitted data.

  In addition to these software-level security measures, take appropriate precautions at the physical level. With WLANs, the placement of antennas and access points directly contributes to the amount of signal leaked beyond the four walls of a building. Orient antennas so they face inwardly, and consider putting WLANs in the center of buildings rather than the outskirts.

- *Application Layer Security*   Applications themselves can incorporate security measures such as encrypting data and validating access. If network security is inadequate and/or the sensitivity of the data being transmitted is heightened, consider incorporating application-level routines to provide an additional layer of security.

9.3.4 Viruses

The potential for contracting viruses on client devices is becoming a reality. With reports of viruses invading smart phones and Palm OS-based PDAs, the need for virus protection at the client level is acute. Any device that can access the Internet or receive e-mail is at risk of catching a virus and passing it on to other nodes within the corporate network. Because of the memory limitations of handheld devices, anti-virus software has typically been hosted on a PC or laptop, with the client device physically connecting to the host to perform virus scanning. Commercial vendors, however, are just starting to release device-resident virus scanning software, although it is limited, at the time of writing, to certain devices and OS versions. As with a wired PC or desktop, the client device must periodically download the latest virus updates to remain protected.

9.3.5 Putting Security in Context

The security risks outlined above, and the measures to mitigate those risks, must be taken in context. Before adopting rigorous security measures, and parting with a hefty amount of cash, consider:

- The *value* of the data you wish to protect
- The potential *damage* that might ensue from loss or interception of the data, including public embarrassment as well as economic loss
- The *risk* that the damage will occur

If you are transmitting financial transactions or medical records over a wireless network, then there is sufficient reason for concern, and it is worthwhile to take advanced measures to ensure against risk of loss or interception. Conversely, if your fleet management application is transmitting arrival and departure times intermittently, the value of the data might not justify the investment to secure it against eavesdropping. Although our society is very security-conscious right now, the truth is that not all data is so valuable, confidential, or sensitive that it justifies going to extreme lengths to protect it.

Wireless Devices

When you conjure up the image of wireless technologies, the first thing that comes to mind is the mobile device. In fact, most people are initially attracted to the way a device or gadget looks; its operational features and wireless capabilities are usually secondary. Everyone has an opinion about devices, and every device has camps of fervent followers and vocal detractors. Ask an owner how they feel about their Palm, Pocket PC, RIM BlackBerry, or Nokia Communicator. Chances are, you'll get an earful.

For other people, the mobile device is a means to an end. The device and wireless experience are inextricably intertwined. Form and function meld into one. An Avis parking lot attendant doesn't spend time contemplating the device and the resident applications sitting in his hand. To him, it is simply another gadget that lets him check-in cars, obtain signatures, and authorize credit card payments. Where the device, software, and network start and end is irrelevant.

Somewhere along the food chain, however, someone *does* have to care about the device and its characteristics. Someone must question the durability of the device used by the Avis attendant—its ability to withstand drops onto concrete and exposure to inclement weather and fluctuating temperatures. Someone must worry about the power consumption of the device and what will happen when the batteries die during a credit card authorization. Someone must be aware of the costs, integrated capabilities, third-party software, and physical characteristics of the device.

When you look beyond the sleek form factors, cool colors, and hip attachments, and try to choose a mobile device as the platform for your enterprise solution, where do you start? How do you pick the right device?

In reality, there is probably more than one "right" device. If a device lets your company accomplish its objectives, facilitates your employees' performance, and gets the job done, then it is the "right" device. Just as there is more than one type of desktop computer that will work within your wired networks, there are also several device types that will let you accomplish what you want to do wirelessly. At the end of the day, it is highly likely that your organization will wind up supporting more than one type of device.

You should also expect to seek and receive plenty of input from the user community regarding device selection. In the wired world, few users have opinions about their desktop computer models and few develop emotional attachments to them. Wireless devices are different. Today, many users are apt to have some level of experience with wireless devices. Many users will own a device or two, and will have biases and preferences that will need to be reconciled with corporate plans and objectives. The device becomes an extension of the person, and understandably so, since it is carried by a person on a daily basis, often close to their body, much like a wallet, set of keys, or clothing accessory.

Before you alienate your work force by making them jettison their favored devices, be sure to evaluate and consider their usage patterns. If most of your users are quite adept at using a particular PDA's handwriting recognition scheme, is it really necessary to make them learn another device's proprietary method? Different user communities may also use devices in disparate ways. Sales staff or other mobile professionals will use their devices differently than a pipeline repairperson or a UPS delivery person. Someone who is primarily working from a vehicle will have different device needs than someone working within an office environment. Being sensitive to how your workers will use their devices to accomplish their work will help you choose a more suitable solution.

If a new device does not offer a positive, or even superior, user experience over existing devices, it will be shunned. If a device is cumbersome, awkward to use, or too delicate for its operating environment, then trouble will ensue. Fortunately, with the wide selection of devices available, and with more on the horizon, you should be able to find an appropriate device, provided that you spend time up front to investigate your options and consult your users.

Lastly, do not underestimate the value of providing a "hook" with your chosen wireless device. As many organizations have discovered, getting users hooked on a particular aspect of the wireless solution can do wonders for the acceptance of the solution as a whole. Often, the device can provide such a "hook." For example, a common and popular hook has proven to be wireless e-mail access. Users quickly become addicted to sending and receiving e-mails on the go. A device that simplifies e-mail access, or even makes the process fun, may well influence the ultimate acceptability and success of the entire wireless solution.

This chapter introduces the topic of wireless devices. It opens by exploring some of the larger device considerations, such as cultivating the right mindset and thinking in terms of "fitness for purpose," and explores other important selection factors and future trends. Next, the chapter provides an overview of important features and characteristics of wireless devices, items that you will want to keep in mind as you evaluate and select device candidates. Following this discussion, the chapter divides the universe of devices into several categories, and looks at some of the commercial offerings in each category. Finally, the chapter briefly touches on device support and management issues, topics that are explored in more detail in Chapter 13.

10.1 Device Considerations _____

Selecting a wireless device, or devices, to deliver application functionality is a trickier proposition than choosing desktop components in the wired world. Many organizations start out having little exposure to or experience with wireless devices and their associated networks, applications, and infrastructures. To come up to speed, both the technical and non-technical sides of the house must invest time in product research and comparisons, and must test-drive a variety of devices. The range of devices is large. The differences between devices can be great. Device capabilities are immature and evolving. Support and maintenance are performed by trial and error. With so many variables to contend with, expect to devote some time to the selection process. Many organizations end up selecting more than one potential device, and have different groups within the company pilot each device. Based on user feedback, a final device choice is then made.

Before you jump into device selection, prepare yourself. Choosing a wireless device is very different than choosing the equipment and components comprising your wired universe. You'll need to adopt a new perspective, think of devices in terms of their "fitness for purpose," and be sensitive to factors like vendor viability, integration, and maturity of development tools.

10.1.1 Cultivate a New Mindset

To understand and eventually settle on the right type of device for the situation at hand, an organization must develop a different mindset. This caveat applies equally to users, executives, mobile workers, and technicians—anyone involved with the wireless solution. Why is a different mindset required? Consider the following points.

- *Device Limitations* The nature of wireless devices requires that we accept limitations that have long since been conquered in the wired world. Physical constraints such as small screens and unwieldy input mechanisms call for a high degree of patience and ingenuity. Performance constraints such as limited memory and low battery life can impede the way we work with devices. Even the way that we physically place devices—upright, on their sides, upside down—can impact their usability. Things that we take for granted in the wired world are conspicuously absent in the wireless one. To avoid becoming frustrated, understand that every device will have its limitations.

- *User Primacy* When was the last time your organization consulted its users prior to selecting hardware or equipment, purchasing servers, or deploying desktop computers? Most organizations are accustomed to making equipment choices completely independent of their users. Selecting a wireless device, however, takes a much more user-centric approach. Users must be involved at all stages of the selection process, since user buy-in is critical to device acceptance. Choosing the wrong device can doom the entire wireless solution, since the device will bear the brunt of user displeasure.

- *Varied Usage and Environments* With stationary users sitting in offices, we can make some pretty safe assumptions about how they will use their computing equipment, and the conditions to which the equipment will be subject. Usage patterns are fairly predictable, and with few exceptions, users generally have the same type of equipment. In the mobile and wireless context, there are many more variables with which to contend. Until we've thought things through, we can make few assumptions about where and how users will actually use their devices, or the environmental conditions that will affect the units. Users may be sitting, reclining, driving, or talking on cell phones at the same time they use their wireless device. The surroundings may be dimly lit or bright, noisy or quiet, hot or cold, dry or wet. Devices can be used in all manner of fashions, in all manner of places, and by all manner of people. The variety of scenarios requires that we discard our old assumptions and try to envision afresh how the device will be used and the environmental conditions that will affect it.

- *Development Austerity*   For developers, wireless devices mean a return to austerity, especially when compared to the rich desktop computing platforms prevalent today. Developing applications that will run on a constrained platform, where resources are precious, requires a consciousness of, and sensitivity to, the operating environment that has long since vanished in a world of unlimited computing resources. Resourceful development techniques become highly prized, as explored more fully in Chapter 12. Even though each generation of device is more powerful than preceding ones, it is unlikely that constraints will disappear completely. As our experience with PCs, servers, and mainframe computers has amply shown, as more processing power or memory appears, developers quickly use it up and begin to clamor for more.

If we approach wireless devices the way that we approach the desktop components found in the wired world, we are in for a shock. No single wireless device will offer all of the capabilities—tangible or intangible—offered by wired desktop components. Nor will these devices be used in the same fashion or types of locales. To assist in developing this new mindset, Appendix A contains a questionnaire designed to remind an organization of the many variables affecting device selection.

10.1.2 Fitness for Purpose Approach

The Uniform Commercial Code (U.C.C.) is a set of model laws that, among other things, governs the sale of goods. These laws contain a concept called "fitness for particular purpose." The concept acknowledges that buyers may sometimes seek a product that is fit for a particular purpose, and legally obligates sellers to make sure the product sold can meet those purposes, if the buyer is relying on the seller's judgment. For example, if a buyer tells a seller that she is looking for a toaster that can handle oversize bagels, then the seller should not sell her a toaster designed for only the thinnest slices of bread.

Legal ramifications aside, this concept of fitness for purpose is also applicable when approaching the selection of a device. The buying organization should develop some idea of the particular purpose it has in mind for the device, and should ensure that it is fit for that purpose. Does the device, in its present state (or perhaps with some minor tweaking), fit the particular purpose your organization has in mind? This "purpose" includes, at a high level, the business objectives of the project, the needs of the users, and the needs of the application. Keeping this purpose in mind will help when it comes time to make choices between competing devices, or make compromises over a single device. For example, a device with a color display may be the most visually appealing of all the candidates, but may not be appropriate for a simple checklist-based application where color is superfluous. Or, the high power consumption of a color device may not be suitable for a situation where conserving battery power is critical.

Conducting a "purpose-based" review of devices will also help keep your organization rooted in the present. As you perform your device research, you will undoubtedly notice that devices are in a state of flux with a set of evolving capabilities. New devices appear frequently, and device manufacturers are constantly announcing the next generation of devices that will be better than ever before. With so much change occurring in the market, it is tempting to delay device selection and wait for the next best gadget to appear. In reality, no one knows what the device market will look like in 1, 3, or even 5 years, despite predictions to the contrary. If you become mired in indecision, sitting on the sidelines waiting for the optimal device to appear, you'll stall your project indefinitely. Instead, re-focus on whether the candidate devices are fit for the purposes you have in mind. If the answer is yes, then choose one. If the answer is no, you have three options. You can engage a device manufacturer to create a device that meets your unique needs. You can adjust your business objectives and solution definition to accord with reality. Or you can forget about the project, and revisit it sometime in the future when devices have evolved.

10.1.3 Other Factors Influencing Device Selection

For some companies, selecting a device will devolve into a price and feature war. While device characteristics and price are important components of the decision, they are not the only ones. A variety of other factors should be weighed when considering the choice of device. Some factors will be more important than others; however, it is worth investigating each one prior to committing to a particular device. These factors include vendor viability, migration strategy, development environment and tools, ease of integration, and network connectivity options.

- *Vendor Viability*   How financially stable is the target vendor? With no obvious market leader, and with no truly entrenched group of users, the device market is in flux. These rapidly changing conditions make it easy for new entrants to appear and garner market share, and for existing vendors to succumb. Early leaders in sales to the consumer market, such as Palm and Handspring, saw their market shares erode as rival Compaq (now merged with Hewlett-Packard) made inroads with its iPAQ device. In this dynamic market, many device manufacturers are rethinking and revamping their business models to remain competitive and viable. Psion, for example, decided to exit the consumer handheld market due to its poor performance vis-à-vis its competitors. In an uncertain market, a risk for organizations purchasing devices is that their vendor will either go out of business or be acquired by another, leaving future upgrades and support up in the air, the situation now confronting Jornada users after Hewlett-Packard's merger with Compaq. In such a case, an organization may be forced to purchase new devices from a different vendor while still supporting the old devices, or may have to write off its entire device investment and outfit its staff with new devices. For these reasons, it is wise to do a little research into the financial stability and future economic outlook of the target vendor prior to committing any capital.

- *Migration Strategy* The project that is prompting your organization to consider wireless devices has its particular purposes and objectives, and these clearly must be satisfied in your choice of device. As pointed out in Section 10.1.2, these considerations are the primary ones when it comes time to choosing a device. If, however, future applications can already be envisioned—like adding expense reporting capabilities or spreadsheet applications to a mobile sales force application or speech recognition capabilities to a wireless ERP application—then it is worth considering how and whether the selected device will support these future add-ons. Some devices are strong in one area but not in others. An e-mail appliance, for example, may not be a good choice if you think that speech recognition features will be needed sometime within the next year. Some devices have an abundance of associated third-party software that can be used to meet future needs; others would require custom software development.

- *Development Environment and Tools* What development tools and environments are available for a particular device line? Many device manufacturers, especially those with some longevity, provide development tools and software development kits (SDKs) to third parties so that they can develop, test, and debug applications that will operate on the device. A strong development environment can greatly ease software development and future maintenance. It will also result in a greater portfolio of third-party applications for the subject device. Other considerations are the languages supported by the platform, the availability of expertise on the market, and the availability of third-party tools and middleware to help put together a complete solution. Many middleware providers offer application development and deployment platforms that cover a range of device types. The advantage of these platforms is that they enable quick deployment of applications; however, they also lock a company into a particular vendor's solution, which may be risky depending on the longevity and viability of the vendor.

- *Ease of Integration* How easy is it to integrate the device with other devices, applications, and software packages? Physically integrating components requires a variety of cradles, docking stations, cables, and connectors. What kinds of proprietary accessories or attachments are needed to physically integrate the device with other equipment? Integrating non-physical components with the device—such as network interfaces, applications, and tools—depends on platforms, language support, protocols, and even the availability of developers skilled in these areas. Does the device use any non-standard or

arcane conventions? Are interfaces and specifications published? Is there sufficient access to underlying utilities and functions?

- *Network Connectivity Options*   What kinds of network connectivity options
  are available for the device? Some devices are optimized to work on certain
  types of networks. Phone handsets, in particular, work on only one or a subset
  of cellular networks. Will the device work with short- and medium-range network solutions, such as Bluetooth, infrared, or WLANs? What about wireless
  WANs, both voice and data? Are attachments or accessories needed to connect
  to a network? Is it a hassle to use them? For example, if your organization was
  looking for a device that could provide access to e-mail without having to initiate a connection, a popular choice would have been the devices like the RIM
  BlackBerry or the Palm i705 with their "always on" network connection.

10.1.4   Looking to the Future

No one can predict with certainty what will happen in the device market in the coming years, especially considering the tremendous amount of change that has
occurred to date. Device capabilities are evolving rapidly. An increasing amount of
packaged software is available for licensing. More powerful functionality is being
integrated into devices.

Despite this uncertainty, a few trends are emerging.

- *Courting the Enterprise Market*   Every device manufacturer is salivating over
  the enterprise market for wireless devices. The consumer market provided
  proof-of-concept and gave capital infusions to device vendors. But the business
  market is where the real money is to be made, a fact of which every device
  manufacturer is keenly aware.

  To that end, and to remain competitive in the enterprise space, device manufacturers are allying themselves with enterprise software vendors to offer stripped
  down business functionality on handheld devices. Some variation of popular
  SFA, CRM, and ERP applications from well-known enterprise vendors will
  make their way onto wireless devices. Handspring, for example, has launched
  enterprise alliance partnerships with companies like Aether Systems, AvantGo,
  and Wireless Knowledge to develop CRM, SFA, e-mail, and ERP applications
  for its Treo and Visor product lines.

  Manufacturers are not only looking at software to attract business users, they
  are also offering more powerful and complete devices. As noted below, each
  subsequent generation of device "borrows" the more successful features of
  competitive devices to woo business users.

- *Melding of Device Types*   Device types will continue to meld and lose their
  sharp distinctions as device manufacturers borrow features and functionality

from different classes and merge them into their own devices. Phone handsets now have PIM- and PDA-like capabilities, while PDAs like the Handspring Treo are adopting voice capabilities. Smart phones now closely resemble PDAs in terms of their functionality, yet they offer superior voice capabilities. Nextel, for example, is working with Motorola and RIM to develop a BlackBerry-like device that will incorporate both voice and data capabilities. Palm launched its i705 PDA with "always on" e-mail capabilities to stem the inroads RIM has made in the enterprise market with its BlackBerry e-mail device. No doubt, other device manufacturers will fire similar salvos in the coming months in an attempt to halt advances made by their competitors.

- *Smaller Is Better, Usually*   In the mobile handheld market, smaller is definitely better. Balancing this desire for small, and even miniaturized, devices is the demand for greater functionality, which generally equates to the addition of more physical components and a resulting larger-sized device. Device manufacturers will continue to explore ways to shrink the size of their devices, yet try to satisfy the demand for larger screen sizes and keyboards. Future devices will fold, slide, and collapse in an effort to minimize their overall dimensions. Users will prefer to carry multiple small devices rather than one large, bulky one.

10.2 Device Features and Issues

Anyone familiar with purchasing computing equipment knows the importance of the product specification or "feature list." In fact, the vast majority of computing equipment is purchased solely on the basis of its features combined with its price. As mentioned above, the selection of a wireless device is a bit more complicated, given the types of usage scenarios and user preferences with which to contend. However, device features remain very important as they influence the usability of the device, and the types of applications that will run on them. Some features will be critical (such as the amount of processing power available to a custom application), while others (such as a color display or audio capabilities) may be superfluous or simply nice extras. Some devices can be expanded and made more powerful by adding peripherals or accessories. Other devices may contain everything you need in one integrated package.

This subsection looks at several features that you will encounter as you research wireless devices, features that may be more or less relevant depending on your users' needs and the needs of the target application. Accordingly, you should have a good understanding of these features, and their nuances, to compare devices on an apples-to-apples basis.

10.2.1 Voice Capabilities

Depending on the application that you have in mind, voice capabilities may be an essential feature. If your application will not use voice processing, and you expect that your users will continue to carry and use cell phones, then voice capabilities are largely irrelevant to your inquiry.

A few different scenarios may lead you to prefer a device that can handle voice communications. If it is inconvenient or unlikely that your users will carry a cell phone in addition to a handheld device, or if users would prefer a single multi-purpose device that can function as a phone and application platform, then your candidate device must have strong voice capabilities. A field service repairperson that must carry manuals, tools, and a handheld device for data collection and work order processing may not be able to juggle a cell phone.

Speech recognition capabilities may also be warranted if your target application accepts voice commands to navigate or perform transactions. If users spend a lot of time driving, for example, and want the ability to use the application hands-free, then a voice interface may be required. In this case, you need a device that can process voice as well as perform any data-related functions.

As more hybrid devices appear, we can expect many future devices to have voice capabilities. Phone handsets and smart phones will continue to be strong contenders in this category, since they are optimized for voice processing, integrate those capabilities in the device, and can also serve as a platform for data-based applications. The downside to these powerful handsets is that the more they combine voice and data capabilities, the bulkier they become. In contrast, PDAs are optimized to handle data-based functions, and may be converted into a phone via the addition of an external modem, as shown in Figure 10.1. An advantage of this non-integrated approach is that the footprint of the PDA can remain relatively small when used in data mode. The disadvantage is that yet another component—a modem—must be purchased and carried, with the potential for loss or misplacement. With a large deployment, purchasing modems for hundreds of PDAs can be expensive, with current modem prices running at about half the cost of a PDA.

10.2.2 Size, Weight, and Portability

The size and weight of the wireless device will influence its portability and usefulness to the user community. Depending on where the device will primarily be stowed, and the amount of other gear that the user expects to carry, the physical size and weight of the device may or may not be problematic. For a salesperson that frequently travels on airplanes and wants to carry the device on her person, a lighter and smaller device

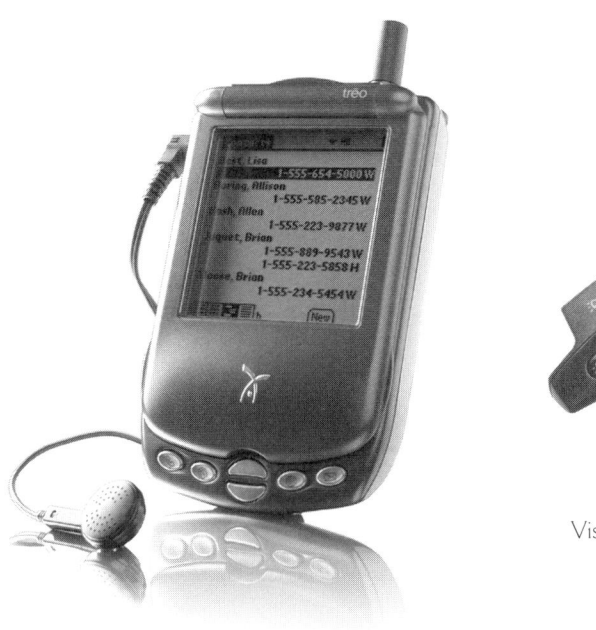

VisorPhone Attachment

Treo with Headset

FIGURE 10.1
PDA with Phone Capability
Source: Handspring

is preferred. For a driver using a dashboard-mounted device in his truck, size and weight are largely irrelevant. A wireless scanning device worn on the finger, or a terminal worn on the hip, must be lightweight and as inconspicuous as possible to avoid burdening the wearer or interfering with physical activities. Figure 10.2 shows a sampling of devices in a range of sizes.

Mobile, wireless devices can be as small as a credit card or as large as a laptop computer. In the handheld category, most commercial devices try to come in less than 1 pound, with the exception of smart phones and communicators which are becoming larger and more bulky from integrating voice and data capabilities in a single device. Manufacturers quote device weights in terms of ounces, and even a one-ounce reduction in weight is considered a marketing coup. Even traditionally weightier devices such as laptops, notebooks, and tablet computers are becoming smaller and lighter. At the time of writing, Toshiba had announced its Protégé 2000 notebook, which is a little over half an inch thick and weighs only 2.6 pounds, much lighter and narrower than its competitor, the Vaio notebook from Sony.

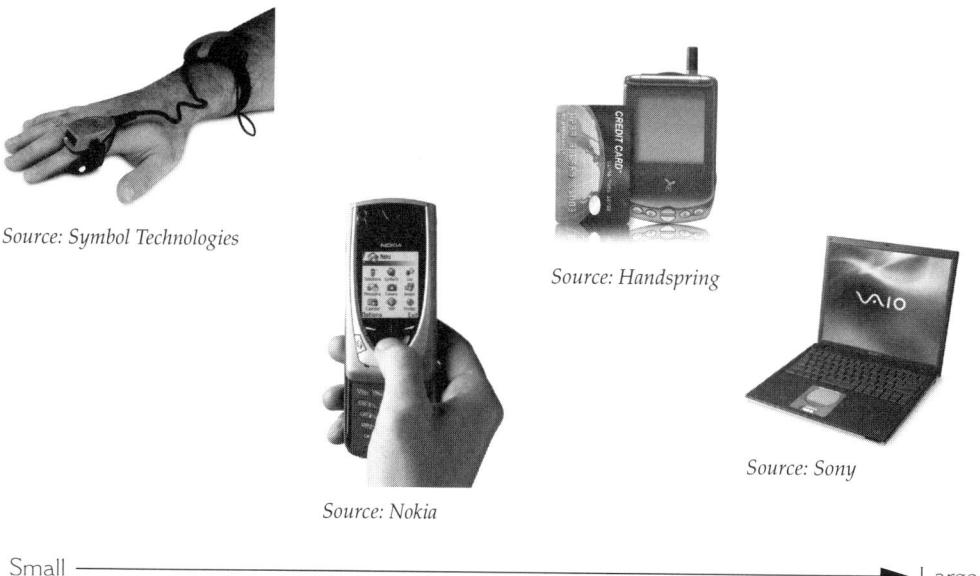

Source: Symbol Technologies

Source: Handspring

Source: Sony

Source: Nokia

Small ────────────────────────────────────▶ Large

FIGURE 10.2
Range of Device Sizes

If size and weight matter for your deployment, remember to include any necessary accessories or peripherals that will need to be carried. The handheld device itself may fit into a shirt or coat pocket, but if the required accessories take up half a briefcase, the solution may not be convenient. For certain uses, the handheld must fit within one hand, a requirement that may be compromised when peripherals are taken into account.

Smaller, lighter units are prized because of their portability and convenience. The more convenient the device, the more likely a user is actually going to carry it with him or her. A device that requires minimal or no extra physical space for setup and operation makes it easy to use in a wide variety of locations. The disadvantage of small devices is that integrated features like displays and keyboards become proportionally smaller too. Components that require physical dexterity, such as keyboards, or that are viewed, like displays, become more difficult to use as their size shrinks. Power sources, such as batteries, are one of the limiting factors on device size. Batteries can only be shrunk so much, and as they become smaller, they supply less power to the device. Smaller devices must also be constructed with durable materials; the more delicate they are, the more prone they are to damage. And, smaller-sized devices are more susceptible to loss (easy to overlook) and theft (easy to conceal).

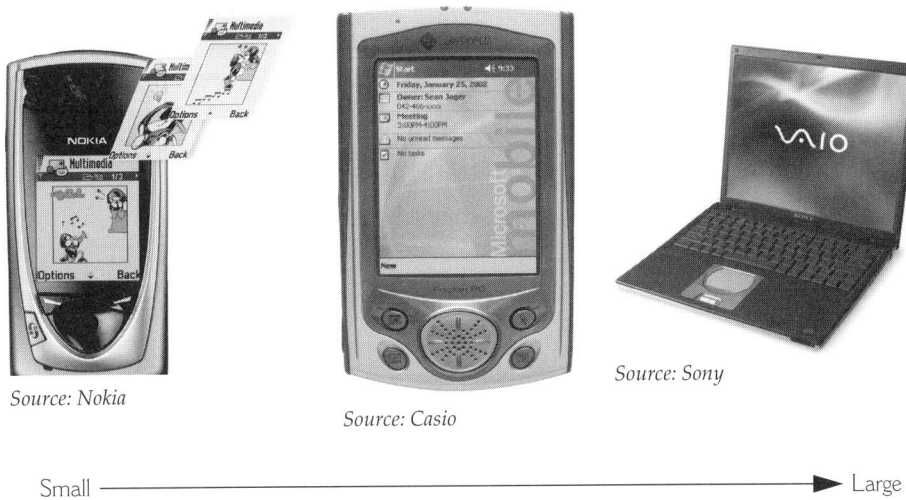

Source: Nokia

Source: Casio

Source: Sony

Small ──▶ Large

FIGURE 10.3
Range of Screen Sizes

10.2.3 Display

Handheld devices come with a variety of displays from monochrome to gray-scale to color. Active matrix and passive matrix displays are available. Active matrix displays are faster and easier to see, while passive matrix displays consume less battery power. Displays, especially color displays, can use quite a bit of battery power, so make sure any battery-saving features are activated, such as dimming the backlight when the display is not in use.

While many screens will perform well in a variety of lighting conditions, backlighting is an important feature in low-light conditions. If users are apt to work in dark, dim, or variable environments, a device may need backlit displays and keyboards. Bright light conditions may also interfere with viewing, so be sure to test candidate devices in a range of lighting, both indoors and outdoors.

With the exception of the larger, mobile form factors such as laptops, notebooks, and tablets, display sizes on wireless devices are notoriously small, as shown in Figure 10.3. Some fit as little as 4 to 6 lines of text; others will allow as many as 10 to 12 lines. The lines themselves are fairly short, ranging from 12 to 30 characters or slightly more. This limited amount of screen real estate is problematic for users and developers alike. For users accustomed to the rich color, graphics, and multimedia capabilities offered by desktop computers, the visual experience of a handheld device may seem paltry. Just imagine trying to read a large document using the small screen on a typical cell phone. Developers must use clever design techniques to cope with small display sizes yet still provide users with an acceptable, positive experience. Wireless business applications tend to eschew graphics and text-heavy screen displays, and favor terse sentences, checklists, and other skeletal types of displays. They avoid deeply layered information that requires navigating through multiple screens. Chapter 12 provides more detail on how developers must adapt programming techniques to deal with device constraints.

10.2.4 Data Input Mechanism

How will users enter data using the device? Standard size QWERTY keyboards are obviously too large to incorporate in a device that may measure just a few inches. The challenge for device manufacturers is figuring out how to overcome this physical constraint and still give users a workable and convenient method of entering text.

Three basic data input schemes have emerged for handheld devices. The first is based on a physical keyboard embedded in the device, as illustrated in Figure 10.4. The second consists of a virtual keyboard or keypad that is displayed on a touch-screen for users to select letters. The third is based on a handwriting recognition scheme used in combination with a touchscreen and stylus. Many devices offer all or some portion of these mechanisms. Several models of attachable, portable QWERTY keyboards are available that expand to a fairly normal size for power typing. Another method of overcoming physical data input constraints is to allow users to save canned messages and text that can be inserted in e-mails or forms to avoid typing.

- *Physical Keyboard*   The physical keyboards on most handheld devices are quite small, making it difficult, if not impossible, to position the fingers and type as one would on a full-size keyboard. Instead, users learn to adapt their typing, in many cases using only their thumbs to press the keys. Anecdotal evidence indicates that users can become proficient quickly in all-thumbs typing, and that this style of input is not an impediment to usage. RIM was an early proponent of this design technique, and their BlackBerry e-mail devices have enjoyed tremendous popularity. This type of "all thumbs" keyboard is now available as an accessory for many types of PDAs. Sharp recently introduced a device with a built-in miniature QWERTY keyboard that can slide up into the body of the device when not in use. Other manufacturers offer various keyboard sizes, such as 28-key boards for accounting professionals, and 37- and 47-key alphanumeric ones. Still other manufacturers are investigating the use of pressure-sensitive fabrics and cloth in keyboards, which can be easily folded or rolled for stowing.

Phone handsets have numeric keypads that double for alphanumeric data entry. To differentiate between letters, some phone handsets have predictive, word-completion algorithms to intelligently guess the word the user is typing

Source: Sony

Source: Handspring

Source: Symbol Technologies

FIGURE 10.4
Device Keyboards

after just a few characters have been entered. In reality, the numeric keypads on most phone handsets are suitable only for limited data entry such as short messages. The newer smart phones have larger, more usable keyboards, a factor accounting for their greater size.

- *Virtual Keypad*   Many handheld devices use touchscreen technology to allow users to enter data. With a touch-sensitive screen, users can apply their fingers or a stylus to the keypad to select letters or numbers. Using a "hunt and peck" approach, users can tap out words and enter data. This approach is not conducive, however, to entering any volume of data and is best limited to brief bits of text. An innovative approach being pioneered by an Israeli company uses a red laser keyboard projected onto a surface to create a virtual keyboard for data entry. The module has optical infrared sensors that are able to detect keystrokes or finger movements and convert them into letters or numbers for the device to process.

- *Handwriting Recognition Schemes*   Relying on a touchscreen and a stylus or pen, handwriting recognition schemes allow users to enter "shorthand" for characters and words. These schemes are mostly proprietary, and users must learn a different method depending on the device they use, whether it's a phone or a PDA. Natural handwriting recognition is still in its infancy.

The shorthand used by these schemes consists of a set of specialized characters. Each letter in the alphabet has its own representation—a unique stroke or partial stroke of the stylus, maintaining contact with the screen throughout the stroke. For example, Palm's handwriting recognition scheme, Graffiti, recognizes the letter "A" as up and down diagonal strokes omitting the horizontal connecting line between them. Pocket PCs also use a form of character recognition that attempts to identify the letter the user is writing by the strokes entered.

Many users claim that these handwriting schemes are not as daunting as they first appear, that the learning curve is relatively short and that the end result is a quick writing technique. Users must still master a cryptic handwriting scheme, however, or even two if they own more than one device type. Don't be surprised if users balk, at least initially, at learning a new writing style that is not immediately intuitive.

10.2.5 Processing Power

Every device must have enough horsepower to fuel the operating system, drivers, routines, agent software, and applications needed to get work done. Handheld devices may not have the processing power of a desktop computer, but they are closing the gap with each new release. This increase in processing power comes at a price, however. Greater processing power generally means more bloated operating systems. An operating system that monopolizes 85% of the CPU capacity will leave little room for your custom application. Moreover, the greater the processing power, the more battery power needed to fuel the device. Applications that require a great deal of processing power will burn through batteries much more quickly. Color applications and wireless communications, in particular, are battery hogs. See Section 10.2.8 for more information on battery life.

Not surprisingly, entry-level PDAs have less processing power than higher-end models targeted at the enterprise market. You can get a Handspring Visor Edge with a 33 MHz Motorola processor or a Compaq iPAQ with a 206 MHz Intel CPU. Intel's StrongARM processors are favored by many device manufacturers, and its XScale technology promises to offer more horsepower while consuming less battery power, precisely what many of the more robust wireless and mobile applications need.

When used in conjunction with a wireless network, the apparent processing speed of a handheld device can be affected by the network's transmission rates. A device may be perceived as performing slowly when the real culprit is a bottleneck on the network. Or the network may be operating just fine, but the device's processor may be overloaded.

10.2.6 Memory

Like processing power, handheld memory has been steadily increasing. The amount of memory that a user will need on his or her PDA depends on the type of work they do, the type of applications that will be running, and the type of data that will be stored. Until recently, 2 MB to 8 MB of memory was sufficient for most users. As processing power has increased, however, and as application functionality has improved, 16 MB to 64 MB of memory is becoming the norm. Pocket PCs, with their more robust bundled application software, regularly come with 32 MB of memory.

The trick with memory is making sure there is enough on the device to support current and foreseeable needs. Even if 16 MB of memory seems exorbitant right now, chances are it won't in a couple of years. If you expect to use the devices for several years before upgrading them, it is better to err on the side of too much memory. Many devices have expansion slots that allow users to add memory cards to store files and programs, as discussed in Sections 10.2.10 and 10.3. These cards provide extra storage, but can't do anything to alleviate "out of memory" conditions due to insufficient RAM.

Note that some device management tools, as discussed further in Chapter 13, will monitor the hardware status of deployed devices and warn about "out of memory" conditions so that network administrators will know when upgrades are needed.

10.2.7 Durability

In everyday use, handheld devices are subject to a lot of manipulation. As their name implies, they are handled much more than other types of computing equipment. As such, their design must be rugged enough to withstand constant use, exposure, and even touching. Touchscreens in particular are roughed up by fingertips and pens, and often require screen protectors to keep them from getting scratched over time. Even something as mundane as the storage slot for a stylus, if poorly designed, will result in the loss of styluses on a regular basis.

For handheld devices, "ordinary use" encompasses a far more diverse set of circumstances than for desktop computers. Since most users are unlikely to drop their desktop computers, use them outdoors, or store them in their cars, the durability of these items is seldom an issue.

Durability is crucial for handheld devices. They are exposed to a range of conditions—indoors, outdoors, and temperature extremes—and subject to a lot of abuse—drops, heat, and even immersion in water. A device must be durable, in its design and construction, to support the different environmental conditions that confront handheld device users from the office to the warehouse to outdoor locales. Field service personnel may work in hostile conditions—in the snow and ice or the heat of the desert. Construction area workers must worry about dust and grime. Shipping firms must deal with water and corrosion. Doctors that stow devices in the breast pocket of their lab coats must worry about them falling out when they bend over to treat and examine patients. Some industry analysts go so far as to advise organizations to purchase a variety of devices and subject them to the "drop test" to see which ones fall apart when dropped onto concrete.

Device manufacturers tend to specify "drop, sealing, and temperature" characteristics of their devices. These descriptions provide an indication of the device's ability to withstand ordinary wear and tear in many types of environments, from office to industrial to outdoors locations. An organization should review these specifications, particularly the drop specifications, to see whether they fit anticipated operating environments. Sealing specifications indicate the ability to withstand dust and rain. Humidity specifications indicate the ability to withstand condensation. Drop specifications indicate an ability to withstand a drop from X number of feet to concrete. Operating temperatures provide a range of degrees at which the device will continue to function properly.

10.2.8 Battery Life

Batteries are the lifeblood of handheld devices. A mobile or wireless device simply will not work without sufficient battery power. Handheld devices may use permanent rechargeable batteries, replaceable rechargeable batteries, or standard alkaline batteries. Make sure that your users understand what to do when the battery runs out, especially if it is during a customer presentation or a credit card authorization.

The amount of time that users can use their devices in between charges depends on the characteristics of the device and the types of applications they are running. Over time, users will typically get a sense of when their devices will need a recharge. Every user, however, must be sensitive from the outset to the device's reliance on battery power. A failure to do so could mean loss of data stored on the device if battery power is allowed to deplete completely. Fortunately, most devices are equipped with a battery indicator that signals the amount of charge remaining in the battery.

The amount of time it takes to recharge a device will vary depending on the battery, but could take a number of hours. Many handheld devices recharge while they are stored in their cradles, and can even be used while they are recharging.

Many things cause a battery to run down. Color displays and backlighting use a lot of power, as do multimedia applications. Wireless communications draw a lot of battery power, especially satellite communications. Satellite-capable devices are quite bulky precisely due to their large batteries. Many accessories do not come with their own power source, and rely on the power of their host device. While an attachable keyboard will not draw much power from the device's battery, a wireless modem will.

Users should develop the habit of conserving battery power—turning off the backlighting feature when not needed, turning off the device (or setting the auto-off feature) when not in use, setting the volume to low or off, and removing accessories after use.

10.2.9 Integration

Depending on the target audience for the device, manufacturers will integrate capabilities that appeal to that audience. For younger consumers, an MP3 audio player may clinch the sale. For an enterprise user, wireless connectivity options like Bluetooth, 802.11x, and infrared may be required. Integrated modules are appealing because they are easier to troubleshoot and don't require carrying around separate components that are easily lost or forgotten. The downside to integration is that the device is often more bulky, and it can be necessary to upgrade the entire device when a component becomes obsolete or deficient in some way.

Integrated capabilities depend on the device and the target market, but may include: wireless connectivity, security, voice recording, audio players, electronic book reading programs, bar code scanners, magnetic strip readers, and more. Palm, Compaq, and Ericsson offer integrated Bluetooth support in their most recent devices. The Handspring Treo has integrated voice capabilities. Sony and Toshiba are integrating 802.11b wireless capabilities in their notebook computers.

10.2.10 Extensibility

Just about every type of handheld device is expandable or extensible through a range of accessories and peripherals. Devices are equipped with expansion slots, ports, and jacks to permit the addition of various hardware components and gadgets from printers to modems to games and memory cards. Devices may also have sleeves that slip onto the device to add certain capabilities. Figure 10.5 shows some of the common expansion options available for PDAs.

For a buying organization, it is important to know how and if the device can be extended, the number of extensions that can be added and used simultaneously, and the price tag for the whole package. For example, many devices come with two expansion slots that take CompactFlash (or PCMCIA) cards for additional memory, games, program content, or other types of attachments. If a device has only one slot, then a user cannot concurrently add memory and a game card for example.

Many of the accessories and peripherals available for use with handheld devices are described more fully in Section 10.3.

Digital Camera
Source: Handspring

Keyboard Attachment
Source: Handspring

Wireless Network Card
Source: Symbol Technologies

Wireless Modem
Source: Handspring

FIGURE 10.5
PDA Extensions

10.2.11 Bundled Software

Similar to the integrated capabilities discussed in Section 10.2.9, devices come with a range of bundled software that varies depending on the target market. At the most basic level is the operating system software that resides on the device. The most prevalent operating systems are the Palm OS, which runs on Palm devices as well as devices from Handspring, Sony, Kyocera and Samsung and Microsoft's Pocket PC 2002 and Windows CE that run on Pocket PC devices from manufacturers such as Hewlett-Packard/Compaq, Casio and Symbol Technologies. In the not too distant future, many devices will start to support Linux. Research in Motion has its own proprietary operating system that runs on its line of devices. Phone handsets and communicators also use a variety of operating systems from Symbian and others.

Besides the operating system, most devices come with some sort of PIM software to organize, store and access information. Many devices also contain software that allows Internet and e-mail access, including micro-browsers. Spreadsheet and word processing software may be bundled with devices, as well as synchronization programs, security software, games, electronic book readers, media players, and calculators. Phone handsets may come with programs for SMS and dialing numbers.

Updating or upgrading bundled software can be tricky. Some device manufacturers allow users to apply software patches or fixes by installing them on the device's random access memory (RAM). Fixes applied in this way remain in RAM indefinitely, but are lost when the batteries are removed from the device or a hard reset occurs. Devices that contain flash memory allow software located in read only memory (ROM) such as the operating system or other applications to be overwritten partially or completely, enabling users to upgrade their operating system software to a later version. If a device does not have flash ROM, however, the only way to upgrade the operating system or other bundled software is to purchase a new device. Even with the ability to upgrade the operating system or other software, it is likely that your organization will end up supporting more than one device type (with different software versions) as devices are lost and replaced, and new hires are equipped with the latest models.

10.2.12 Third-Party Applications

In addition to the bundled software mentioned in the preceding section, a variety of third-party applications are available for purchase including travel guides, location guides, and games to CRM- and SFA-type business software. The breadth and depth of these offerings depends greatly on the maturity of the underlying operating system, and the willingness of the device manufacturer to work with software vendors to create workable applications. Many ERP enterprise packages, in particular, already have wireless functionality that is relatively simple to enable. And Microsoft offers handheld versions of many of its popular Office programs including Outlook, Word, Excel, and PowerPoint.

Many device manufacturers, recognizing a lucrative opportunity when they see one, are eagerly expanding into the enterprise market and collaborating with software vendors to create a line of compelling business software. In the coming years, the number of packaged business applications available for handheld devices should skyrocket.

Besides business applications, a variety of management and support applications are starting to appear. These applications aid organizations in operating, administering, managing, and supporting their wireless devices and applications. These tools are discussed in more detail in Chapter 13.

10.2.13 Personal Information Management (PIM) Capabilities

PDAs have their origins in providing PIM capabilities, and almost every handheld device comes with some sort of PIM capabilities. Whether or not these features are integral to your wireless solution, they will likely be part of the device package and relied upon to some extent by your users. PIM capabilities allow users to organize personal information such as contact lists, address books, schedules, appointments, and tasks, as depicted in Figure 10.6. They may offer calculators, notepads, and alarms and permit the creation of to-do lists.

Palm™ m505 Handheld Screen
Source: Palm, Inc.

FIGURE 10.6
PDA with PIM Capabilities

Almost every handheld device can exchange or synchronize information with a desktop or laptop computer. Synchronizing data between different platforms may be as simple as using the cradle or docking station and software provided by the device manufacturer, or as complicated as purchasing accessories and a third-party package to support, manage, and synchronize devices on an enterprise-wide basis. Ensure that the PIM software used on the device is compatible with the tools used in your office environment, and that the synchronization software can support the versions in place.

10.2.14 Internet and E-mail Capabilities

By and large, PDAs and e-mail devices allow users to access the Internet and their e-mail accounts. In non-wireless mode, these devices rely on synchronization technology and cradles to download e-mail or Internet content from the desktop to device (and vice versa) for off-line viewing. In wireless mode, certain "disconnected" devices such as phone handsets and many PDAs, allow a user to initiate a network session (i.e., dial in) to access e-mail or browse the Internet. Other devices, like the RIM BlackBerry, support "always on" wireless network connections so the user does not have to initiate a session or do anything to retrieve e-mails. E-mails are delivered straight to the device in near real time, and the user is notified via some type of indicator on the screen that e-mail has been received. To date, the ability to view e-mail attachments has been problematic without the purchase of additional software or services, and even then, it isn't always a seamless process.

Browsing Internet and web site content is also greatly limited on a handheld device compared to a desktop computer. Audio and video features, games, graphics, and other multimedia programs available on the web are generally not supported in their entirety on a handheld device. Depending on the browser resident on the device, you may be able to browse any URL on the web or only specific web sites coded in a special format. WAP browsers, for example, can view variants of HTML-encoded web pages such those encoded in WML. Palm devices rely on the Palm.net wireless service to convert standard HTML web pages into formats optimized for viewing on the device.

10.2.15 Cost

An important consideration when selecting a device platform is the cost per user and the total cost of ownership. These costs include acquisition costs and lifecycle costs.

Acquisition costs occur when a device is purchased. They include the cost of the device itself and any accessories such as cables, cradles, modems, keyboards, batteries, chargers, cases, etc. Software license fees may also be incurred as a one-time charge, a recurring fee, or a combination of both. Extended warranties may also be purchased at an additional charge.

Lifecycle costs include expenses to operate, support, and maintain the devices. These costs include replacement devices and parts, service fees for airtime (per packet, per message, per minute, or flat rate), and personnel costs to administer and manage the network of devices. Chapter 6 contains more details on estimating the costs of a wireless solution, including device costs.

10.3 Device Categories _____

The device landscape is changing so rapidly, with new products announced weekly, that it would be foolhardy to attempt to categorize every type of handheld device. The number of special purpose devices alone precludes an exhaustive list. An organization that is researching wireless devices should take note of the general categories listed below, and perform its own research into the most current offerings.

The categories covered by this section include: e-mail devices, PDAs, laptops and notebooks, special-purpose devices, gadgets, and leading-edge devices. A sampling of these device types is shown in Figure 10.7. Most organizations will find that an existing device type will meet their needs. In some cases, a unique solution will be required, and an organization will have to engage a manufacturer to create a specialty device. For example, UPS retained Symbol Technologies to create a wireless ring scanner and special software to resolve conflicts between its 802.11b and Bluetooth networks as part of its new wireless distribution system.[1]

---

1. "UPS to deploy Bluetooth, wireless LAN network," Bob Brewin, *Computerworld*, July 23, 2001.

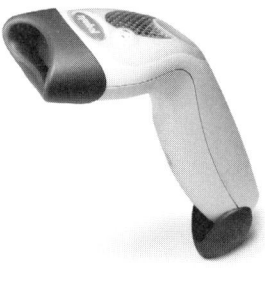

Bar Code Scanner
Source: Symbol Technolog

Vehicle-Mounted Terminal
Source: Symbol Technologies

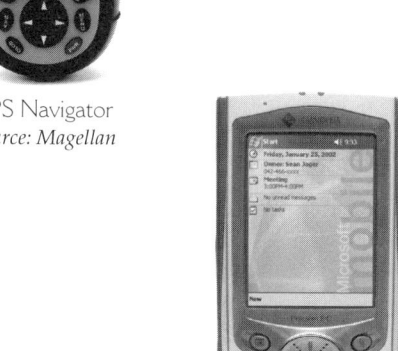

GPS Navigator
Source: Magellan

Communicator
Source: Nokia

Pocket PC PDA
Source: Casio

Ring Scanner
Source: Symbol Technologies

FIGURE 10.7
Sample Device Types

In general, organizations will have to make tradeoffs when choosing a device. E-mail devices excel at providing seamless, "always on" connectivity and messaging but may not offer the types of application functionality found on a Pocket PC, for example. A Pocket PC may appeal to an organization for its fast processing power, larger screen, and flexibility, but if those characteristics are important, perhaps a notebook computer would be a better choice. A Palm device may have an attractive price tag, but if the cost of developing custom software is too high, is the lower price worth it?

Making these tradeoffs, and choosing the best device for your situation, requires some knowledge. To that end, this section will provide a broad categorization of device types available at the time of writing. Be sure to supplement your reading with research into the current state of the device market before making your final choice.

10.3.1 E-mail Devices

While every handheld device is now offering some sort of e-mail capability, there is a class of specialty e-mail devices that rely on "always on" network connections to deliver e-mail straight to the device. Whereas other devices require the user to take some action to establish a network connection and retrieve e-mail, these devices, in combination with their underlying network technologies, bring e-mail directly to the user. A classic example of this type of e-mail device is the BlackBerry offered by RIM.

The RIM BlackBerry devices, shown in Figure 10.8, are based on RIM's proprietary operating system. The hallmark of these devices is their "always on" network connection that provides instantaneous access to incoming e-mail messages, and speedy submission of outgoing e-mail messages. The BlackBerry models also have a small keyboard that can be used to compose messages using an "all thumbs" style of

FIGURE 10.8
RIM BlackBerry Devices
Courtesy of Research in Motion Limited

typing. Like other device manufacturers, RIM is partnering with software providers to offer a set of packaged, enterprise-class applications on its devices. In addition, RIM and Nextel have agreed to partner on a device that will incorporate both voice and data capabilities in a form factor that will appeal to the enterprise market.

The platform and features offered by the RIM BlackBerry can be leveraged to provide more than e-mail. For example, as cited in Chapter 3, Atlantic Envelope Company designed a custom CRM application to run on the RIM BlackBerry that allows its salespeople to view order, inventory, and customer history information. The application consists of device-resident and server-resident modules that use the messaging and e-mail capabilities of the device and network as a conduit to send information back and forth, but rely on their own formatting and processing capabilities to present, interpret, retrieve, and manipulate the data.

Related somewhat to e-mail devices are two-way alphanumeric pagers. These devices also rely on "always on" network connections to send and receive short messages. They are not suited, however, for enterprise-class applications due to their limited functionality and anticipated dwindling market share as third generation wireless networks take hold.

We can expect that more devices will include "always on" e-mail capabilities in the future, beginning with Palm's i705 PDA. The wireless data services offered by 2.5G wide area networks, combined with their more extensive coverage, will prompt device manufacturers to regularly integrate connected or "always on" features into their products. As a result, more single-purpose devices like the RIM BlackBerry will lose their edge, and will have to expand and offer a more diverse set of capabilities to maintain and grow their audience.

10.3.2 PDAs

By far the most common platform for delivering wireless application functionality, PDAs encompass a wide range of device types with differing sets of features. PDAs obtained their name—Personal Digital Assistant—because of their original PIM capabilities. Today, they serve as much more than a personal organizer, and have become the platform of choice for organizations to deliver wireless application functionality to their mobile workforces. To illustrate the variation of PDAs on the market, listed below are several commercial, enterprise-grade PDAs.

Palm OS-Based Devices   Palm developed one of the earliest commercial PDA devices based on its Palm OS operating system. Palm licenses this operating system to several other PDA and phone manufacturers such as Handspring, Sony, Kyocera, and Samsung. In the PDA market, Palm, Handspring, and Sony have some of the most well-known device types.

FIGURE 10.9
Palm™ i705 Handheld
Source: Palm, Inc.

- *Palm*   Palm offers a range of PDAs, from low-end to high-end, based on its proprietary operating system, Palm OS®, which it also licenses to other third parties. In early 2002, Palm split into two operating entities, one to manufacture devices and the other to develop software for the platform, including the Palm OS®.

Market research shows that the Palm OS®-based devices from Palm, Handspring, and Sony accounted for just over 80% of the market in the latter part of 2001.[2] With this high market penetration, Palm devices have thousands of third-party applications and hardware accessories. When compared to similar class Pocket PC devices, Palm devices generally have less powerful CPUs and memory. These drawbacks are outweighed, however, by Palm's user-friendly device design, intuitive interface, robust development environment, established user base, and broad appeal to consumers and business users alike.

As part of its campaign to capture the hearts and minds of the corporate market, Palm debuted its i705 handheld model (Figure 10.9) in early 2002. Designed to appeal to the enterprise user, the i705 device has a persistent, "always on" network connection, e-mail service including behind-the-firewall e-mail, Internet access, PIM capabilities, and built-in infrared and wide area wireless communications. A clever feature of the i705 handheld is the light at the top of the device that blinks red to notify a user of incoming e-mail, even when the device is turned off. The i705 device comes with a monochrome LCD display, has one expansion slot, weighs a mere 5.9 ounces (battery and wireless

---

2. "Palm Completes OS Subsidiary Spin-off," David Haskin, *www.internetnews.com*, January 22, 2002.

capabilities included), measures 3.1 x 4.7 x .6 inches, has 8 MB RAM, and 4 MB Flash ROM, sports a 33 MHz processor and is powered by a rechargeable lithium battery. A Universal Connector at the base of the device allows for a range of hardware peripherals to be attached, including a keyboard, GPS device, or cradle. The features and functionality incorporated in the i705 handheld will compete head on with the RIM BlackBerry e-mail device. And for users that prefer to type, Palm also offers its Mini Keyboard, an accessory that slides on to the base of the device.

- *Handspring*   Handspring also offers a range of PDAs based on the Palm OS. Handspring has historically targeted the consumer market, particularly with many of its low-end Visor models, but has moved aggressively to court the enterprise market with its Edge and Treo models. Handspring is also working with enterprise partners like Aether Systems, AvantGo, and Wireless Knowledge to develop a line of CRM, SFA, ERP, and e-mail applications for its PDA lines.

An example of a Handspring PDA is the Treo model, as depicted in Figure 10.10. The Treo has integrated GSM phone capabilities as well as Palm OS PDA features. It has a built-in QWERTY keyboard and also offers Graffiti handwriting recognition for stylus-based input. The Treo comes with monochrome and color displays, 16 MB of memory, a 33 MHz processor, and a rechargeable lithium battery. It features wireless e-mail, short message service, and Web browsing, and comes loaded with Palm desktop software. Primary applications are Phone Book, Date Book, web browsing, and messaging. Web browsing relies on Handspring's proprietary Blazer browser. Phone capabilities include one-handed dialing, flip lid speakerphone, and three-way calling.

Source: Handspring

FIGURE 10.10
Handspring Treo

- *Sony* Sony offers its Clié PDA line along with a line of notebook and pen-based computers. The Clié has been targeted at the consumer market to date, and has many built-in entertainment and multimedia capabilities including an MP3 audio player as well as support for many third-party entertainment applications. The Clié device uses the Palm OS, and comes with 8 MB to 16 MB of memory, rechargeable batteries, a cradle, and synchronization software. Sony's notebook computers—the Vaio line—are mentioned in Section 10.3.4.

Pocket PC Devices Pocket PC devices are platforms that run versions of Microsoft operating systems, such as Windows CE and Pocket PC 2002, developed specifically for handheld devices. The Pocket PC operating system is full featured, and the hardware platforms are generally quite robust. Due to their later entrance into the PDA market, Pocket PCs do not have quite the breadth of third-party applications that Palm devices enjoy, with the notable exception of Microsoft office product such as Word, Excel, and Internet Explorer. High-end Pocket PCs are generally more expensive than high-end Palm devices, but since pricing is widely variable, the reader is advised to compare current prices. In the Pocket PC market, Compaq/Hewlett-Packard and Casio are well-known brands, but other manufacturers such as Toshiba, Audiovox, Symbol Technologies, Melard, and Intermec also offer Pocket PC devices.

- *Hewlett-Packard/Compaq* After concluding its merger with Compaq in early 2002, Hewlett-Packard is expected to discontinue its Jornada line of PDAs in favor of Compaq's more popular and much-publicized Pocket PC, the iPAQ line. The iPAQ model comes with varying amounts of memory, color and monochrome displays, a heavyweight 206 MHz Intel processor, and weighs in at about 6 ounces. The device ships with either 32 MB or 64 MB of memory, a docking cradle, and battery charger, and a slide-on expansion sleeve as opposed to built-in slots. Some models also offer integrated Bluetooth support. Bundled with the iPAQ are several Microsoft programs including Pocket Outlook, Pocket Internet Explorer, Pocket Word, Pocket Excel, Windows Media Player, and Microsoft Reader. Also included is ActiveSync software to synchronize the device with the desktop.

- *Hewlett-Packard* Hewlett-Packard has offered a line of PDAs based on its Jornada model, although the future of this line is in doubt after the company's merger with Compaq. Like the iPAQ, the Jornada comes with varying amounts of memory from 32 MB to 64 MB, and either color and monochrome displays. The Jornada has expansion slots for memory cards and other accessories, and some models offer electronic book reading, MP3, and voice recorder capabilities.

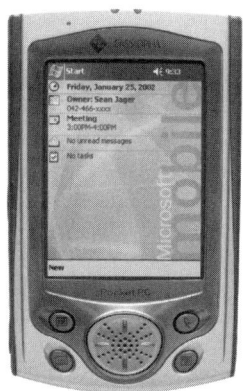

FIGURE 10.11
Casio E200
Source: Casio

- *Casio*   Casio's Pocket PC line is the Cassiopeia. Its enterprise E200 model, shown in Figure 10.11, comes with a color display, 64 MB of memory, and two expansion slots.

Other PDA Devices   Besides the Palm OS-based and Pocket PC PDA models cited above, many other successful brands are available on the market. Some of these use either the Palm or Microsoft operating systems, while others rely on proprietary versions. The up-and-coming operating system is Linux, which is just starting to appear on PDA devices.

- *Symbol Technologies*   Symbol Technologies sells a range of Pocket PC and Palm OS-based data collection devices, as well as special purpose scanners, pen-based computers, vehicle-mounted devices, and other handheld units. Symbol's PPT 2800 series, for example, is a Pocket PC-based touchscreen device that combines PDA capabilities with bar code scanning and real-time wireless communications technology. The PDT 8100 series is a Pocket PC-based device with multiple keyboard options for tactile data entry. In general, Symbol's devices are intended to be sold in volume—to retailers, manufacturers, field service organizations, warehousers, and logistics and distribution companies.

  Symbol devices excel at data collection through bar code scanning and magnetic strip reading technologies, which are integrated in their devices. In addition, Symbol devices are famously rugged and comply with demanding "drop, sealing, and temperature" specifications. By embracing the Pocket PC and Palm platforms, Symbol is opening up its device line to other types of business applications. Symbol also offers a SDK to encourage enterprises and

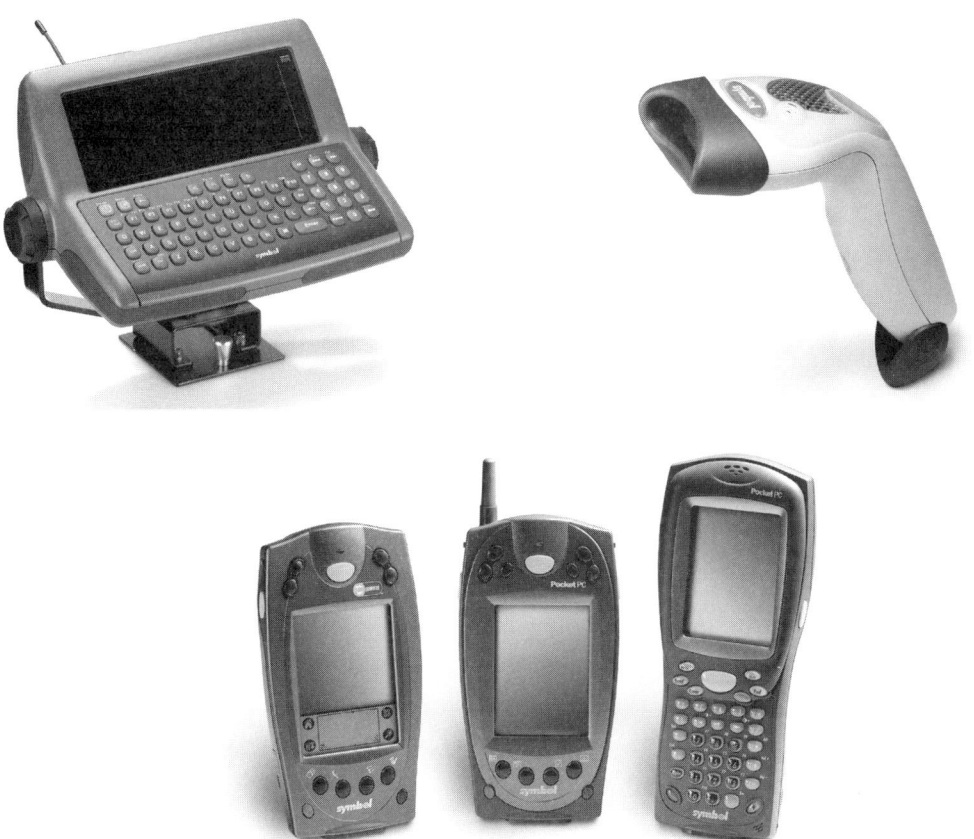

FIGURE 10.12
Symbol Technologies Devices
Source: Symbol Technologies

third parties to develop their own business applications based on Symbol's superior data collection technologies. Shown in Figure 10.12 are a variety of Symbol handheld devices.

- *Sharp*   Sharp is also entering the enterprise device market with gusto, offering a PDA based on the Linux operating system. The Zaurus model features an integrated miniature QWERTY keyboard for all-thumbs typing, which can slide away when not in use, as well as a handwriting recognition scheme. The device also has 2 expansion slots for external memory cards, wireless connections, and other accessories, and weighs in at about 7 ounces.

10.3.3 Phones

Across the world, there are dozens of phone handset manufacturers. Because phones are manufactured to work with specific network technologies that vary around the world, some phones are available only in certain geographic locations and not others. In general, far fewer phone handsets are produced for the U.S. market compared to the rest of the world.

Phone handsets encompass simple cell phones with voice-only capabilities; data-enabled handsets with limited Internet, e-mail, and short message services, aimed mainly at consumers; and smart phones or communicators that combine voice and PDA capabilities.

Communicator devices are very powerful compared to a basic phone handset. Their downside, however, is their bulk and size due to a combination of larger displays, keyboards, and voice capabilities in a single device. For a user that primarily wants to use the device as a phone, it would be more convenient to carry a smaller and lighter cell phone than the communicator device. In addition, most devices do not allow concurrent voice and data sessions, and battery life is relatively short. These devices are best suited for power users and field workers who are looking for integrated voice and data capabilities, and for whom the convenience of a single device outweighs its somewhat awkward physical size.

Manufacturers such as Nokia and Kyocera offer smart phones and communicator devices for the U.S. market. The Nokia 9210/9290 communicator, depicted in Figure 10.13, offers voice and data capabilities, Internet and e-mail access, and has a keyboard for data input and a larger display size.

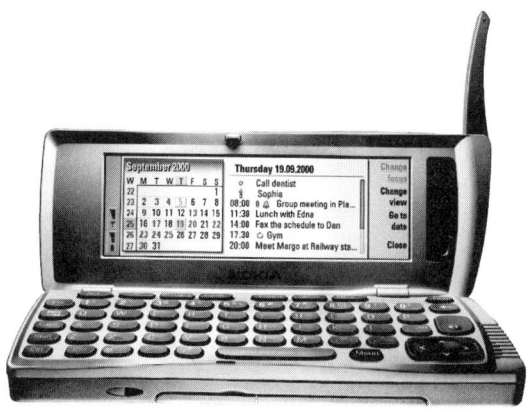

FIGURE 10.13
Nokia 9210/9290 Communicator
Source: Nokia

Another example of a smart phone is the Kyocera model QCP 6035 that merges a conventional PDA (based on the Palm OS) with a digital wireless phone. Unlike Nokia's bulkier communicator devices, the Kyocera phone is relatively small at 5.6 inches high and 2.5 inches wide and relatively light at 7.34 ounces. Running on standard lithium rechargeable batteries, the device can be used for only a few hours between charges. To give the reader a flavor of the features on a communicator device, consider those found on Kyocera's QCP 6035 model.

- PDA capabilities
  - Software tools such as an address book, date book, memo pad, and to-do list
  - E-mail access
  - Web browser with support for secure on-line transactions
  - Support for any Palm OS-compatible software
  - Support for multiple wireless data technologies such as SMS, WAP, and HTML, as well as the Palm OS web clipping service
  - 8 MB of memory
- Wireless phone capabilities
  - Supports three modes of network coverage: CDMA digital PCS, CDMA digital cellular, and analog
  - Voice services like speakerphone, voice memo, and voice dialing
  - Multiple ring tones and vibrate alert

Phone manufacturers are also moving aggressively to capture corporate market share, and we can expect to see a variety of device types emerging from this sector in the coming years. Of some significance, Nextel inked an agreement with RIM to develop a "smart" BlackBerry PDA device with both voice and data capabilities. The device is anticipated to have standard PDA features, "always on" network and Internet access, text paging, and two-way messaging. If the device works as advertised, and the form factor is an acceptable size, the device may well give other PDA manufacturers stiff competition.

10.3.4 Laptops, Notebooks, and Tablet Computers

Although they certainly do not fall into the "handheld" device category, laptop and notebook devices provide mobile, wireless connectivity just as PDAs and phones do. In addition, they offer a wide range of application functionality identical to the software and tools that appear on users' desktop computers. Although these devices are becoming smaller and lighter, they remain far more cumbersome, and in some cases simply too large, to be used by a widely dispersed mobile workforce. Many of these

FIGURE 10.14
Sony Vaio
Source: Sony

larger format mobile devices are used for special-purpose applications in vertical market solutions such as field service or health care.

Laptops, notebooks, and tablet computers have advantages, including bigger displays, a greater range of connectivity options (both wired and wireless), more memory and full-size keyboards, which allow them to overcome some of the inherent device constraints that plague handheld devices. For wireless connectivity, many of these devices contain integrated Bluetooth and 802.11x WLAN support, and a range of commercial wireless modems is available to access WANs. They are especially appropriate for applications that require a lot of memory or processing power, high-volume data input, or multimedia capabilities to display high-quality graphics or complex images. The downside to these devices is that they have low battery life, much higher prices than handheld devices, larger size and weight, and a more complex operating environment that may befuddle novice users.

Device manufacturers such as Sony, Panasonic, Hewlett-Packard, Hitachi, Toshiba, IBM, Symbol Technologies, Intermec, and Psion offer laptop, notebook, and tablet devices. Sony offers the Vaio line of notebook computers, as depicted in Figure 10.14, which includes integrated 802.11b capabilities for connection to a workplace WLAN or a public WLAN. The notebook is only 1 inch thick, weighs about 2.7 pounds, and contains a battery that will power the device for up to 6 hours. Hewlett-Packard offers a Pavilion notebook PC with integrated 802.11b and Bluetooth wireless capabilities. Most of these devices are constructed of durable materials to withstand both normal wear-and-tear and abuse from outdoors use. Manufacturers like Panasonic, Psion, and Symbol Technologies offer particularly rugged versions of these large form factors.

10.3.5 Special-Purpose Devices

The preceding devices help untether mobile professionals from their wired, desktop environments. They are used to power applications that are often similar to those found in the office environment, or to gain access to data that is easily retrieved by users located in the workplace.

Special-purpose devices, in contrast, help to automate specific processes or perform specific tasks. They come equipped with vertical applications for fields like health care, financial services, manufacturing, retail, field service, and shipping to name a few. For example, data collection and inventory management tasks are aided by devices with integrated bar code scanning technology such as those offered by Symbol Technologies. Credit card processing can be accomplished using devices with integrated magnetic card reading technology and wireless connectivity, such as those offered by Apriva, Nurit, and Thales.

Telemetry applications for tracking, monitoring, and locating assets will generally use other classes of equipment and hardware including passive and active tags, transceivers, and terminals. Sierra Wireless, for example, offers a telemetry solution based on its Dart line. Solutions that rely on satellite technology to perform tracking and locating will use various forms of GPS devices, attachments or fixed-mount equipment from vendors such as Magellan, Garmin, and Lowrance.

Many manufacturers of standalone special-purpose devices also make special-purpose accessories that can attach to PDAs and laptop computers through their expansion slots and other physical connections. For example, Apriva's credit card scanning accessory can be attached to a RIM BlackBerry device. Many types of GPS kits are offered for Palm and other PDAs. Bar code scanners may also be snapped onto various handheld devices.

10.3.6 Gadgets

While enterprises will be most interested in the types of devices and equipment mentioned in the preceding section, a whole other category of gadgets is available that may be useful in limited circumstances. These gadgets include items such as:

- Casio's wearable wrist camera that connects to a PC via an infrared link
- Seiko Instrument's SmartPad2, a transmitter-enabled pen and a notepad with sensors that allows handwritten notes to be transmitted to a PC via an infrared link
- MicroOptical Corporation's VGA Clip-On Monitor that attaches to glasses, enabling the wearer to view the contents of a video display when not within line-of-sight of the screen

Chances are that if your wireless solution requires some specialized gadget, there is a manufacturer who can meet your specifications. Often, technologies used within other industries can serve as the basis for a solution in your line of business. The VGA Clip-On Monitor, for example, was developed to assist engineers and repair personnel when fixing computer equipment—however, any worker that needs to read the contents of a video display from a distance might find the device useful.

10.3.7 Futuristic Technologies

In the race to make devices smaller and lighter, and with greater power sources, several companies are trying to bring new, leading-edge technologies to market. Although some of these technologies may appear a bit far-fetched, their underlying principles may be adapted and applied to more common, pressing problems. Consider the following:

- Using plastic transistor technology, Royal Philips Electronics has created a prototype video screen that is flexible, pliable, and lightweight, and that can be rolled up like a newspaper.[3]

- U.K.-based ElekSen has created a prototype washable, fabric keyboard that can be rolled up and stowed.[4]

- Applied Digital Solutions has created a miniature thermo-electric generator that converts body heat into a source of power to avoid the use of batteries.[5]

- Motorola, with Freeplay Energy Group, has developed an accessory, a hand crank, that connects via cable to a cell phone which the user can wind up to power the phone for 3 to 6 minutes of talking time.[6]

10.3.8 Accessories and Peripherals

Last but not least, there is a great body of accessories and peripherals to use with the devices categorized above. These accessories include cameras, modems, MP3 players, GPS receivers, credit card scanners, keyboards, headsets, screen protectors, cards, monitors, battery chargers, printers, projectors, and docking cradles as well as cards containing software, games or additional memory. Figure 10.15 shows a few types of accessories. Accessories can be inexpensive or quite pricey. External modems, for example, were averaging about 50% of the cost of the device at the time of writing, a hefty amount for an attachment.

---

3. "Philips Electronics: Rolling Up Your Computer," Tim McDonald, *WirelessNewsFactor.com*, December 6, 2001.
4. "Mobile Users Travel Light with Washable Fabric Keyboards," Lou Hirsh, *WirelessNewsFactor.com*, January 7, 2002.
5. "Body Heat—The New Wireless Power Source," Brian McDonough, *WirelessNewsFactor.com*, October 2, 2001.
6. "Motorola To Take Phone Crank for a Spin," Brian McDonough, *WirelessNewsFactor.com*, September 27, 2001.

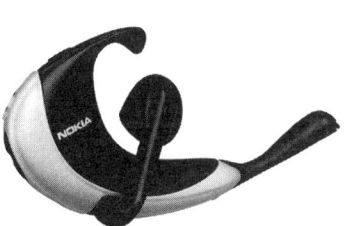

Bluetooth Headset
Source: Nokia

MP3 Player from InnoGear
Source: Handspring

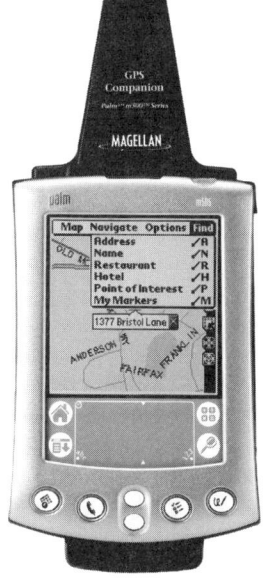

GPS Attachment
Source: Magellan

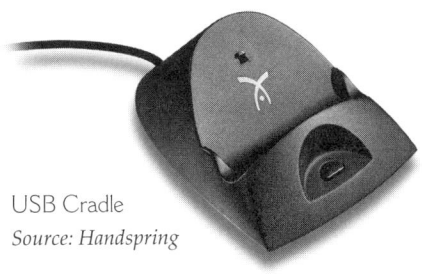

USB Cradle
Source: Handspring

Eyemodule Digital Camera Attachment
Source: Handspring

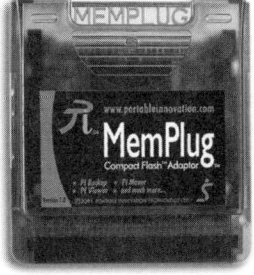

VisorPhone Attachment
Source: Handspring

CompactFlash Adapter
Source: Handspring

FIGURE 10.15
Accessories for Mobile Devices

Accessories connect to a device through physical connections such as expansion slots on the device, sleeves that slip onto the device, USB ports, and jacks. Devices with infrared or Bluetooth capabilities can also use these features to connect to similarly equipped devices and peripherals, such as a printer or desktop computer. The number of accessories that can be attached to a device at any one time depends on the number of slots, connections, or ports available. Many PDAs come with two expansion slots to allow users to extend their devices in more than one way, at the same time. For example, a user could insert a CompactFlash card to the Palm m505 for additional memory and at the same time use an MP3 player attached to the second expansion slot.

Some accessories come with their own power source or batteries, but many draw power from the device to which they are attached. Although some accessories draw minimal power, others such as audio or video, or wireless connections can consume a hefty amount of battery power, so it is wise to inactivate and detach these accessories when not in use.

To illustrate a few types of accessories, consider the examples listed below.

- *Wireless Connectivity* If wireless support is not integrated into a device, there are several options for acquiring it. For wide area wireless network access, add an external modem to the device or an air card with antenna that fits into one of the expansion slots. For access to a WLAN, or even Bluetooth capability, add an appropriate air card complete with antenna. Wide area wireless connectivity can also be obtained by connecting a PDA, via a cable or a Bluetooth link, to a data-enabled phone that is effectively used as a modem. Figure 10.16 shows some of these wireless connectivity options.

- *Keyboards* For devices lacking keyboards, or with barely usable ones, a variety of attachable, portable keyboards are available. These keyboards fold up when not in use, and expand to full-size versions when needed. They draw their power from the device itself, although power requirements are usually minimal. Targus offers its Stowaway portable keyboard that opens to a full-size QWERTY version and folds into the size of a PDA. Nextel also sells a folding, portable keyboard, manufactured by Think Outside, for use with its wireless phones. Handspring devices use a Snap N Type keyboard, shown in Figure 10.17, that is about the size of the keyboard found on a RIM BlackBerry device. Figure 10.18 shows an example of a portable, stowable keyboard.

Reproduced with permission of
3Com Corporation

Source: Nokia

Source: Symbol Technologies

Source: Handspring

Source: Symbol Technologies

FIGURE 10.16
Wireless Connectivity Accessories

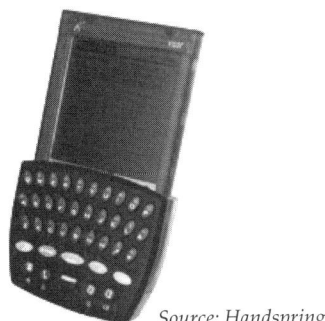

Source: Handspring

FIGURE 10.17
Handspring Visor with Snap N Type Keyboard

Folded keyboard

Open keyboard

FIGURE 10.18
Think Outside Stowaway Keyboard
Source: Think Outside

- *Health Monitors*   While this accessory is limited in its application, it is a good example of the inroads that wireless technology is making in the healthcare industry. Active Corporation sells an electrocardiogram (ECG) heart monitoring device that attaches to a Palm handheld via a cable. The monitoring device records, displays, and reviews ECG signals to analyze cardiac rhythms. Home health care providers, emergency medical technicians, nursing homes, and even individuals are using the device to monitor heart function.[7]

- *WLAN Adapters*   Several manufacturers, including Symbol, Xircom, Cisco, and Sierra Wireless offer 802.11x WLAN air cards and adapters that allow a device to connect wirelessly to a workplace or public WLAN. The size and convenience of these accessories varies. The Xircom (an Intel company) adapters are somewhat bulky, partly because they come with their own power source. On the Handspring Visor, for example, the Xircom adapter adds a "hump," and the WLAN module for the Palm m500 is about the same size as the device itself. In contrast, Symbol's adapter is just slightly larger than a CompactFlash card, but does not have its own built-in battery. Sierra Wireless offers air cards that are slightly larger and that include a built-in antenna.

---

7. "How's Your Heart? Check with Palm," Jay Lyman, WirelessNewsFactor.com, December 28, 2001.

- *GPS Navigator Kits*   In addition to standalone handheld GPS navigation devices, GPS manufacturers also sell a series of GPS attachments or kits that consist of a receiver that snaps or slides onto the device, plus mapping software for a city, state, region or even the entire U.S.
- *CompactFlash Cards*   These accessories are small, external cards that fit into a device's expansion slots. They are used to add memory to a device, and to back up and store files and programs. They may also hold third-party software, games, travel guides, or location-based guides.

10.4 Device Management and Support

Device management and support is a thorny issue for most companies. Whereas the desktop and enterprise markets have mature and rich software tools to administer, manage, and upgrade wired networks and components, fewer offerings are aimed at the mobile market. In addition, the nature of the devices themselves makes management a tricky proposition. These devices are meant to travel, and they do—often right out the hands of their users. Loss, theft, and misplacement can lead to constant device turnover. Mobile devices are also subject to greater physical abuse, and a range of environmental conditions, leading to breakage and malfunctions. As a result, devices are constantly shuttling back and forth between users, support groups, and repair centers, making it difficult to track and account for the inventory of devices.

The range of devices that an organization will end up supporting is also apt to be large. Even if an organization standardizes on a single device type at the outset of its implementation, it will likely end up supporting more device types, and several models of a single device type, over time. Users may acquire new devices. Replacement devices may come in new models. More powerful options may appear.

Upgrading the software and applications contained on a device is also problematic once those devices are out in the field. Downloading or "pushing" software or information upgrades to devices is a complicated process. For operating system upgrades, devices with flash ROM can be upgraded via software means while others must be replaced wholesale to obtain the latest software versions. Many companies will take the path of least resistance and purchase new devices when it comes time to migrate to new software versions.

For more information on device management and support issues, consult Chapter 13.

10.5 Wireless Device Resources

Information on specific devices is located on the manufacturers' web sites, as referenced in Appendix A. In addition, please visit *www.justenoughwireless.com* for updated information on wireless devices. Other wireless device resources include:

www.anywhereyougo.com

www.allnetdevices.com

www.pcmall.com

www.pdastreet.com

Wireless Networks

Wireless networks underpin every wireless application. Their job is to transport information, whether voice or data, from point A to point B. A device such as a laptop, PDA, or phone handset can use a wireless network to connect with another device or any number of traditional wired networks such as the Internet, a private corporate network, or the public telephone system.

There are a variety of wireless networks that differ by their characteristics. Having a basic understanding of these network types is important when first conceiving and defining a wireless application. The wireless application envisioned will affect network choices, but the opposite is also true. The traits and capabilities of wireless networks will influence the design and functionality of the ultimate wireless application. Knowing these traits and capabilities helps one avoid unrealistic expectations and potentially fatal design flaws during the strategy and definition stages discussed in Chapter 5.

In an ideal world, end users, executives, and business people should not have to be concerned about, or even aware of, the type of wireless network used or its idiosyncrasies. The wireless network should simply work when called upon to do so, and otherwise be invisible, creating no problems or distractions. Whether one, two, or a dozen wireless networks are at play, or two, five, or ten vendors involved, the network should perform reliably and seamlessly. We expect no less from the telephone system, the wired computing networks in the workplace, and even the cable networks that deliver television and broadband Internet access to our homes.

In the real world, wireless networks have not yet reached such a state of perfection. There is no "one size fits all" network ideal for every type of wireless application. Some networks are strong in certain areas, such as throughput, but fall short in other areas, such as geographic coverage. Instead of treating wireless networks as invisible conduits, managers, designers, and developers must know just enough about network choices, strengths, and weaknesses to make informed choices. Mastering network details, protocols, and other operational minutiae should not be required. That level of study is best left to network specialists and engineers. Rather, managers and designers must be aware of the wireless network landscape so they can maneuver successfully, and make good judgment calls, when it comes time to create and use wireless solutions.

This chapter provides a high-level introduction to wireless networks. It begins by explaining what wireless networks are and how they differ from wired networks; breaks current network choices down into two broad categories; and discusses the relevance of data networks. Next, the chapter describes the important features and issues to pay attention to when investigating wireless network solutions, from coverage to reliability. These features are also critical to application design and functionality. Following this discussion, the chapter examines the different wireless network options available today, concentrating on those that facilitate data transmission—the predominant need of business applications. Lastly, the chapter presents considerations that organizations must weigh when evaluating their wireless network choices, and offers a table of network types with associated features and their pros and cons. The chapter wraps up with a list of resources that the reader can use to learn more about wireless networks.

11.1 What Is a Wireless Network? _____

A number of wireless networks exist, but they are not all equal. Although similar in their basic approach—using light waves or airwaves to transmit information—they differ in their use of hardware and software, standards and technologies. They also differ in coverage and throughput, and how consumers use them. The wireless network market is populated by hundreds of competing vendors, carriers, software companies, equipment manufacturers, and service providers. As you might guess, the market is rapidly evolving and expanding, with new offerings appearing almost daily.

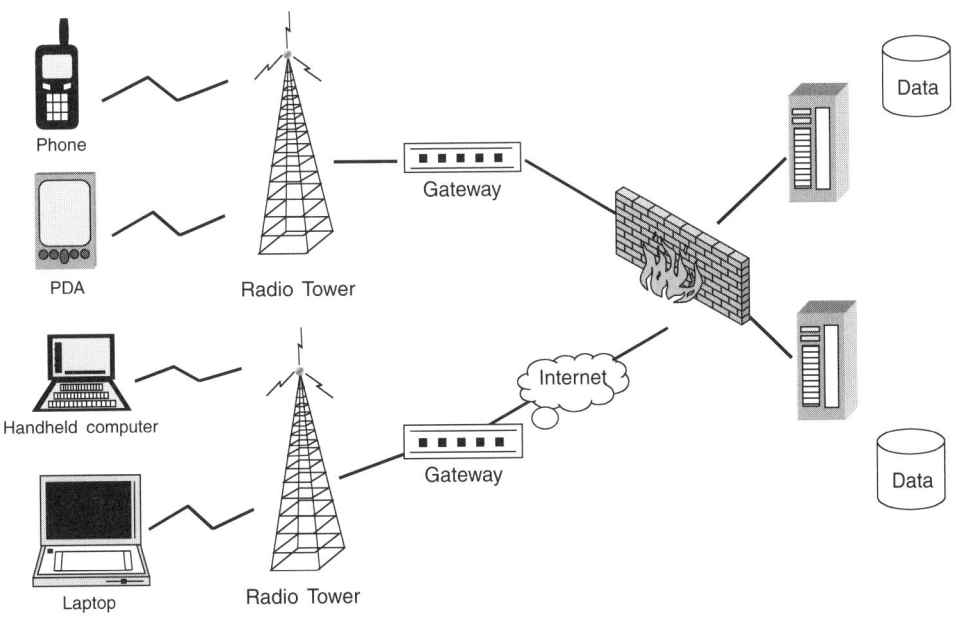

FIGURE 11.1
Basic Wireless Network

Examples of wireless networks surround us. Some are quite familiar. Dispatch services used by taxicab companies, police departments, and field service organizations are one example. FedEx's two-way radio system, instituted in the 1980s to dispatch and communicate with drivers is another example.[1] The cellular telephone network, tapped into by every mobile phone user, is an obvious and pervasive wireless network. Paging networks, popular for years with mobile professionals of all types and more recently with young adults, are also prime examples of wireless networks. A host of other, private wireless networks exist, used by corporations, government organizations, schools, and even homes.

11.1.1   Wireless Versus Wired Networks

The most notable and obvious difference between a wireless and wired network is the lack of cables or wiring. Wireless networks rely on technology that does not require physical connections or proprietary cables to transmit information from one point to another. Instead, information is sent and received over radio frequencies or even light waves (Figure 11.1). This conspicuous lack of hard-wired connections is

---

1. "Wireless Early Adopters Survival Tips," *CIO* Magazine, March 15, 2001.

precisely what enables workers and employees to leave their offices and desktop computers behind, and yet remain connected to vital information while mobile.

By eliminating the need for physical cables, wireless networks offer many tangible benefits. At a macro level, large public wireless networks provide:

- superior convenience and ease-of-use for end users who can rely on portable, personal devices rather than stationary, sometimes shared, devices or terminals

- quicker initiation of service

- affordable service and low total-cost-of-ownership for subscribers and enterprises, particularly in developing countries where wired connections are scarce and costly

- cost effective and safe alternative to extending wired, landline service into remote or hazardous areas

At an enterprise level, private wireless networks afford similar financial and time-saving benefits. Within an office or university campus, for example, an IT organization can quickly establish a local, short-range wireless network without the expense, time, or disruption of installing and pulling cables through offices, walls, ceilings, and floors. These networks can handle a variety of mobile and stationary devices, from desktops to PDAs, without the need to purchase and deploy proprietary cradles, cables, and other connectors on a per-user basis. They can also be reconfigured quickly when needs change or users move. The resulting wireless networks not only extend or replace the functionality offered by wired networks, they also enable a host of new wireless applications.

Despite their physical differences, wireless and wired networks are alike in many respects. Both must be reasonably speedy, reliable, secure, and available for the applications using them to function properly. From a business process perspective, both types of networks must facilitate secure information access and exchange in the most user-friendly and unassuming manner possible. Deficiencies in any of these areas will undermine even the most well designed applications regardless of the type of network selected.

11.1.2   Do It Yourself Wireless Networks Versus Purchased Wireless Network Services

To a beginner, making sense of the different types of wireless networks is a vexing proposition. Wireless networks vary greatly by attribute. Some may perform well over long distances while others work only at short distances. Some may excel at transmitting high quality voice, but offer poor transmission rates for data. Wireless networks may vary based on their underlying technologies and standards, and by their acceptance in different commercial and geographic markets. Because of these varying attributes, an organization may find that a complete wireless solution requires knitting together several different types of wireless networks to gain the level of service and coverage desired.

Thankfully, when it comes to wireless networks, managers need only know what type or types of network are useful in the situation at hand. They do not need to memorize every wireless network option, its throughput rates, and cost. Instead, they need a simple framework for evaluating wireless networks. The next section offers a framework that breaks wireless networks into two broad categories— private networks that a company owns, installs, and manages itself, and the large-scale, commercial networks owned by third parties that can be "time-shared" for a subscription fee.

Do It Yourself Wireless Networks
Just like the wired networks found in corporate offices and buildings, companies can purchase and deploy various kinds of private, wireless networks. These do it yourself ("DIY") networks rely on internal resources, or hired consultants, for implementation, administration, and management. The company owns the resulting private network and controls access to it. These wireless networks are used to tap into the company's wired ones, and provide access to enterprise data and applications housed on servers and mainframe computers, corporate intranets, and the Internet. They facilitate information exchange and messaging among co-workers, and synchronization of information among different device types. They eliminate the need for proprietary cables and cradles. And, they open the door to new types of wireless applications that encourage mobility and allow workers to be more productive. Specific types of DIY wireless networks are discussed in Section 11.3. Figure 11.2 shows a simple example of a DIY wireless network.

Companies commonly use DIY wireless networks in the following circumstances.

- *Local, In-Building Scope* If users are located or roam within a relatively contained area such as a building, facility, warehouse, manufacturing site, office complex, or campus, and will primarily use the network in-building, then a private, DIY network may make the most sense. Some special-purpose places, like hospitals and airport facilities, may restrict use of DIY networks that rely on radio frequencies for fear that they will interfere with the operation of important equipment.

- *Skill Availability* Although some of the actual wireless network technology products are new, their concepts are familiar. Expertise for installing, administering, and maintaining DIY wireless networks is generally available among IT professionals. If a company lacks knowledgeable internal resources, experienced consultants can perform the work.

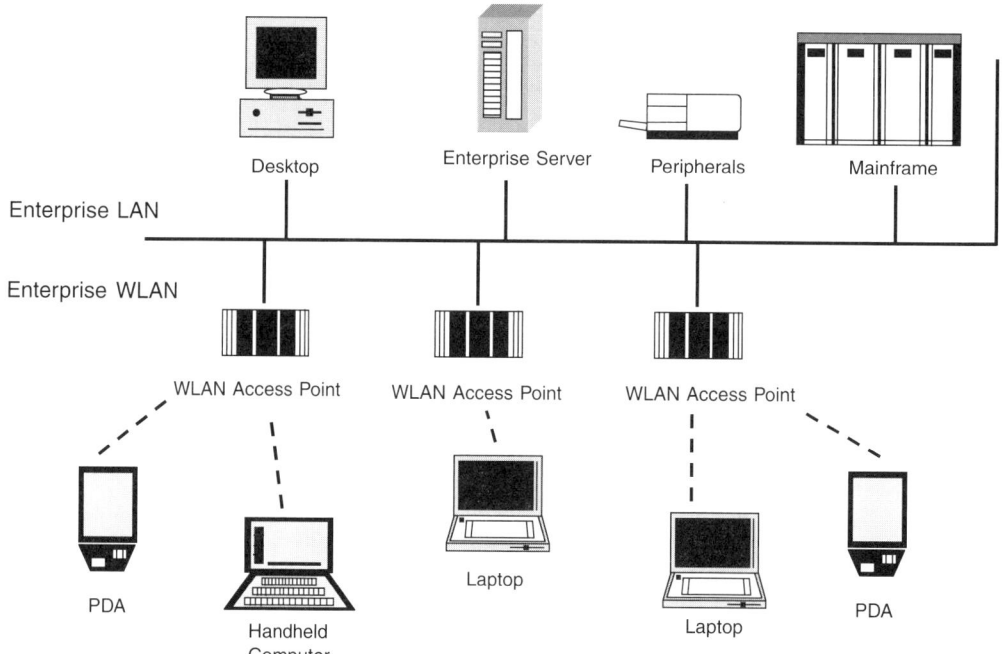

FIGURE 11.2
DIY Wireless Network

- *Convenience*   DIY wireless networks are an attractive option when it is inconvenient or infeasible to outfit users with cables or to run wiring through a building. Users that must be mobile while they work—in an assembly line or on a manufacturing or warehouse floor—cannot trail cables along with them. Cables are easily snagged, or may even restrict movement. Certain buildings may also pose obstacles to physical wiring. This situation may arise when the building is physically large; the age, construction, or historic value of the premises preclude inexpensive types of installations; existing wiring is already unreliable or problematic; or employees move offices or equipment is swapped frequently. In these cases, DIY networks can avoid the constraints and limitations imposed by physical surroundings, and accommodate changing physical setups.

- *Cost-Effectiveness*   Depending on the scope of the rollout, DIY wireless networks are generally cost-effective and affordable. In some cases, network components, such as infrared ports, are already built into existing devices requiring no additional investment or upgrades. The network cards, access point equipment and associated software comprising a DIY wireless network may require a higher initial investment than their wired counterparts; however, they are easily cost-justified in the long run when the time, labor, and equipment involved in maintaining an equivalent wired network are taken into account.

- *Maintenance and Support*   As with other computing resources, the burden of maintaining and supporting the DIY wireless network falls squarely on the IT organization, unless it is outsourced to a third party. The company must have sufficient IT resources to perform ongoing administrative and upgrade activities and to ensure that security is adequate for the network.

Purchased Wireless Network Services   In contrast to a DIY network, national wireless networks are owned and operated by third parties. These networks are much more sophisticated in their operation and underlying technologies than DIY wireless networks, and rely on equipment and software installed throughout the country or even around the planet in the case of satellites. Essentially open to the public, these networks are "timeshared" by subscribers that pay fees for purchased services. Large telecommunications carriers such as AT&T Wireless, Cingular Wireless, Sprint PCS, and Motient own and operate terrestrial, cellular networks. Companies like Motient and Iridium also own and operate extraterrestrial satellite-based networks.

When would a company purchase wireless network services rather than install a DIY wireless network? The most typical case is when a company needs *wide area* coverage, extending beyond the confines of its headquarters or office park. A wireless solution may require a company to extend service and provide information access to a group of workers who may roam far from a corporate office, data center, classroom, or other "wired" haven. Field service workers, for example, visit customers in locations that may be remote from corporate headquarters or dispatch facilities. Salespeople have meetings with prospects and customers at widespread locations.

A company may also opt to purchase wireless network services even if the anticipated scope of coverage is local. If users are primarily located outside, like security personnel on a campus or parking area attendants, then wide area services work better at sending and receiving data than DIY networks. Wide area networks are also preferable if the installation or operation of a DIY network would present problems.

Coverage Considerations	Network Choice
Local office or campus	DIY
Primarily in-building use	DIY
Wide geographic area	Purchased services
In-building interference concerns	Purchased services

FIGURE 11.3
DIY Versus Purchased Wireless Network Services

Within the operatories of a hospital emergency room, for example, a DIY network might cause interference with medical equipment and devices that happen to use the same radio frequencies as the network.

When more extensive wireless coverage is needed or DIY networks are just not feasible, companies will purchase network services from entities that offer coverage in the necessary geographic areas. If mobile users will roam only in the southeastern U.S., a provider like Cingular Wireless may suffice. If operations will occur throughout the U.S., then services will have to be obtained from multiple providers. All wireless carriers post coverage maps on their web sites, making it fairly easy to find out which ones have adequate coverage in a given area. Depending on how far afield users are expected to roam, a company may need to purchase network services from more than one provider to obtain seamless coverage unless its primary provider (or a third-party aggregator) has already negotiated roaming agreements with competitors.

If a company needs to purchase wireless network services from a third party to run its wireless application, it should be prepared to:

- Execute agreements with carriers to purchase airtime on their networks. Depending on the uniqueness and needs of the wireless application, these agreements may be heavily negotiated or simply a standard contract. As wireless applications become more prevalent, carriers will offer pre-defined and pre-priced offerings, translating into less negotiating and legal wrangling. Back in 1990, when UPS first deployed its wireless application, it had to negotiate separately with its top four cellular carriers to obtain the coverage and pricing desired in a process that stretched out over a year.[2]

- Determine appropriate billing and chargeback mechanisms. The uniqueness of the wireless application, the extent of coverage, the number of users, and the amount of data transmitted, will influence the total cost of service, which may be expensive. Carriers have typically charged a fixed monthly fee and a variable fee based upon usage (number of kilobytes of data transmitted, for example). Some carriers may offer tiered or preferred rates of service, and may even offer higher service level commitments for additional fees. With the rollout of 2.5G data services, Verizon Wireless, for example, is offering a variety of flat rate and variable pricing plans.

- Consider purchasing services through an intermediary such as a WISP (wireless Internet service provider) or WASP (wireless application service provider). These entities do not own any networks, but have negotiated airtime agreements with carriers, often at favorable pricing due to their economies of scale. Not only can WASPs stitch together the airtime and coverage needed, they also provide access to packaged or customized wireless applications and functionality.

---

2. "Wireless Early Adopters Survival Tips," *CIO* Magazine, March 15, 2001.

11.1.3   The Data Versus Voice Distinction

When considering current generation wireless cellular networks, there is an important distinction between voice and data capabilities. This distinction is especially important to companies that are thinking of purchasing wireless network services from one of the large carriers.

Not surprising, large commercial wireless networks, with their roots in telephony, have been historically geared toward efficient voice rather than data communication. These networks were spawned precisely to offer wireless voice services as an alternative to landline-based telephone services. Indeed, as analysts and commentators often note, voice remains the primary application of cellular wireless networks for now and into the foreseeable future. Although support for data is improving on these networks with the advent of 2.5G technologies, it remains limited when compared to voice.

Why does this voice versus data distinction matter? Because just like every other business application, wireless applications primarily send, receive, manipulate, display, and process data. Few involve voice processing, except as a method of overcoming device limitations. Wireless cellular networks do not yet have the bandwidth to offer wireless business applications the type of performance, response time, or throughput that wired, or even DIY wireless networks can offer. Data transmission rates over cellular networks vary, because the carriers are in the throes of a massive upgrade to improve their data capabilities. In many areas of the country that do not yet enjoy 2.5G network coverage, data transmission rates still max out at 19.2 kbps. In other areas where 2.5G networks have been rolled out, speeds comparable to dial-up services (40 to 60 kbps) are available, and 144 kbps capabilities theoretically exist. Throughputs of current wireless networks are listed in Figure 11.6. Network performance in general is further discussed in Section 11.2.

Besides bandwidth constraints, data networks have been marked by sparse geographic coverage generally limited to larger metropolitan areas and strongest in the northeastern U.S. In fact, for many current wireless applications, coverage is more important, and a greater limiting factor, than bandwidth.

When it comes to optimizing wireless networks for data transfer, the carriers are not idle. They are engaged in an extensive effort to upgrade, and build out, their existing network infrastructures. Their goal is to deliver broadband data transmission rates with a theoretical high of 2 mbps, but with 384 kbps being more realistic in the vast majority of mobile situations. The upgrade also promises to provide seamless national coverage, and is slated for the 2004 to 2005 timeframe, although these dates keep moving out.

Technology	Carrier
CDMA, CDMA2000	Sprint PCS, Verizon Wireless
TDMA, GPRS	AT&T Wireless, Cingular Wireless
GSM, GPRS	VoiceStream
iDEN*	Nextel

* Similar to GSM and TDMA

FIGURE 11.4
Underlying Wireless Network Technologies and Carriers

Upgrading these cellular systems is complex, expensive, and time-consuming. It also requires more complex client devices because high data rate transmissions demand resource-intensive modem chipsets and multimode capabilities. Exacerbating the upgrade chore are the three different and competing network technologies used by the U.S. carriers as depicted in Figure 11.4. Moreover, there is no single path to upgrade these network technologies. Each carrier must choose its upgrade path, and then retrofit its infrastructure with new equipment, hardware, and software. At the end of the day, the carriers will still use different, and competing, technologies.

The size of this undertaking is so huge that the carriers are targeting two major milestones with interim delivery capabilities at each stage. Figure 11.5 depicts these milestones, termed 2.5G and 3G in telecommunications parlance. These milestones represent different generations of wireless networks just as Windows 98, Windows 2000, and Windows XP denote different versions of Microsoft's operating system. Each subsequent generation of technology will offer greater data transmission rates, coverage, and responsiveness.

The first target, 2.5G networks with data transmission rates as high as 144 kbps, was reached in late 2001/early 2002. Although these 2.5G networks are capable of reaching 144 kbps rates, the carriers are being very careful not to over-promise and under-deliver. In general, carriers are conservatively quoting transmission rates of 40 to 60 kbps, similar to dial-up service, until they see how the networks perform under a range of conditions.

What's next after 2.5G networks? The roll-out of 2.75G and 3G networks, naturally. Targeted for arrival sometime after 2004, 3G networks will offer broadband transmission rates that, if they work as claimed, will change the wireless computing landscape tremendously. The timing of these upgrades will depend, in large part, on acceptance and demand within the corporate community in the U.S. Once the carriers gauge the level of interest for 2.5G data services, they will determine how soon they need to roll out next generation networks.

Generation	Introduction	Characteristics
1G	1980s	Analog
2G	1990s	Digital, voice (circuit-switched)
2.5G	Circa 2002	Digital, voice, and data (packet-switched)
3G	Circa 2005	Digital, broadband voice, and data (packet-switched)

FIGURE 11.5
Wireless Network Milestones and Generations

Until higher bandwidth wide area networks are available, companies can "design around" data transmission limitations. Developers can create wireless applications with efficient data exchange, enabling them to work within the constraints of current wireless networks. This design approach favors transmission of short streams of data like alerts, notifications, one or two-line messages, e-mail messages (without attachments), and brief strings of data. Transmission of videos, rich graphics, and lengthy documents will remain problematic until 2.75G and 3G networks materialize, and even then, pricing will likely be an issue.

11.2 Wireless Network Features and Issues _____

When sampling the various wireless network options, a common set of features emerges. These features help define the networks themselves, but also influence the kinds of applications which will run on them. Some features may be critical to an application—such as the ability to transmit data in near real time—while others are less so. Some features can be expanded or augmented by combining network solutions in a "best of breed" approach. Other features may be inadequate in current generation wireless networks, but are expected to be more robust in future ones.

This subsection examines seven aspects of wireless networks that a user must understand to select a wireless network and properly design a wireless application. These seven features include the following.

11.2.1   Coverage

Coverage is, by far, the most influential factor when choosing a wireless network. The geographic area in which users will roam—a building, office park, city, state, region, or country—will determine the network used. Some users may work within a confined space, at a station on a factory floor, or in their own office. A narrow area of coverage is sufficient to support them. Other users may roam within a larger, but still defined area, such as a building, hospital, car dealership, or campus. For these users, coverage will have to span several hundred feet. Still other users may be very mobile, traveling within a city, state, or region. They will require the widest area of coverage.

Networks offer different coverage. Some transmit data only over short-ranges, while others operate well at medium- or very wide-ranges. The range of a given network must support users in the environments where they expect to operate. Even the widest of networks—the national cellular network—has spotty coverage in many areas of the U.S. Few carriers have a nationwide footprint. Urban areas have the best coverage when it comes to WANs, except that satellite networks often cannot cope with obstacles such as tall buildings. Rural or remote areas generally have poor terrestrial coverage, but have much stronger satellite coverage.

Coverage is also tempered by various other application needs. Even though a mobile user may range within a two-state area, for example, it does not mean that a WAN is required. If the user does not require real-time access to information, but can live with periodic synching of data, then a WAN may be overkill. Similarly, a satellite network may offer the best coverage in the furthest reaches of Nebraska, but be difficult to cost-justify for the application.

11.2.2 Bandwidth

Bandwidth refers to the volume of data that can be transmitted over a network in a given period of time. With speedy wired network connections such as Ethernet, cable modems, and DSL lines, bandwidth is seldom an issue. As a result, user expectations have been set very high.

Many current generation wireless networks do not come close to the bandwidth of these wired options, severely constricting their ability to support robust, data-intensive business applications. For example, as shown in Figure 11.6, Bluetooth networks operate at a maximum 1 mbps bandwidth, while many wide area data networks are only just starting to operate at speeds approximating a dial-up modem. Even when a publicized bandwidth rate appears sufficient for user needs, expect actual bandwidth to be half as high or less. In practice, overhead on the network, traffic load, signal obstructions, the actual mobility of users, and even the placement of antennas and other transmission equipment can negatively impact bandwidth rates.

Bandwidth may or may not be an issue depending on the type of applications that will run on the network. An application that captures and transmits only small amounts of data, such as bar code information or telemetry data, can work with a relatively low-bandwidth network. Conversely, an application that performs database updates or supports full web browsing requires a high bandwidth network. Keep in mind that application throughput can increase the cost of a solution, especially in the case of purchased network services that impose charges based on the amount of data transmitted. To put the bandwidth discussion into perspective, Figure 11.7 shows bandwidth requirements for certain common functions.

Network Type	Max Bandwidth*
Local	
Infrared	4 mbps
Bluetooth	1 mbps
WLAN	
802.11b	11 mbps
802.11a	54 mbps
802.11g**	54 mbps
Wide Area	
Data Networks***	19.2 kbps
Digital Cellular	
2G	19.2 kbps
2.5G	up to 144 kbps
3G	up to 2 mbps
In motion	up to 384 kbps
In a moving vehicle	up to 144 kbps
Satellite	9.6 kbps up to 64 kbps

* Actual bandwidth is generally lower due to network traffic, overhead, obstructions, etc.

** Approved by the IEEE in November 2001

*** Includes CDPD (AT&T and Verizon), Mobitex (Cingular) and ARDIS (Motient)

FIGURE 11.6
Wireless Network Bandwidths

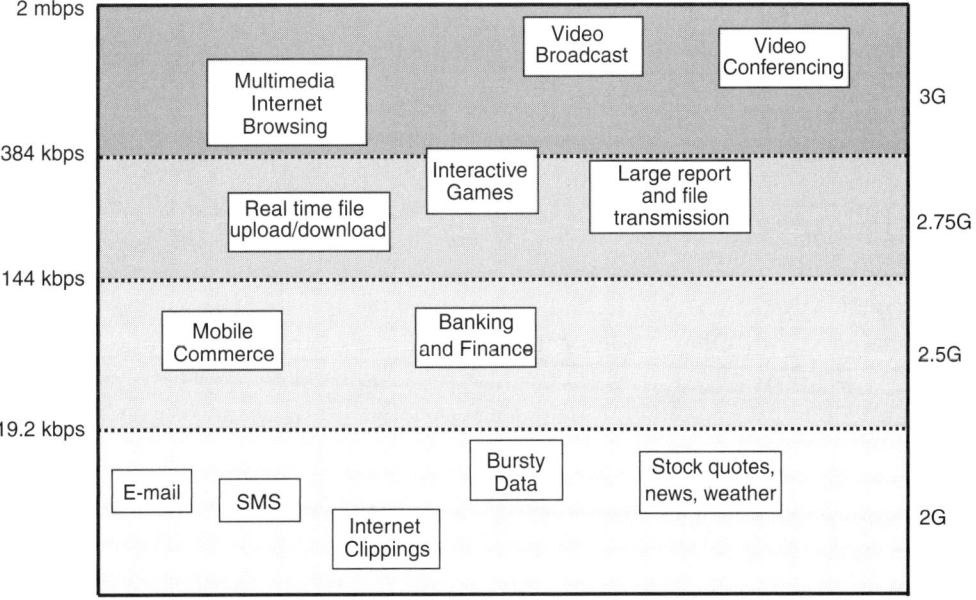

FIGURE 11.7
Application Bandwidth Requirements

Vendors and standards committees are busily at work increasing the bandwidth of every wireless network option, and the table shown in Figure 11.6 will undoubtedly be outdated shortly after publication. Restrictive bandwidths will become a thing of the past, and high-speed networks will become the norm. Wide area networks in particular will see bandwidth increase by a factor of 50 in the coming 2G to 3G migration.

In the interim, however, present networks are best suited to the transmission of short text messages such as e-mails, alerts, notifications, and simple form content. They are incapable of transmitting very large documents, graphics, or videos, or matching the Internet browsing experience offered by broadband wired connections. Even when high bandwidth networks are a reality, application functionality may remain "low bandwidth" for some time until constrained device features—tiny screen sizes and limited data entry capabilities—are overcome.

11.2.3 Latency

Latency is concerned with the ability of the network to convey information in a timely manner. Delays in transmission, or the ability to contact a user, affect the freshness of information. In certain fields, such as disaster management, stock trading, or the delivery of emergency health care, the "shelf life" of information is very short, and it is vitally important to convey it instantaneously to field or mobile workers. For a salesperson in the field, the ability to lock or reserve inventory in a timely manner is critical for filling a pending order. While there is always some degree of latency with every network type, most networks strive toward a zero latency model. As latency declines, networks approach real-time transmission of information.

Latency is affected by factors such as coverage and network technology. Clearly, if coverage is lacking in a given area, there is maximum latency. As a convenience, some wireless networks have a store and forward capability that effectively stores transmissions until a user re-enters an area with coverage and then pipes things through, but this feature does not eliminate latency concerns for communications that must happen in real time. Network technology also impacts latency. If network technology is such that it takes considerable time to establish a connection with a user in the field (the case with present circuit-switched cellular networks) or vice versa, then latency is high. If a network supports an "always on" connection with a mobile user, latency declines dramatically. Some present day data networks offer "always on" connections, which is also one of the goals of the 2G to 3G migration. An "always on" connection allows organizations to "push" information out to the field in the most time-efficient manner possible. Of course, the desired recipient must be in an area of coverage to receive the information in near real time.

11.2.4 Reliability

Before moving business applications to wireless networks, those networks must perform reliably. Anyone who has used a cellular network to make voice calls will agree that reliability problems abound. Difficulties with roaming, dropped or disconnected calls, noise, interference, and signal weakness are annoying when it comes to voice calls, and even more problematic with data transmissions.

Reliability covers areas such as the ability to connect to a network in different locations; the availability of the network; the ability to complete a transaction without interruption, dropped data or errors; and the ability of the network to perform error checking and validation and to recover from or correct errors gracefully in stationary and mobile situations. Bluetooth networks, for example, use a built-in scheme to detect and retransmit lost packets of data. Packets are re-sent whenever a device does not receive an acknowledgement that a packet has been safely received. Other wireless packet data technologies, such as the 2.5G GPRS technology, employ this same scheme. In dense traffic areas, when the subscriber moves at increasing speed, or where strong signal interference is likely, packet re-transmission is essential. If a network has enough bandwidth, it can re-send lost packets without any significant delay.

When network reliability is suspect, application developers must take extra steps to ensure the integrity of transactions. At the software level, developers can add code to detect errors and either roll back or resume transactions without loss of data.

11.2.5 Security

Like wired networks, wireless networks must have sufficient security to keep intruders at bay and ensure the integrity and privacy of information. Security is usually accomplished through classic means: a combination of passwords, authorizations, authentication, and encryption schemes.

Some wireless networks, however, pose security risks that are not shared by their wired counterparts. If radio frequencies are used to transmit data over longer distances, hackers intent on eavesdropping or intercepting data can breach data in transit without the need to be physically close to the communicating devices or network. And, if wireless networks provide an entrée into wired ones, these hackers can also gain access to entire corporate systems.

Still other wireless networks, like infrared, are inherently secure from eavesdropping due to their native design (line-of-sight light transmissions). The remaining types of wireless networks rely on traditional security mechanisms, which may be built-in, to keep information secure. Researchers and specialists regularly test the security of wireless networks, and publicly report on their vulnerabilities, as has been the case with the security protocol for 802.11 wireless LANs (WEP) and certain gateways (WAP) used on the cellular network. The intent of these researchers is to prod vendors and corporations into creating and using more bullet-proof wireless networks.

Although wireless networks offer varying levels of security, it is incumbent that companies fully understand the security offered, and enhance it with other commercial products, operating policies and procedures, just as they do with wired networks. As a rule of thumb, companies should determine the value of the information traveling over the wireless network, and take commensurate measures to ensure its security. In some cases, built-in security schemes will need to be augmented with third-party software and tools that provide stronger encryption, establish virtual private networks (VPNs), or monitor system access and usage.

11.2.6 Interoperability and Standards

Unless a company intends to build its own private wireless network, it will have to ensure that the network and the products that operate on it adhere to some set of industry standards. If not, a company may find itself dependent on proprietary software and hardware that cannot be easily upgraded, and that do not integrate and operate with other products. Adherence to standards helps reduce the risk that software and hardware will not interoperate.

Independent bodies regularly publish and upgrade wireless network protocols and operating standards. Various associations will also test and certify compliance to these standards. Before investing in a wireless network solution, companies are well advised to research the standards followed and compliance attained by the products at hand, and their ability to interoperate with other products and/or standards. Sometimes, however, a company will be forced to accept a solution that is not interoperable. A classic example is the wireless cellular networks. In the U.S., second generation cellular networks rely on four different network technologies—CDMA, TDMA, GSM, and iDEN. Phone handsets are manufactured to work with a particular technology and cannot be used on a network that uses a different technology unless they are pricier "multimode" units. And a phone handset that works on a GSM network in the U.S. must be "multiband" to operate on networks using GSM technology in Europe, for example. Application code developed to run on these networks must be "network agnostic" by design or else companies may wind up supporting three different versions of software, one for each network technology.

Interoperability is also an issue where more than one type of wireless network is used, with each operating in the same radio frequency. Such is the case with Bluetooth and 802.11x WLANs. Unless this spectrum clash is taken into account and explicitly addressed, network interference is a distinct possibility.

11.2.7 Cost

For the end user, network costs may include device, network airtime, gateway, software, and support components. Pricing for wireless network components, as with most technology products, has declined and is predicted to fall even lower in the future due to fierce competition. This price competition is advantageous not only for consumers, but also for the health of the overall wireless market. Declining technology prices, especially for items like Bluetooth chips, will help propel categories of wireless technology into widespread acceptance and use.

Companies that purchase, install, and operate their own wireless networks face two categories of costs: initial and ongoing. Initial investments in software, hardware, and client devices may be higher than traditional wired components. In the long run, however, maintenance fees and labor are significantly less due to the ease with which wireless networks are reconfigured.

Companies that subscribe to network services will see a wide range of pricing options. As wireless networks are upgraded, and increasing numbers of commercial data-based applications come on-line, carriers will become more sophisticated in pricing their data services, and will likely offer a variety of plans similar to those offered for voice services. The use of software to track and monitor network usage and traffic will allow carriers to determine precise chargebacks and develop tiered pricing schemes tied to different levels of service. Fee structures will likely combine a fixed fee and a variable fee based upon packets of data sent and subscribed service levels.

11.3 Wireless Network Options _____

As the contours of a wireless application take shape, it is time to think about wireless network options. The preceding section talked about two types of wireless networks—DIY for local, defined coverage, and purchased network services on large-scale national networks for wide area coverage. Business process needs and the target wireless application will steer a company toward one or the other of these network types, or possibly a combination of both.

Each wireless network type—DIY or wide area—offers several options for enterprise use. These options are generally referred to by the underlying technology used. Each option has its pros and cons, which are summarized in Section 11.4. Some options overlap in their capabilities; some provide complementary coverage. Many of these wireless network options are developed, defined, and monitored by standards bodies that ensure interoperability between different vendors' products. Certifying bodies also test products for compliance with these standards, and stamp those that comply with logos to indicate that they meet standards.

When reviewing the options below, keep in mind that it may make most sense to combine several options to achieve the level of coverage, performance, and convenience desired. Be advised that some technologies, such as fixed wireless, are not covered in this book, and that others, such as paging networks and satellite networks, are mentioned only briefly.

11.3.1 Wireless Network Types

Rather than digress into too much technical detail, this part of the book provides an overview of different wireless network types. For managers and executives that need to get up to speed quickly on wireless networks, an exhaustive survey of network options is overkill. Rather than inundating managers with too much detail and too little practical advice, this section focuses on supplying just enough information about network options to differentiate between categories and hone in on the most suitable options for a given situation. Some technical discussion is inevitable, however, this section deliberately seeks to limit the details and relies on an overview format instead. Technicians interested in a more detailed discussion should supplement their reading with the many reputable texts and guides available on the market.

Wireless networks are constantly evolving. This section offers a snapshot of capabilities that exist at the time of writing. Cost and performance data are particularly susceptible to variation over time, and companies are advised to perform their own research on current performance claims and vendor pricing before pursuing a given avenue. In general, prices are declining and performance is improving, as subsequent generations of technology become publicly available. Bandwidth, in particular, is expected to improve demonstrably over the next few years; however, it is important to verify actual performance in the intended environment as high-end ranges quoted by vendors are seldom achievable except in controlled situations.

The first three network types discussed below—infrared, Bluetooth, and WLAN—are all DIY network solutions. These DIY networks are short- and medium-range solutions. The short-range solutions—infrared and Bluetooth—are ideally suited for setting up wireless, personal area networks (WPANs) that allow an individual to connect to the devices—phone handset, PDA, desktop, printer, etc.—found within his or her personal workspace. Short-range solutions are also useful for establishing ad hoc, dynamic networks between two or more devices to share documents or data. Medium-range solutions, such as WLANs, work well as an extension or replacement to wired networks and operate over longer distances than those supported by the short-range solutions. WLANs also appear in purchased network service solutions, such as the public hotspots that are cropping up in places like airports, convention centers, hotels, and other business-friendly venues as a means to connect to the Internet.

Once an organization strays beyond short- and medium-range network solutions, by necessity it moves into the purchased network services category. Organizations purchase wireless network services when they are interested in highly mobile applications that require coverage beyond the confines of an office, building, or campus. While local, medium-range WLAN solutions can span several miles between buildings using special equipment, their range is inadequate to support workers or customers at remote locations. In those circumstances, the only choice is to turn to a provider that offers coverage over a much larger geographic area, such as a city, state, or region, through a wireless wide area network (WAN). These types of networks are covered in Section 10.3.5.

11.3.2 Infrared

An infrared network is a short-range, DIY wireless network option. Its characteristics are as follows.

- *Description*   Infrared (IR) networks are cordless, high-speed, short-range networks. They enable spontaneous, point-to-point communications—two devices talking to each other—using infrared light beams as shown in Figure 11.8. To establish a connection, two devices are pointed at each other and one device sends a signal to initiate the conversation. A familiar example of IR technology in the home is the remote control device used to operate a television set. In the work environment, a user holding a PDA might point the device at another PDA to transfer items such as contacts and calendars, or at an IR-enabled printer to print a document.

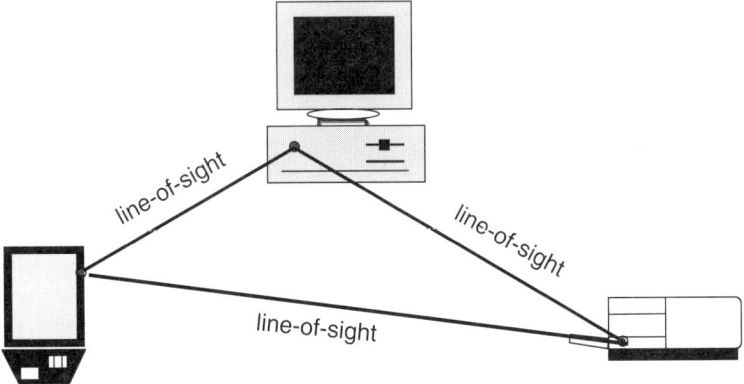

FIGURE 11.8
Infrared Network

- *Range* With IR, devices must be within line-of-sight (a 30 degree angle) to communicate. If either device is moved out of this zone, the connection is disrupted. To meet this line-of-sight constraint, devices must remain relatively stationary while communicating. Devices can be a maximum of 3 feet apart, although 6 feet is purportedly acceptable if the transmitting and receiving devices are perfectly aligned. These directional and distance requirements mean that IR networks are suitable only when devices are within close proximity, are aligned within a narrow angle, and are kept in a comparatively fixed position during connection and transmission.

  IR connections must be line-of-sight because the light waves used cannot penetrate objects such as floors, ceilings, walls, doors, or even people. For example, if an IR-enabled device was in the midst of transmitting a presentation to a projector for screen display, the transmission would be interrupted if someone walked between the two devices.

  In an office setting, IR technology is not an acceptable alternative to wired cabling for access to a corporate network. The line-of-sight feature would require companies to install IR access points in each office to allow employees to access the LAN.

- *Throughput* Within the business world, data transfer rates for IR range from a low of 512 kbps, to 1 mbps in some PDAs, to a high of 4 mbps in some notebook computers. Speeds do not vary based on the distance between devices, but remain constant. Several organizations are working on upgrades that will boost transfer rates to 16 mbps sometime in the future, with 50 mbps speeds rumored down the road.

- *Adoption* iR-enabled devices are widespread and include desktop computers, notebook computers, PDAs, printers, digital cameras, kiosks, cellular pagers, phone handsets, instruments, machinery, keyboards, and other types of mobile devices. Analysts estimate that over 100 million of these devices have been sold, although few buyers are aware of or take advantage of the built-in feature. Most popular operating systems support IR, including Windows, Windows CE, Palm OS, and Linux. A low-power technology, IR draws upon the power supply of the host device in which it is embedded.

- *Security* IR networks are inherently secure due to the line-of-sight requirement between transmitting and receiving device. Moreover, IR-enabled devices do not automatically communicate with each other. The person using an IR device makes a conscious decision to point and shoot the device at another, leaving little room for doubt about which devices are communicating. Eavesdropping is not really feasible. A snoop would have to pick up a reflected light beam and filter out any noise.

- *Interference*   The directional nature of IR networks not only aids security, it also makes connections less prone to interference. The sun or another bright light source can affect an IR device, but they generally do not suffer from interference concerns. Unlike other wireless technologies that rely on radio frequencies, IR is not subject to FCC oversight or regulations. Manufacturers and users can simply enable IR technology and use it at will, license free, without worrying about federal compliance or conflicting technologies operating in the same air space.

- *Standards*   A standards body, called IrDA (Infrared Data Association), seeks to ensure interoperability between IR devices of all types. IrDA publishes specifications for manufacturers and vendors to ensure proper operation of hardware and software, and has issued profiles for certain common usages. The IrDa also administers a certification program called IrReady to test vendor products and certify them as compliant with the standards.

- *Cost*   IR solutions are less expensive than the other DIY wireless options—Bluetooth and 802.11x. Incorporating IR technology in devices translates to a $1 to $2 difference in the price tag. If IR support (the port and software drivers) is not native to a device, owners can purchase an IR adapter and software for minimal cost.

- *Application*   IR networks excel at simple, high-speed data exchange between two devices that are within close proximity. The inherent security of the technology makes it attractive for financial transactions where buyer and seller are physically present, like retail transactions in stores, restaurants, and parking lots. A pilot is underway using IR for financial, point-of-sale (POS) purchases performed with a credit card. With IR-enabled credit cards, purchasers can beam data to and from their financial accounts to IR-enabled POS terminals, or even vending machines.

  IR networks, combined with LAN technology, are also popular at universities as an alternative to cabled networks. The University of South Dakota is piloting an IR LAN within its campus.[3] First-year medical and law students and faculty were given IR-enabled Palm PDAs. IR access points were hard-wired into the university's cabled network, and installed in obvious places like classrooms, libraries, student centers, laboratories, and even hallways. Using the IR network, students and faculty are able to access the university's intranet to share files, synchronize information, browse web files, view work assignments, and send and receive e-mail. An important consideration for the university was to deliver these capabilities at a low cost. IR-enabled PDAs meant that students and faculty would not need to purchase cradles and cables, and IR access

---

3. "Wireless Is Academic," Susana Schwartz, *Field Force Automation*, November 2001.

points were a low-cost way to continue leveraging the university's wired network. The vendor on the project, Clarinet, quoted a price per student of approximately $40 for the IR solution, compared to a $100 price tag for a cradle or a $170 investment to purchase Bluetooth or 802.11x capabilities.

11.3.3 Bluetooth

Like an infrared network, a Bluetooth network is a short-range, DIY option (Figure 11.9). Its characteristics are as follows.

- *Description*   Bluetooth, like IR, allows for the creation of ad hoc networks that rely on radio frequency technology. Commentators often refer to Bluetooth wireless networks as an "alternative to cables," effectively allowing users to unwire their offices. An office outfitted with Bluetooth devices allows a user to create his or her own WPAN and jettison proprietary cables, synching cradles, and other connectors. As devices get in range, tune in, and recognize each other, communications can commence with very little human intervention, allowing the user to synchronize devices, transfer files, and access local peripherals. Users can also create networks, called piconets, with other Bluetooth-enabled devices (up to 8 altogether) to exchange and synchronize data.

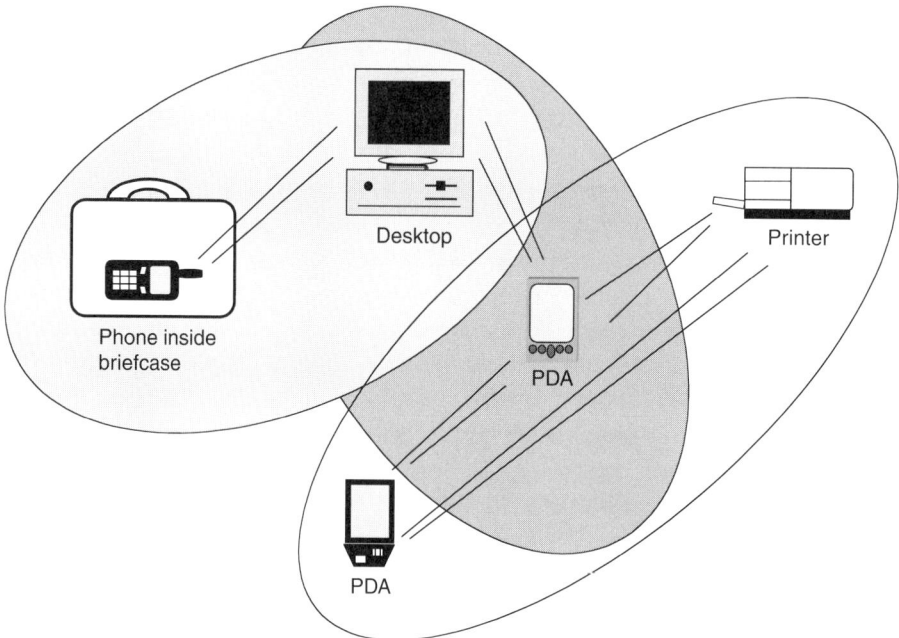

FIGURE 11.9
Bluetooth Network

- *Range*   Bluetooth is omni-directional, meaning it can establish a network with any device within its zone of operation, whether those devices are above, below, in front of, behind, or beside the others. Bluetooth-enabled devices are equipped with mini antennae, and constantly test their surroundings to detect, and possibly connect, to other Bluetooth devices.

  Bluetooth radio signals travel through non-metallic obstacles like walls, cement floors, wooden desks, clothing, and briefcases. Whereas IR is not recommended for use in a crowded room—people and furniture can interrupt transmissions—Bluetooth adapts well to this situation because of its omni-directional and penetrating nature. Bluetooth has a range of 30 feet in low-power mode to 330 feet at high power. The lower powered version is favored because it costs less to implement and experiences less interference.

  Its ability to penetrate solid objects allows Bluetooth to perform better than IR as part of a wireless LAN solution. Bluetooth access points—bridges to the wired network—can be situated in strategic locations (every 30 feet or so) giving users access to corporate intranets and the Internet. This configuration requires more access points, however, than a similar medium-range WLAN implementation.

- *Throughput*   The maximum data transfer rate for Bluetooth is 1 mbps. In reality, throughput lies in the 700–800 kbps range. Actual throughput may be even less, depending on the number of Bluetooth devices sharing the same channel.

- *Adoption*   Bluetooth is a relatively recent technology, developed by Ericsson, the Swedish mobile phone maker, in 1994. Over 1,800 companies support Bluetooth, including IBM, Nokia, Intel, and Toshiba, and have formed the Bluetooth Special Interest Group (SIG).

  Manufacturers have incorporated Bluetooth chips in some, but not all, mobile devices. Although the chip is physically small enough to fit in almost any device, the cost per chip has historically been high, making it economically feasible to use only in more expensive laptop and notebook computers. The longer term goal is to embed Bluetooth chips in every mobile device imaginable—phone handsets, computers, PDAs, printers, digital cameras, fax machines, headsets, projectors, and other units. This goal is already coming to fruition, as Palm is offering an expansion card that allows its PDAs to talk to other Bluetooth devices and Ericsson is offering a Bluetooth-equipped phone handset. Over the next few years, Bluetooth technology should go mainstream as chip prices fall further. A host of big name vendors have pledged their support to Bluetooth, including Microsoft who announced that Bluetooth support would be integrated with its Windows XP operating system beginning in 2002.

- *Security* Bluetooth comes with built-in, lower-level security that authenticates users participating in the network via passwords and that can also encrypt transmitted data. Because Bluetooth is omni-directional, it is feasible that a hacker could fortuitously place himself or herself in range to intercept data during transmission, but the odds of this happening are slight.

  Bluetooth devices will try to establish networks with other connectable devices automatically; however, a device does not have to accede to requests for connection. A device that wishes to communicate with another identifies itself first, and is either accepted or rejected based upon its identification. By using the built-in security capabilities, Bluetooth users can ensure that their devices do not surreptitiously communicate with others. Identifying, classifying, and authenticating devices, and defining options for passwords and encryption, are accomplished using separate software to manage and administer the Bluetooth environment.

- *Interference* As a radio frequency technology, Bluetooth operates within the 2.4 GHz band, the same one used by 802.11x WLANs and a variety of home devices such as microwave ovens, cordless phones, baby monitors, and garage door openers. While an ordinary business user is unlikely to encounter these potentially interfering devices in the work environment, some locales such as hospitals and airports use special-purpose equipment that relies on radio frequency technology that could conceivably clash with Bluetooth. As a result, Bluetooth may be circumscribed in these sensitive locations to avoid any potential interference. The FCC does regulate the use of airwaves, including Bluetooth's 2.4 GHz band, but allows the public to use this particular spectrum license-free.

- *Standards* Members of the Bluetooth SIG develop standards for Bluetooth software and interoperability requirements for devices. Similar to the IrDA's efforts with IR technology, the Bluetooth SIG is in the midst of defining Bluetooth usage profiles for common scenarios like faxing and data synchronizing. They also administer the Bluetooth SIG Qualification Program. Qualification is a prerequisite for licensing Bluetooth intellectual property and applying the Bluetooth trademark to a product. At the time of writing, the Bluetooth SIG web site, *www.bluetooth.org*, listed over 400 vendor products that had passed the qualification program.

- *Cost* At publication time, Bluetooth is more expensive to implement than IR. Bluetooth chips cost between $15 and $20 compared to IR's $1 to $2 price tag. This imbalance will not last long. Chipmakers expect Bluetooth prices to fall to $5 per chip once production fully ramps up, eliminating any financial impediments to implementing Bluetooth. Adapter kits are available for non-native Bluetooth devices at prices comparable to those for 802.11x WLAN functionality.

- *Application*   Like IR, Bluetooth works best for short-range data exchange. It is very effective in creating WPANs, allowing users to eliminate cables, interact with computers, PDAs, and peripherals with ease, synchronize data effortlessly, and boost their productivity. In situations where IR would work, but greater distance is required, Bluetooth is a good alternative.

In addition to the office environment, Bluetooth is appearing in automobiles. Motorola has introduced a Bluetooth Car Kit to enable hands-free use of a mobile phone while driving. A Bluetooth-enabled phone handset can connect to built-in speakers and a microphone installed in the automobile. Speech recognition software allows the occupant to tell the phone what to do. Daimler-Chrysler is also offering a Bluetooth-enabled hands-free calling system based on Mercedes-Benz's Tele Aid telematics application.

Bluetooth is also effective in warehouse and factory environments, where data collection occurs in real time, and where hauling cords around is a nuisance. UPS plans on outfitting sorters at its distribution centers with Bluetooth-enabled rings to scan bar codes of packages moving on a conveyor belt, precisely the kind of place where cords can get in the way.[4] The rings will send information to a terminal worn on the sorters' hips, for eventual transmission to a WLAN. FedEx is looking into a similar application. Bluetooth devices can also be used on warehouse or factory floors to work as locator devices. By situating Bluetooth receivers at strategic places, companies can track the whereabouts of a person wearing a Bluetooth-enabled device throughout the networked area.

11.3.4 802.11x WLAN

In contrast to infrared and Bluetooth networks, a WLAN provides wider coverage. It is a type of medium-range DIY network, and its characteristics are as follows.

- *Description*   WLANs are primarily used to extend or replace a traditional Ethernet wired LAN. They offer all the functionality of wired LANs, but without the headaches of the wires themselves. They also open the door to a host of new, mobile applications. WLANs are commonly found within office buildings, hospital grounds, or university campuses, the same places where wired LANs appear. They can use either IR or radio frequencies to communicate, but business implementations favor radio frequencies for their power and range. Microsoft, for example, has used a WLAN to "unwire" its entire Redmond campus and various branch locations in Seattle.[5] WLAN products are almost

---

4. "UPS to deploy Bluetooth, wireless LAN network," Bob Brewin, *Computerworld*, July 2, 2001.
5. *http://www.microsoft.com/business/mobility*

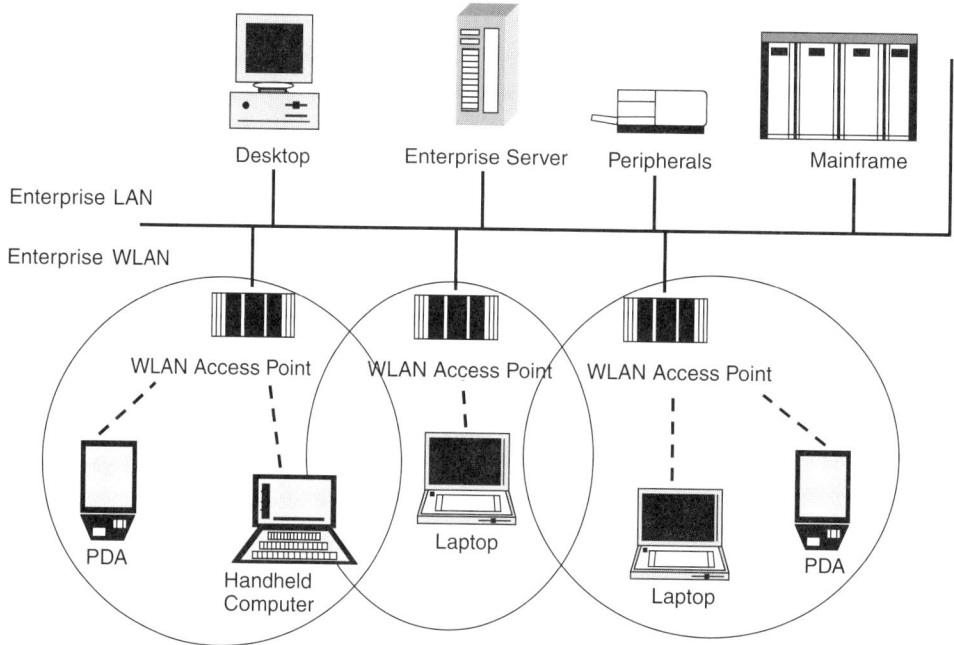

FIGURE 11.10
WLAN

universally based on standards published by the Institute of Electrical and Electronics Engineers (IEEE) under the designation 802.11.

WLANs, as shown in Figure 11.10, are comprised of client devices and access points. Client devices include desktop computers, laptops, and PDAs containing WLAN cards or fully integrated WLAN components. Access points receive, buffer, and transmit data between the WLAN and the wired LAN, and let client devices access enterprise applications, data, intranets, and the Internet. They are installed at strategic locations and connect to the wired LAN via standard cabling. WLAN technology also allows client devices to connect directly to each other to form independent, peer-to-peer networks to exchange data like documents, contacts, or files.

- *Range*　Like Bluetooth, 802.11 WLANs are omni-directional. Their radio signals penetrate non-metallic obstructions such as walls and floors. The range of a WLAN is less than 100 feet to over 500 feet, making it ideal for in-building or campus use. The type and number of obstructions along a signal path affect the range of the WLAN. Sometimes merely moving a computer or an antenna can improve the range and throughput. Installing multiple and overlapping access points provides seamless wireless coverage to an entire building or campus, allowing users to roam and remain connected. Rooftop antennas and equipment can provide "bridges" between WLANs in separate buildings that are miles apart.

- *Throughput* Actual WLAN throughput depends on the product and its design, the number of users per access point, the types of applications used, signal obstructions, and responsiveness of the wired LAN. Data transmission rates range from 1 mbps to a high of 11 mbps with the 802.11b IEEE standard. A newer standard, 802.11a operating in the 5 GHz range, achieves maximum rates of 54 mbps. The successor to 802.11b, called 802.11g, will operate in the same 2.4 GHz range and reportedly achieve throughput of 54 mbps. Future standards are rumored to approach Ethernet-like speeds of up to 100 mbps.

  Careful siting of access points makes a tremendous difference in actual throughput. Since access points are shared among users, the number of users and the bandwidth demanded by their applications will affect overall throughput. Organizations typically perform a site survey prior to installing access points to determine how best to balance the load.

- *Adoption* WLANs were originally used as specialty applications in vertical industries like retail and warehousing for inventory management and data collection. Broad market acceptance ensued when the IEEE published its 802.11 standards for WLANs in 1997 and major manufacturers signed on to the standard. All of the large PC and device manufacturers offer built-in WLAN support, and WLAN cards are available for most other mobile devices. Microsoft has built WLAN support into its most recent operating system, Windows XP.

- *Security* The radio signals sent by WLANs can leak outside buildings, making it possible for an eavesdropper to pluck transmissions out of the air from an outdoor location. Moreover, an intruder who gains access to a WLAN can attempt to use it as a conduit to the wired LAN. Careful placement of antennas can help limit the amount of radio signal that escapes beyond a building's walls.

  802.11b WLAN products come with built-in security called WEP (Wired Equivalent Protocol). Most products ship with the security turned off, however, and unless organizations remember to turn it on, they leave their WLANs exposed to eavesdropping. WEP has a history of much-publicized weaknesses, which several groups are determined to address. In the meantime, companies are advised to choose a security approach commensurate with the value of the information traveling over the WLAN.

- *Interference* 802.11b and 802.11g WLANs use radio frequencies in the FCC-regulated but license-free 2.4 GHz band, like Bluetooth and any number of home appliances such as microwave ovens and cordless phones. Most WLANs are designed to avoid interference with microwave ovens, but interference with co-located Bluetooth and 802.11x networks must be separately addressed. The 2.4 GHz band is also populated by special-purpose equipment used in hospitals and airports, so care must be taken in those locales to avoid interference by WLANs. The 802.11a standard operates in the less populous 5 GHz range, and raises fewer interference concerns.

- *Standards*   The 802.11 family of standards published by the IEEE is one of the primary reasons for WLAN growth. Although non-802.11 WLANs do exist, manufacturers have converged on the IEEE standards to ensure interoperability between products. The IEEE continues to issue new standards, the most recent of which are 802.11a and the still-in-progress 802.11g. From an enduser point of view, the standards are similar except that later versions offer increased bandwidth, which may or may not be required depending on application needs.

  A separate body, the Wireless Ethernet Compatibility Alliance (WECA) has created interoperability test suites based on the IEEE standards, and will certify compliance to the standards. The logo "Wi-Fi" indicates WECA's stamp of approval, and is used to signal that a product is 802.11b compliant. WECA also plans to certify subsequent IEEE standards under other logo names.

- *Cost*   WLAN costs have two components—device costs for WLAN adapters and access point costs. The total cost of a WLAN implementation will vary depending on the number of access points needed and the number of client devices that must be outfitted with WLAN adapters. Market pressures have caused WLAN prices to drop precipitously, but they still remain higher than those of IR or Bluetooth. Enterprise-grade access points cost several hundred dollars, and WLAN adapter cards can be had for under $100. Although the initial investment in WLAN equipment may be higher than the direct costs for equivalent wired components, lifecycle costs are lower. The labor costs of installing and maintaining a WLAN are less than those associated with pulling cables and changing setups for wired networks. Indirect costs, such as user downtime during moves and administrative overhead, are also avoided with WLANs.

- *Application*   Most organizations deploy WLANs to extend or replace their wired LANs, and to capture the efficiencies and savings that result when physical cables no longer need to be installed. Organizations industry-wide use WLANs including hospitals, universities, transportation companies, and businesses of all kinds. Harvard Medical School, Stanford University, and the Massachusetts Institute of Technology use WLANs to give students and faculty access to course descriptions, work assignments, and reference materials. Hospitals use WLANs in emergency rooms to collect patient information and transmit it to patient-management systems. Car dealerships use WLANs to give sales reps and repair technicians access to inventory and service applications from the car lot and service bays. UPS is using WLANs in its distribution centers to collect tracking information transmitted from Bluetooth ring scanners worn by package sorters.[6]

WLANs not only offer untethered access to existing wired data and applications, they permit the creation of new mobile applications. Buildings and grounds personnel can receive alerts, maps, and schematics on handheld devices. Executives can dynamically access and update forecasts and plans during meetings.

WLANs are also appearing in public places, like hotels, airport lounges, and coffee shops, to provide Internet access to guests and patrons. Starbucks is building out an 802.11b network in its coffee shops that customers can use to browse the Internet while they eat and drink. Many hotels, convention centers, and airport lounges have public WLANs that subscribers can use to obtain network access.

WLANs are also complementary to short-range solutions, such as IR or Bluetooth, which excel at spontaneous, point-to-point data exchange. Bluetooth can provide a user with wireless links between mobile devices, computers, and peripherals, while a WLAN can offer continuous connectivity to enterprise data, applications, and the Internet.

11.3.5 Wireless Wide Area Networks

Wireless wide area networks (WANs) provide extensive geographic coverage, much wider than that offered by short- or medium-range solutions. Why wouldn't an organization simply opt for using a WAN for all situations? Cost, reliability, and in-building reception are factors that may give an organization pause, but the biggest culprit is the low bandwidth inherent in the underlying network technology. WANs excel at providing wireless *voice* services, but have traditionally fallen short when it comes to data services—precisely what data-intensive wireless business applications need. That situation is changing rapidly, however, with the rush to upgrade cellular networks to next-generation capabilities. Once those capabilities are widely available, and the carriers can attractively package and market data services to businesses, the wireless network landscape will change irrevocably. Of the network types mentioned below, some will disappear completely and others will become subsumed into the 3G networks.

WANs use radio frequencies to transmit information. The FCC regulates the spectrum in which the WANs operate and licenses network operators. Spectrum has traditionally been in short supply and network operators vie against each other to snap

---

6. *http://www.microsoft.com/business/mobility.*

up additional spectrum whenever it is offered. As new spectrum becomes available, the FCC auctions it to the network operators for hefty prices. The FCC is expected to allocate additional spectrum for wireless networks in the coming years, although aspects of its auction approach may change. In addition, the FCC intends to allocate spectrum to companies that will develop non-cellular types of technologies, to encourage competition in the wireless arena. While some of these technologies may provide breakthrough capabilities, they are years away from practical, widespread commercial use.

The next sections describe four different WAN options: dedicated data networks, digital cellular networks, paging networks, and satellite networks. To keep the discussion oriented to wireless business applications, the section will focus on data as opposed to voice capabilities.

Dedicated Data Networks

Dedicated data networks are the precursor to next generation, 2.5G networks in that they offer *data* services. For organizations that required wide area data services, these networks were the only way to go. They differ from first and second generation cellular networks in their underlying technology, which is packet-based. Of the national telecommunications carriers, AT&T Wireless and Cingular Wireless operate wide-scale packet data networks. AT&T's data network is called CDPD and functions as an overlay on its existing cellular voice network. It is available in few geographic markets, and purportedly covers only about 55% of the U.S. population. Investments in expanding the CDPD network, and other types of data packet networks, has come to a halt as carriers focus instead on migrating their digital cellular networks to offer high-speed, packet-switched data services. As businesses begin to rely on 2.5G networks, use of these dedicated data networks will wane, although they may remain popular for certain special-purpose applications.

These dedicated data networks are characterized by spotty coverage, very low bandwidth in the 19.2 kbps range, and poor indoor reception (although smart indoor antenna technology is helping to overcome this issue). They work best transmitting small amounts of data—short messages and alerts like weather forecasts, e-mails without attachments, and stock quotes. Other companies have forged deals with AT&T and Cingular to offer their own branded services over these data networks. For example, Research in Motion, manufacturer of the BlackBerry e-mail device, offers a wireless e-mail service that runs on the Cingular and Motient networks. The Palm.net service also runs on the Cingular network.

Motient, an independent data services company, operates a high-bandwidth data network, supplemented by satellite capabilities, to provide national coverage. In contrast to AT&T's and Cingular's networks, the Motient network has wider geographic coverage and deep in-building penetration. Most of the Motient network operates in the 19.2 kbps range, although it is capable of delivering 192 kbps in the New York City area.

In the long term, it is questionable whether secondary data service providers like Motient will be able to compete successfully with the large telecommunications carriers. At the time of writing, Motient was struggling financially, following in the footsteps of other regional providers who tried, but failed, to offer high-speed data networks in major metropolitan areas. The technology is viable, but the ability of providers to attract sufficient subscribers to underwrite their large infrastructure investments is in doubt. The promised arrival of high-bandwidth cellular data networks, as described below, may squelch the efforts of these data service provider.

Right now, companies are using existing data networks for a range of applications. Workers use RIM BlackBerry devices to send and receive e-mail. Companies like Atlantic Envelope Company (see case study in Chapter 3) use data networks to power their wireless CRM application allowing salespeople to respond to customer queries about orders and purchase histories and inventory on-the-spot. The Illinois State Police use a CDPD network to connect officers in the field with a criminal justice system to check vehicle registrations, licenses, and other important information.

Digital Cellular Networks   Digital cellular networks are owned by the large telecommunications companies like AT&T Wireless, Verizon Wireless, Cingular Wireless, QUALCOMM, Sprint PCS, Nextel, and VoiceStream. These networks support wireless voice services, and are being upgraded from circuit-switched technologies to packet-switched technologies, as described in Figure 11.11, to accommodate high-speed data traffic.

Circuit-Switched Versus Packet-Switched Networks

Second generation digital cellular networks (2G) are circuit-switched. When a caller dials a number, a circuit is established that lasts for the duration of the call. The circuit is dedicated to that call, even if the allocated bandwidth is underutilized during the course of the conversation. This design is wasteful, tying up more bandwidth than is actually needed.

Packet-switched networks do not rely on dedicated circuits. In effect, they are "always on" and ready to transmit. To make maximum use of bandwidth, packet-switched systems break data streams into small "packets" of information that are reassembled at the receiving end, similar to the method used by the Internet. This method is ideal for transmitting large amounts of data.

FIGURE 11.11
Circuit-Switched Versus Packet-Switched Networks

With the debut of 2.5G networks in late 2001 and 2002, carriers are just beginning to offer enterprise-specific data services, although they are downplaying network capabilities until they see what kind of reception they will receive among the business community. Verizon Wireless, for example, offers "Express Network" services that deliver "always on" high-speed wireless Internet, intranet, and e-mail access to data-enabled devices such as PDAs, laptops, and smart phones at transmission rates ranging from 40 to 60 kbps, similar to dial-up service. Although 2.5G networks will theoretically achieve 144 kbps transmission rates, the carriers are loath to over-promise until they are able to assess how the networks perform under varying loads and conditions. If your company or employees operate in an area without 2.5G network coverage, you will have to resort to a dedicated data network with a low throughput of 19.2 kbps.

Once the 2.5G and 3G wireless network upgrades are complete, and high bandwidth wireless networks are widely available, the networking landscape will change dramatically. If 2.5G and 3G networks perform as promised, and are attractively priced, cable modems, DSL lines, and other types of high-speed wired connections may become obsolete. Only device limitations, and not network limitations, will constrain the kinds of application functionality that can be deployed. Performance characteristics for next-generation cellular networks are noted in Figure 11.12.

Even with 3G networks, coverage will remain a concern. In rural or remote areas, companies will have to rely on satellite networks to provide "last mile" coverage. As noted below, satellite networks have their downsides, primarily in terms of cost and bandwidth.

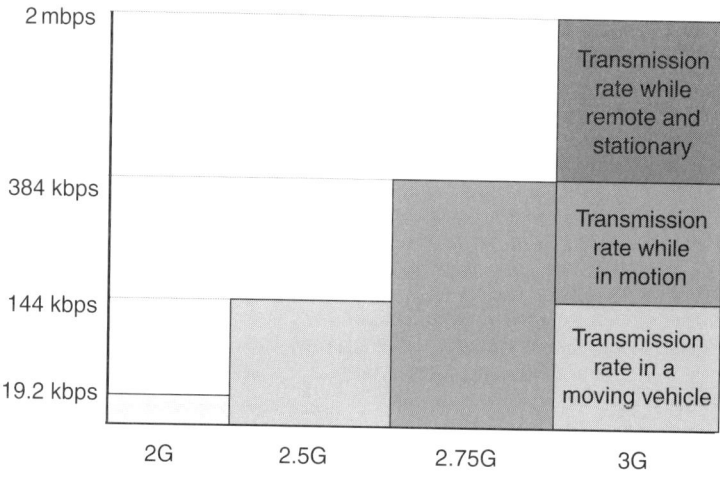

FIGURE 11.12

Next Generation Cellular Networks

Paging Networks   Paging data networks allow for two-way wireless messaging services. These services are limited to short messages including e-mails without attachments, alerts, and the delivery of stock quotes and other time sensitive information. Companies like SkyTel and Metrocall own and operate paging networks.

Paging networks offer perhaps the best in-building performance of all the WANs; however, limitations on coverage, responsiveness, and throughput make them unsuitable for running data-intensive mobile applications. Moreover, while they perform well in their target area—short, two-way messaging—they are facing increasingly stiff competition from services offered by the telecommunications carriers. In the long run, once 2.5G and 3G services are in place, the market for paging data services is expected to contract severely.

Satellite Networks   Satellite networks offer data services as well as one-way and two-way paging services. Companies like Globalstar, Inmarsat, Iridium, TMI, and Motient offer satellite data services in the U.S.

Satellite networks are the only choice for coverage in remote or rural areas, but typically lack extensive urban coverage and operate poorly, if at all, inside buildings. Data transmission rates are in flux, but generally range from 9.6 kbps to 64 kbps, too slow for data-intensive mobile applications. To date, satellite data services have been aimed at industries with wide geographic reach such as oil and gas, trucking and transportation, and at low-bandwidth applications like tracking and monitoring.

Mobile devices using satellite networks need a lot of power and must have large antennae, which excludes most small, handheld devices. Satellite devices are frequently located within vehicles, where they can draw upon the abundant power supply of their host. For all of these reasons, it is unlikely that satellite operators will be able to provide enterprise-grade, two-way mobile data services at attractive prices anytime in the near future.

Nevertheless, satellite networks may be the only option for a company that needs to roll out a mobile application across a very wide geographic area. Penske Logistics, a trucking and shipping company, outfits its delivery trucks with wireless devices and applications so that truckers can report on their arrivals and departures in real time. Anderson Trucking Services relies on a GPS-enabled application and satellite networks to let customers track their loads in real time as they are hauled on Anderson's trucks. The Federal Aviation Authority (FAA) and airlines are pondering proposals by Iridium and others to use satellites to receive real-time flight data from in-transit airplanes, and to transmit the data to ground control stations.[7]

---

7. "New Technology May Make 'Black Boxes' Obsolete," Timothy C. Barman, *WirelessNewsFactor.com*, November 5, 2001.

Various tracking, telematics, mapping, traffic, and weather applications rely on location-based information provided by Global Positioning System (GPS) locator devices working in conjunction with a satellite network operated by the U.S. Department of Defense. A variety of GPS adapters allow mobile devices, including handhelds and automobiles, to obtain location information from the satellites and pass it to applications that can intelligently interpret and leverage the information. Telemetry applications combine radio frequency identification (RFID) tags, GPS, and cellular technologies to allow airline freight and road transport businesses to track and locate individual packages in transit or within warehouse environments.[8]

Less interesting from a business perspective, but showing some of the promise of the technology, satellite operators now provide broadband Internet access to consumers. In a partnership with satellite operator Hughes Network Systems, Internet Service Provider EarthLink has begun offering consumers two-way satellite service for broadband Internet connections.[9]

11.4 Selecting a Wireless Network

Constructing a wireless solution does not begin with the selection of a wireless network. Rather, as explained in Chapters 4 and 5, it begins with recognizing an opportunity or identifying a business process that could benefit from application of wireless technologies. These opportunities and processes are investigated and, if they are worthy candidates, are translated into a set of requirements for a wireless application. These requirements are then used to drive the definition of a solution that includes components such as devices, networks, and infrastructure products.

11.4.1 Considerations When Selecting a Wireless Network

Application requirements clearly drive the selection of a wireless network. Network capabilities, in turn, influence application design. Often, features that are deemed "requirements" quickly become "optional" when real-life solution constraints are encountered. It is always helpful to differentiate "must have" features from "nice to have" features, if only in the back of your mind, to avoid forcing an inappropriate solution.

---

8. "Mobile Package Tracking Takes Off to Zap Theft Loss," Jay Wrolstad, *WirelessNewsFactor.com*, August 31, 2001.

9. "EarthLink Gives Satellite Broadband Service Another Boost," Dan McDonough, Jr., *WirelessNewsFactor.com*, October 9, 2001.

As a first step, take the application requirements derived in Chapter 5 and categorize those requirements according to the list of features outlined in Section 11.2. For example, estimate application bandwidth requirements, coverage, and the latency or timing of information updates. Second, if the desired area of coverage is local, discard purchased network service options. Conversely, if the desired area of coverage is wide, ignore the local, DIY types of wireless networks. Third, examine the network summary in the next section to get a quick handle on the pros and cons, and applicability, of the potential network. Fourth, once a target network has been identified, refer to the more detailed description contained in Section 11.3 above. Lastly, when a preferred network emerges, perform some additional research to see if planned network upgrades, newly announced standards, security issues, or supported products or devices fit with the anticipated current and future needs of your organization.

11.4.2 Wireless Network Summary

The data in Table 11.1 provides a quick synopsis of the pros and cons and applicability of various wireless network options. Sections 11.2 and 11.3 provide more detailed information. Please note that bandwidth figures and other characteristics for these networks are in flux as evolving standards are being introduced with some frequency. It is advisable to perform a quick check as to current performance capabilities, especially if the choice of network hinges on a particular quoted figure.

TABLE 11.1
Wireless Network Summary

Network Type	Pros	Cons	Applicability
Do It Yourself Wireless Networks			
Infrared (See Section 11.3.2)	mature technology inherently secure low cost low power no interference concerns point-to-point data exchange spontaneous; no setup required governed by standards body	line of sight requirement close proximity, 3 foot range cannot penetrate obstacles up to 4 mbps bandwidth devices must remain stationary impedes true mobility impractical for building-wide use	best for exchanging data with other devices best for in-room coverage where very close device proximity feasible not useful for data-intensive business apps suited for transfers of transactional information secure nature good for 2-party financial transactions works well for retail and point-of-sale transactions works well as a public WLAN solution (conference areas, libraries, etc.)
Bluetooth (See Section 11.3.3)	connects personal computing devices spontaneous; little setup required replaces cables and cradles omni-directional user and device can be mobile built in security low power easily embedded, small sized chips governed by standards body	short range, up to 30 feet 1 mbps bandwidth immature technology slow but growing adoption interference possible	best for synchronizing data between personal devices commonly used as cable replacement option to create WPAN best for in-room coverage good for exchanging data with others works well for low-bandwidth, data capture-type apps with prior setup, connects easily with pre-identified Bluetooth devices possible as WLAN solution, but 802.11x preferable complements medium-range 802.11x WLANs
802.11x WLAN (See Section 11.3.4)	superior range - hundreds of feet superior throughput - 11 to 54 mbps extends or replaces wired network alternative where wiring impractical allows wider roaming adds mobility to processes and users lower lifetime costs than wired network governed by standards body	range limited to local area initial investment higher than wired native security has weaknesses eavesdropping possible interference possible new standards create confusion	best for seamless access to wired network apps, data and functionality best for in-building or office park coverage allows creation of new mobile apps works well with data-intensive apps supplement native security if high-value or confidential data cost-effective option where pulling cables is expensive or infeasible complements short-range IR and Bluetooth solutions prevalent in office buildings, universities, hospitals, and building parks

Network Type	Pros	Cons	Applicability
Purchased Wireless Network Services			
Wireless Wide Area Networks (data and cellular) (See Section 11.3.5)	national urban range high bandwidth migration underway 2.5G bandwidth targets hit in 2002 secure medium excellent voice capabilities	low bandwidth (144 kbps with 2.5G) competing standards disconnections lead to integrity issues spotty rural coverage spotty in-building penetration poor latency, with exceptions varied pricing can add up	only choice for wide area coverage where maximum mobility required best coverage in urban areas outside performance generally more reliable than in-building not suited for data-intensive apps; best for short text and bursty data "push" capabilities improving with 2.5G but latency still an issue variable pricing favors low-bandwidth apps seamless coverage may require negotiating airtime with several carriers competing standards puts premium on network agnostic app design prevalent in transportation, field service, and distribution industries
Paging (See Section 11.3.5)	national range two-way messaging superior in-building penetration	low bandwidth competition from 2.5G and 3G network not extensible	best choice for two-way messaging apps best in-building penetration not suited for data-intensive apps not suited for long term reliance due to emerging 3G cellular networks
Satellite (See Section 11.3.5)	global range only choice for very rural or remote areas	very low bandwidth expensive high power large antennas outdoor use only spotty urban coverage	only choice for wide area coverage in remote areas good for GPS or other location-based apps cumbersome equipment not suited for small, handheld devices host device must have a large battery or power supply suited only for low-bandwidth apps that can operate outdoors requires high-value app or data to justify expense

TABLE 11.1 — Continued
Wireless Network Summary

11.5 Wireless Network Resources

Information on specific network technology and hardware is also located on the manufacturers' web sites, as referenced in Appendix A. In addition, please visit *www.justenoughwireless.com* for updates on wireless network topics. Other wireless network resources include:

www.wlana.com

www.irda.org

www.wirelesslan.com

www.bluetooth.com

www.bluetooth.org

www.ericsson.com/bluetooth

www.networkcomputing.com

www.computerworld.com   select the Mobile/Wireless Knowledge Center

www.mobilemarketplace.com

www.attws.com   AT&T Wireless

www.sprintpcs.com   Sprint PCS

www.nextel.com   Nextel

www.cingular.com   Cingular Wireless

www.verizonwireless.com   Verizon

www.voicestream.com   VoiceStream

Wireless Applications

As you may recall from Chapter 5, the four major components comprising a wireless solution are devices, networks, applications, and information architecture. Chapters 10 and 11 discussed devices and networks, respectively. In this chapter, we consider wireless applications and their closely related information architectures. As a caveat, this chapter does not explain *how* to develop wireless applications. Detailed technical books abound to assist you with that task. Rather, this chapter, like the others before it, aims to provide just enough detail to get you started, in this case with wireless applications.

After referring to them numerous times throughout this book, what exactly *are* wireless applications? In a nutshell, a wireless application consists of functionality, distributed to some extent between a client device such as a PDA, Internet-enabled phone, or laptop, and a server. Moreover, it relies on a wireless network connection to exchange data. These wireless networks use a variety of technologies including satellite, wide area cellular or data packet, medium-range radio frequency (WLAN), short-range radio frequency (Bluetooth), or short-range light waves (infrared).

Regardless of the specific network type, the hallmark of a wireless application is that it depends upon a wireless connection to accomplish some facet of its operations.

If the user is accessing the Internet from a handheld device, for example, then a constant network connection is required. With a field service application, technicians might use the network connection only intermittently when they want to update service records back at the home office.

Purely mobile applications, in contrast, do not rely on wireless network connections, even though they are operated remotely, far from wired offices. These applications reside locally on the client device and operate in disconnected or off-line modes only. Many data collection applications, such as those used by inspectors, meter readers, and delivery personnel, are prime examples of mobile applications. These applications communicate with the host or server periodically, through a wired connection like a dial-up telephone line or through a physically connected synching cradle. The data used by the application is loaded onto the device in the same manner, allowing the user to work independent of network connection. In the future, as network services become less expensive, coverage more pervasive, bandwidth more robust, and networks more reliable, many organizations will choose to wireless-enable these mobile applications.

The second aspect of any wireless application is the "application" itself. An application refers to some type or set of functionality enabled by software. Applications are incredibly diverse, covering manufacturing operations, sales, accounting, CRM, and ERP. For a real-life sampling of wireless applications, refer to Chapter 3.

The users of wireless applications are just as diverse as the applications themselves. They include corporations and their employees. Mobile professionals like doctors and lawyers, and even students, consumers, and machines. Users span vertical industries such as insurance, manufacturing, and transportation, and horizontal functions like field service and sales. The unique needs and demands of these users greatly influence and shape the strategy and design of the wireless application. For example, emergency room doctors cannot rely on mobile applications that use "outdated" information; they need to have a wireless application that receives time-sensitive data pushed to it at short intervals. Moreover, they don't have the time to navigate through long menus or lists, so the wireless application must present the information in a terse, readily digested manner.

When designing and developing wireless applications, companies must deal with a host of issues and considerations, both familiar and new. Issues like understanding user needs, following coding standards, figuring out which data sources to use, defining integration requirements, and designing for efficiency and scalability are similar no matter what type of application is at hand. But some design and development issues are unique to wireless applications, stemming from the fact that:

- Wireless and mobile form factors—PDAs, Internet-enabled phones, handheld devices, pagers, etc.—are fundamentally different from traditional desktop computers and more limited in terms of display sizes, data input mechanisms, memory, and processing power.

• The ways in which users interact with wireless applications are strikingly different from how they use PC-based applications.

• Wireless networks are manifested by low bandwidth, incomplete coverage, relatively high costs for data transmission, especially with satellites, and high session disconnect rates requiring clever, careful coding techniques.

The actual wireless application that your company chooses to deploy is decided in large part by its business objectives as discussed in Chapters 4 and 5. Bear in mind, however, that the selection of wireless application is critically important from a proof-of-concept perspective, a funding perspective, and a usability perspective. While opting for a grandiose, cutting edge wireless application on your first attempt may promise compelling competitive advantages, the approach is also fraught with risks. A wiser course of action may be to choose a simpler and more straightforward wireless application on the first try, to gain expertise and establish the foundation for using wireless technologies in your company.

This chapter provides an overview of the more important considerations that will affect the design and development of your wireless application. To begin, the chapter looks at some characteristics of the wireless application development environment. Next, it briefly explores some basic application design concepts, including application components and the partitioning of functionality among them. Specific application design considerations are then noted, from developing usage scenarios to designing screens and navigation, determining information requirements and writing the software. The chapter closes with a look at four separate development paths available to companies to build their wireless applications.

12.1 Application Development Issues _____

What should you expect when you develop applications in a wireless environment? For starters, relatively austere platforms when compared to the rich desktop computers, wired PCs, and even laptops found in the typical organization. The physical and operating limitations of most client devices necessitate a different development mindset, one focused on efficient resource utilization and small, compact coding techniques.

The development environment itself will consist of a mixture of standard development tools, specific kits provided by device manufacturers and special mobile platforms offered by middleware vendors. The quality of software development kits also varies, with some, such as Palm's, being quite robust and others more limited.

Vendor support is generally stronger for more widely deployed, horizontal devices like PDAs and phones than for niche, vertical devices with a smaller audience of developers. Solutions supporting more than one device type will have more complicated development environments with several types of tools and utilities. With increasing demand for enterprise-grade wireless applications, third-party wireless development tools are beginning to appear. These tools generally take the form of "mobile application platforms" that speed development and deployment of applications using components and routines created by the vendor. While they may shorten development cycles, these platforms lock companies into a particular provider, and often focus on specific types of application functionality such as SFA, CRM, or financial services.

When a developer sits down to create a wireless application, what kind of environment does he or she work in? What types of platforms and networks are used, and what are their physical and operating characteristics? What types of tools and utilities does the developer have at his or her disposal to create and test the application?

- *Device Platforms*   From any perspective, wireless client devices are much more austere than their wired counterparts. Although each new generation is more powerful than the last, the devices are still quite constrained when it comes to processing power, memory, battery power, screen size, input mechanisms, and overall usability. Furthermore, current generation networks have constraints that affect transmission speeds, message length, and connectivity. As a result, developers must hone programming skills and practices that address these limitations, especially resource-efficient coding techniques.

  Platforms may appear similar superficially, but have significant differences in how they act. For example, the familiar refrain, "write once, run anywhere," for Java is not necessarily true with wireless devices. In a study comparing the performance, speed, and memory consumption of several Java platforms, including PDAs and cell phones, Fourbit Group found significant variation among the platforms.[1] Choose platforms whose strengths match your application's needs.

- *Development Tools*   The development tools and utilities available to a developer vary greatly depending on the target device. The more widely used the device, the better the vendor support, tools, and utilities. With most platforms supporting popular languages such as C, C++, or Java, developers can often continue to use their existing development environments, such as Microsoft Developer Studio for example, to author the software.

  Vendors supplement these environments with Software Developer Kits (SDKs) consisting of utilities, testing tools, application programming interfaces (APIs), documentation, and even sample application source code to jumpstart coding

---

1. "Had Too Much Java?," Phillip Britt, *Field Force Automation*, February/March 2002.

efforts. Through vendor web sites, developers also have access to FAQs, developer discussion groups, and SDK updates.

The availability of mobile application development platforms is on the rise, but these are seldom offered by independent third parties. Rather, they tend to reflect the vendor's application development expertise and experience, and are geared toward areas like SFA, CRM, or ERP. As mentioned above, these platforms can speed development, but they also tie a company to a particular provider, a risky proposition with some of the smaller, early stage providers.

Programming is typically done on a desktop PC, and software is heavily peppered with references to vendor-supplied APIs. APIs provide functions and routines for applications to exploit or control the target device, for example, to use the timer, messaging capabilities, or user interface elements native to the device. Their goal is to simplify low-level tasks such as file system access, memory management, and multi-tasking. On the RIM BlackBerry, for example, APIs let a developer use the device's file system, message handling, and other network-related functions, and manipulate the user interface (screen, keypad, thumbwheel).

Once an application is coded, most SDKs provide an emulator to test and debug the application on the PC as opposed to the device itself. Full-scale testing of the application is deferred until the prototype and pilot phase, as discussed in Chapter 9.

Although many IT organizations are quite skilled at developing application software, few have real-world experience creating wireless applications. This lack of expertise, combined with the austerity of client devices and hodgepodge of development tools, creates a more volatile development environment. Consequently, development cycles may take longer than anticipated, coding may proceed in a more trial-and-error fashion, and many issues won't be discovered until a prototype is available to pilot in real-life situations.

Performing more than rudimentary testing early in the development cycle is difficult to accomplish. Although SDKs contain simulators that allow developers to "test" their code on the target device, they do not support end-to-end application testing. That level of testing is usually left until the beta or pilot stage. End-to-end testing of wireless applications is quite complex, involving a number of components. Furthermore, few tools exist to help in the effort, and it is difficult for organizations to envision and create sufficient scripts or scenarios to test wireless applications thoroughly. When a problem arises, tracking it down and resolving the root cause can be extremely tricky. Does the problem originate in the device itself? In the device-resident application code? In the network or gateway? In the application server software? In the data?

The wireless development environment may be constrained and evolving, but it is not insurmountable. As Chapter 3 amply demonstrates, many companies have managed to meet and master these development challenges to produce successful wireless applications. The key is to adopt a new frame of mind and a willingness to work within constraints rather than to let them become overwhelming.

12.2 Application Components

Building a wireless application is similar to building other types of applications. Your organization will need the supporting infrastructure, tools, and expertise to design and develop the application. Of those three elements, the one that is probably most lacking is expertise in wireless technologies, device platforms, and wireless networks, and familiarity with the issues and constraints affecting these items. Gaining an understanding of the building blocks that comprise a typical wireless solution, and options for distributing functionality across those components, is a good place to start.

At heart, wireless applications embody core client/server design principles. At one end is a client device, access device, or terminal. At the other end is a host or server. In between the two is some kind of data transport mechanism and gateways that translate data from one format or protocol into another so that it is usable by the recipient. Figure 12.1 illustrates this simplified architecture. Many vendors offer packaged solutions addressing both server and device components. For example, Research in Motion packages the handheld device, airtime over its partner's nationwide network, server software, and a SDK to assist developers in creating applications.

Every wireless application will blend some of the device and server components noted in the sections below. The types of components that are bundled into the solution depends greatly on what the solution is designed to accomplish, the organization's existing infrastructure and platforms, and the types of third-party packages used.

12.2.1 Device Components

Devices, and their associated software, are perhaps the most visible components of a wireless application. While the hardware itself is the facet that most excites users, it is the guts inside the hardware that application developers care most about. Three aspects of client devices are important to developers: the low-level operating system and routines that control the device, the application software that will run on the device, and the data that is stored or processed on the device.

- *Low-Level Operating System and Routines* Device manufacturers are responsible for the operating system code and routines, either creating them (in the case of Palm) or licensing them (in the case of Pocket PCs running

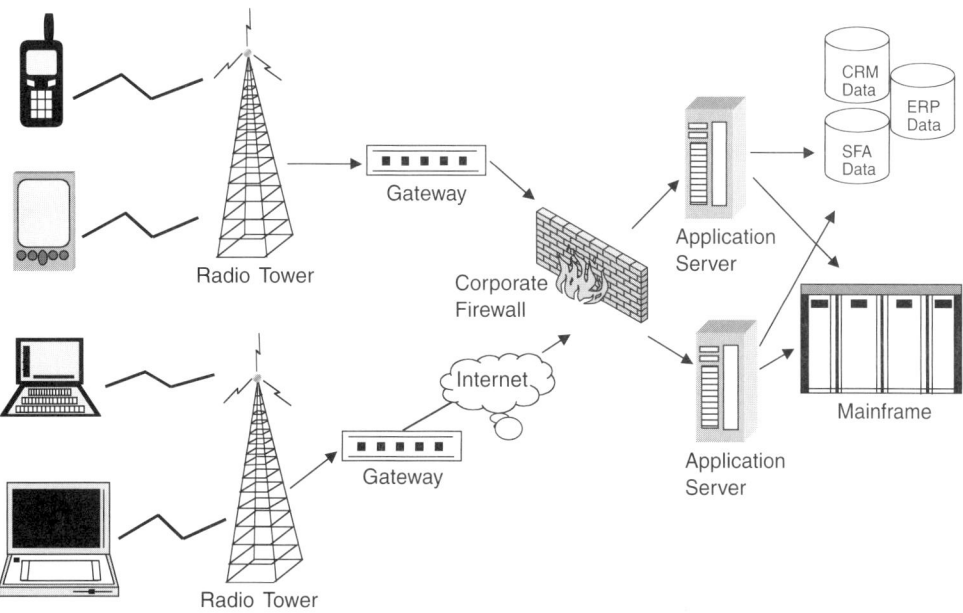

FIGURE 12.1
Simple Wireless Application Architecture

Microsoft's operating system). Developers access these routines by inserting APIs supplied by the device manufacturer in their application code. Using these routines, for example, developers can access the device's native file system, radio and messaging capabilities, and user interface elements. They can also interact with accessories attached to a serial port, such as a bar code scanner, printer, GPS navigator, or credit card reader.

- *Application Software*   On a client device, wireless applications are delivered in three basic ways: through the device's micro-browser, using a voice Internet application, or through a custom-developed, device-resident application.

Perhaps the quickest way to create a wireless application is to take advantage of the device's micro-browser. On cell phones, this means using the Wireless Application Protocol (WAP); on Palm devices, it means using the Palm Query Application (PQA); and on Pocket PCs, it means using Microsoft's Pocket Internet Explorer. These browsers use the native presentation capabilities of the device, web-based protocols and HTML, XML, or WML variants to supply application functionality. Similar to web-based development in a server environment, these browsers provide a quick-and-dirty way to put application functionality on a device.

Voice applications rely primarily on Voice XML, a markup language designed to support voice Internet applications. Using these applications, a user can access enterprise data sources from a client device, using voice commands, in hands-free mode. Despite vendor claims to the contrary, at the time of writing, Voice XML is an immature technology and problematic to implement. Most Voice XML applications are still in the experimental stage, and it will take some well publicized successes for it to move from the bleeding edge to wide deployment.

Custom applications take a bit longer, and cost a bit more, to develop. If users will work in disconnected mode, if the application relies on peripherals or attachments such as scanners and printers, or if significant amounts of data input are anticipated, then a custom application is generally the most appropriate solution. The application executes on the device, uses local data storage to some degree, and transmits data to or synchronizes with back-end systems at various points.

Multiple applications can run on a device at the same time, although only one is in the foreground at any single time. These applications are event-driven. Applications remain "dormant" until some event occurs (user presses a key, data is received, etc.) at which point they awaken. Applications that run in wireless mode must deal with events such as message received, message sent, signal level, network started, radio turned off, and others.

Device-resident applications also manipulate, through APIs, the user interface elements associated with the device to create custom screens, menus, dialog boxes, status boxes, and input fields.

• **Data**   Each device has a native file system and memory. To varying degrees, device-resident applications will manipulate files, write and read records, and store data on the device. Developers must fully understand the nuances of handling data on the device. For instance, if data is stored in persistent memory (ROM), it is not lost when the device is powered off or the battery runs out. Depending on the complexity of the data model, a separately licensed, device-resident database may be required. Once data is stored on a device, questions arise as to its retention period, latency, timing of uploads, and security. Applications that run in disconnected mode generally require some amount of data to be resident on the device to support the user.

12.2.2 Server Components

Servers are the workhorses in any wireless application architecture. Through software, application logic, and middleware, they are responsible for a range of activities including one or more of the following:

- Sending messages to, and receiving messages from, client devices using push/pull technologies

- Performing session management with clients

- Enforcing security through authentication, authorization, encryption, and decryption

- Interacting with web-based and legacy applications using various adapters

- Interacting with e-mail servers and other groupware programs

- Running application software

- Providing location-based services

- Providing a range of file and content management and delivery services, including formatting, transcoding, web re-purposing, and screen scraping

- Performing configuration management

- Managing and administering client devices

- Performing database and groupware synchronization

- Performing software distribution, back-ups, and restores

- Providing mobile platforms for the development and deployment of mobile and wireless applications, including software utilities, tools, etc.

The software, tools, and utilities that carry out these activities are often termed middleware. Middleware provides the services and business and presentation logic that links back-end solutions (web, legacy, etc.) across different networks (wired and wireless) to client devices. These services involve everything from synchronization to security, to location services, to transcoding and more. Some middleware solutions provide the entire gamut of services from nearly any type of back-end system to nearly any type of device. For example, AvantGo offers a product called the M-Business Server that gives various device platforms, including the RIM BlackBerry, Palm OS devices, and Pocket PCs, access to legacy business applications and data. It also provides connectivity to groupware products like Lotus Notes and Microsoft Exchange, performs authentication and security functions, allows for remote setup and administration of devices, and centralizes access control. Other types of middleware products offer only a portion of a solution, such as providing access to a single database, and some offer support for only particular types of back-end systems or device types.

Most middleware solutions claim to be device-agnostic and network-agnostic, supporting a variety of device and network types. In reality, some middleware solutions do require native, client applications, which waters down claims of device-independence. Even if a middleware product supports a broad range of devices, it will need to be updated as new device types appear.

Middleware products also differ in how they integrate with legacy or back-end applications, with some supporting many standards (COM, CORBA, ODBC, J2EE) and others centered on XML. They also vary between the amount and ease of customization permitted, programming languages or scripts supported, availability and usefulness of pre-defined templates, and architecture.

12.2.3 Partitioning Functionality Among Components

In every wireless application, functionality is distributed to some extent between client devices and host servers. On the server side, functionality handles incoming messages, manages sessions, routes requests, and transmits data back to remote devices. On the device side, functionality requests or pulls data from the server and presents it to the user, or sends data to the server to perform updates or transactions.

Determining how to divide functionality between server and device isn't always clear-cut. For example, when data is entered or collected on a handheld device and then transmitted to the server, where should verification or validation of the data take place? Should it be performed at the device level or at the server level? Performing verification at the device level seems to make the most sense. After all, why waste the time and expense of transmitting incorrect information to the server? On the other hand, situating the functionality on the client adds overhead to a device that may already face processing, memory, and battery life constraints.

When considering how to partition functionality between server and device, keep these points in mind.

- *Disconnected Versus Connected Mode*   If network coverage is lacking in a given area, do you want users to be able to work with the application? A purely mobile, non-wireless application avoids coverage issues by allowing the user to work only in off-line or disconnected mode, synchronizing data with the server at a convenient later time through a wired connection. In contrast, some wireless applications require a constant network connection to perform back-and-forth exchanges of data with a server and will not work absent such a connection. Applications that depend on a device-resident browser and server-supplied data, for example, work only when connected to a network. When connectivity lapses, the browser is effectively rendered useless.

  Other wireless applications fall somewhere between the two extremes, working in a combination of connected and disconnected modes. Some applications allow a user to perform work locally on the device and transmit data to a server either intermittently or whenever coverage is available. Many data collection applications work in this manner, capturing and verifying data using local software, storing routine data on the device for later synchronization, and transmitting exceptional conditions to the server as they arise. Some field service applications, such as the one used by Honeywell's ACS division, are designed to work in connected mode, transmitting data in near real time so long as coverage is available.

To deal with gaps in coverage, device and server software are ideally designed to store and forward messages as users move out of and into range. The suitability of this approach depends on the type of data being pushed or pulled, and whether a user is able to continue working without it. For example, if a sales rep using a wireless SFA tool wants to submit an order based upon the contents of a past customer order, but cannot retrieve the information from the server, then he is stuck. Using the application in disconnected mode doesn't help him with placing the order. Likewise, many messages and alerts contain critical or valuable information demanding immediate attention. If the information is delayed, then it loses its value.

Applications that work in disconnected mode depend, to some degree, on data stored locally on the device. While this approach may boost productivity, it heightens security concerns. Depending on the sensitivity of the data, it may be risky to store it on a device that is more easily lost or stolen compared to a desktop computer. And, as the complexity of data stored on the device increases, more sophisticated device-resident databases may be needed, raising the total cost of ownership per device.

- *Agnostic Versus Native Design*   As mentioned above, "agnostic" applications are free of specific device or network conventions, with the advantage of running across the widest possible range of device and network types. In contrast, native applications coded to run on a specific device, and exploiting device features through use of APIs, sacrifice portability for presumably richer functionality.

  Device-resident browsers are a classic example of design agnosticism. They are the antithesis of native applications. A wireless application relying on Microsoft's Pocket Internet Explorer is portable from Pocket PC device to Pocket PC device. The application uses only those features common across Pocket PCs as supported by Pocket Internet Explorer rather than the unique features of the Casio Cassiopeia, for example, or the Compaq iPAQ.

- *Airtime*   The way in which application functionality is partitioned also affects use of network airtime. When a WAN is used, depending on the carrier's pricing plan, charges may accrue based on the amount of data transmitted or airtime used. Minimizing these charges means minimizing the size and number of transmissions from client device to server. Compressing data, or selectively caching data on the device, minimizes the size of transmissions. Putting more self-contained functionality and data on the device minimizes the number of transmissions. Performing high-volume transmissions of routine, low latency data via wired rather than wireless connections is an additional way to reduce both the size and number of wireless transmissions.

- *License Costs*   Cost is an important consideration in how functionality is distributed between components. "Fatter" or more intelligent clients loaded with functionality and data have higher per device costs. They generally depend on third-party application software, databases, agents, and utilities with their associated license fees. Depending on the number of mobile users and client devices, these license costs may be significant.

12.3 Application Design and Development Considerations _____

Whenever a developer begins to work with a new programming language, a new platform or a new technology, he or she needs to become acquainted with a host of new issues and challenges, and will have to master a new set of techniques. Programming guides, product documentation, and training classes help lessen the learning curve. Also invaluable is an experienced co-worker or mentor to alert the novice about the subtleties and intracacies of the new environment. Like such a mentor, the purpose of this section is to alert the reader to certain salient issues involved in wireless application development. Developers are advised to use this section as a starting point, and to perform their own research into the particular items most relevant to their application and technologies.

Every organization will have its own approach to designing and developing wireless applications. This section does not prescribe any particular methodology or lifecycle. It does, however, organize kernels of useful information into five subsections that may or may not correspond to your organization's development phases, but will be useful nonetheless. The first subsection, "Usage Scenarios," describes how the mobile user experience will affect the ultimate design of the applications. The next subsection, "Screen Design," discusses factors affecting the presentation of information and user interactions on client devices. The third subsection, "Navigation," considers issues of screen flow and navigation for applications running on mobile devices. The fourth subsection, "Information Requirements," briefly touches on the data-related aspects of wireless application design. The last section, "Coding the Application," highlights some common issues that developers will need to address in writing their application code.

12.3.1 Usage Scenarios

As Chapter 5 demonstrates, mobile users encounter a variety of environmental, physical, and operational factors rarely present in a wired office situation. For example, whereas most office workers sit at a desktop PC, with fairly constant lighting and low levels of noise and distraction, mobile workers face a gamut of fluid and changing scenarios. A mobile user may stand or sit or even recline while using a wireless application. He or she may operate the device with one or two hands. Lighting conditions may vary from bright to dim, and locales may be noisy or quiet. The questionnaire and cheat sheets contained in the appendices highlight many of these user experience issues.

The conditions confronting mobile users, and the ways in which those users will actually use the application, profoundly affect the design and development of the solution. At a high level, user experience issues have great sway over the selection of devices and networks. At a lower level, they affect the design of the application itself. For example, users that will need to operate the application hands-free while driving may require speech recognition software as part of the solution. A user working in variable lighting conditions will require appropriate font sizes and colors to improve visibility. Users distributed across the world will require an application with multiple language support.

Knowing how users work helps developers optimize interactions with the application, and minimize frustrations with device and network limitations. By studying users, for example, a design team will discover the order in which users generally perform their work, and can give preferential placement to the most commonly used screens or selections. ArcStream Solutions, a wireless system integrator, organizes the functions contained in its mobile sales workbench offering, for example, according to the workflow of a sales rep: pre-visit, visit and post-visit activities. There are many other ways to organize and streamline applications to correspond to user needs and preferences. Combining location-based technologies (i.e., where is the user?) with user preferences and profiles, as shown in Figure 12.2, allows applications to tailor the amount of information displayed to the user, strip out geographically irrelevant details, and order the results based on locale. For example, a salesperson in California might prefer to see customer lists organized geographically, from West Coast to East Coast. A field service technician in Quebec, Canada might want menu choices presented in French, while a technician in Toronto might prefer English.

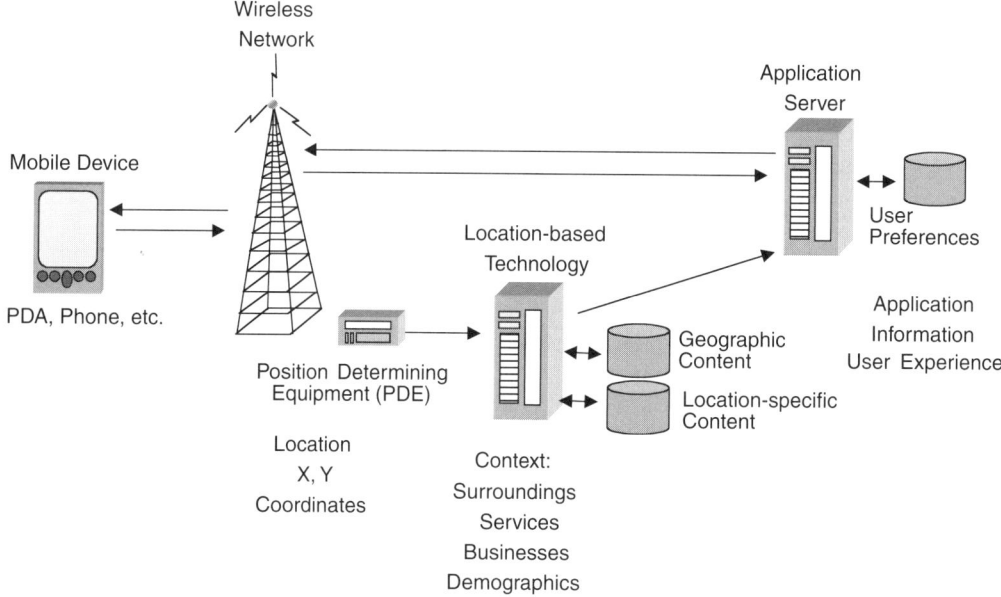

FIGURE 12.2
Location-Based Architecture

12.3.2 Screen Design

Having fleshed out usage scenarios, developers can map out screens to correspond to each one. A mobile sales application like the one offered by ArcStream Solutions, for example, may have a dozen usage scenarios that break down into pre-visit activities (reviewing account background information, product literature, etc.), visit activities (placing orders, looking up order status, etc.), and post-visit activities (writing follow-up letters, updating sales forecasts, expense reporting). Each of these activities will have a set of screens associated with it. Organizing the flow of these screens is addressed in the next section, Navigation.

Screen design determines how information is presented, what interactions are required with users, and the types and formats of data items.

- *Presentation*   The visual presentation of information on mobile devices is challenging, mostly because of the small screen sizes. Many phones are able to display only three to ten lines of twelve to twenty characters apiece. Many PDAs have only ten to eighteen lines with thirty to sixty characters per line. With screen real estate at a premium, developers must carefully plot out the placement of screen elements, and exploit every line and pixel. Size constraints also call for boiling down information to only what is essential and stripping out extraneous matter.

Presentation involves the selection of visual elements such as text or graphics, choosing attributes such as font type, size, and color, determining the placement of these elements on the screen, and ensuring stylistic consistency among all screens. When determining screen layout, select an orientation—horizontal or vertical—that best suits the physical dimensions of the screen. Conserve screen space and avoid waste; try to left-justify and single space lines of text. Opt for terse yet meaningful phrasing. Use consistent, recognizable icons, and buttons. Limit the use of logos and graphics that are used for purely visual rather than functional purposes. Some devices, such as phones, may not support graphics and others may have such poor resolution that it's best to do without the images. If the screen is divided into different logical areas, stick with the theme throughout the application. Wherever possible, provide the user with a status line showing where they are in the application and other pertinent data such as time, signal strength, battery life, etc.

If your application will rely on a native device browser, research its presentation capabilities. Some earlier browsers presented information in an inelegant manner, dividing input fields into separate screens, for example, making it impossible to see earlier inputs or captions. Others did not show the content entered into input fields, leaving the user to guess whether the correct information was provided.

- *Interactions*   Users interact with applications and devices in multiple ways, entering data on keyboards, writing characters using a stylus, selecting menu items by touch, pushing buttons, and scrolling via thumbwheels. Your application should permit the user to interact in his or her chosen way, within reason. Unless the application absolutely depends on an accessory, such as a scanner, do not assume that the user has one on hand. Allow for alternate ways to capture and enter data. When supporting multiple device types, know that a button or key on one device may not exist on another.

Most wireless devices, with the exclusion of laptops, have constrained data input capabilities. Carefully consider how much data the user must enter and via what means. Generally, it is better to present users with a list of selections and let them choose one, than requiring them to enter a string of characters or press a bunch of buttons. For example, do not force the user to type in the current date, present a list of dates for selection. This approach favors forms, checklists, and menus to minimize typing. Try to limit menus to three or five items so that they fit on a single screen.

When users do enter data, preserve it. Do not make users re-enter the same data from screen to screen. If a user is forced to return to an earlier screen, re-display the previously entered data, do not make them re-enter everything from scratch.

12.3.3 Navigation

The limited screen sizes of mobile devices and the bandwidths of wireless networks affect user navigation in several ways. First, with tiny displays, it is difficult to present users with sufficient context to let them infer where they are in the flow of the application. Second, with little space to spare, the number of links and instructions per screen must be kept to a minimum. Third, the time it takes to format and transmit information to a mobile device may make it incredibly time-consuming for a user to travel through even a few screens. In this environment, good navigation techniques are essential.

Once the screens are designed, it is time to organize them. Logically related screens, by activity or function, are often grouped together into sets, and users are allowed to move directly from one screen to another by making a selection from a drop down menu. Minimizing the number of screens that a user must page through before arriving at the desired location is quite important. To accomplish this objective, consider the following techniques.

- *Prioritize Screens*   Present screens and tasks in the order most commonly performed and present menu items and links in descending order of popularity.

- *Consult User Preferences*   Allow users or groups of users to establish preferences or profiles that will help simplify navigation. For example, insurance claims adjusters located in cities might prefer to capture damage information first at an accident site before vehicles are towed away. An adjuster in rural areas might prefer workflow that begins with obtaining information from the insured.

- *Apply Context*   Use contextual information such as location or time to determine the correct order for links or screens and to eliminate meaningless clutter.

Status lines or dialog boxes are also valuable aids to tell the user where they are in an application and what they've accomplished so far. Equally important is to give users a way to escape or return to a known starting point. MAIN, HOME, and BACK buttons typically provide this capability. With network coverage spotty in certain areas, the likelihood that the user will get interrupted or disconnected is high. To allow the user to resume operations, the application should supply sufficient context as to *where* the user is in the application or function, and *what* has been accomplished. Supply feedback ("message sent," "order submitted," etc.) to inform users about completed tasks or transactions so that they know what to do next.

12.3.4 Information Requirements

Every wireless application is dependent on information exchange. The kind and type of information is largely defined by the objectives of the application. An insurance claims adjustment application will obviously require information about the insured, the policy and repair estimates, and may generate information about vehicle damage and driving conditions. A field service application will deal with customer histories, repair records, parts inventory, and time and expense data.

Keep in mind that information may flow bi-directionally between client device and server. Data may originate in the field, entered by users, captured using scanners, or generated by device-resident applications. Data also originates at the host, contained in databases, files, web sites, back-end systems, messaging systems, or other content stores. Mapping information requirements is not simply a one-to-one proposition. Data may need to be combined, processed, calculated, summarized, or reconciled before a device or server can use it, operations typically performed by the application software or middleware.

The particular information requirements for your application will likely be quite varied depending on what the application is trying to accomplish, who will need the information and when, and where they will ask for it. Since it is impossible to define generic information requirements for every type of wireless application, consider the following items as they apply to your specific situation. In addition, many of the coding considerations covered in the next subsection are influenced by the information requirements of the application.

- *Type of Data Access*   Some data is "read only," intended purely for presentation. Client devices query the server for information, but do not process or update the data. E-mail messages are in this category as is much of the data presented via a micro-browser. Other data will be manipulated or processed, used to update records or to trigger transactions, calling for a more complicated data architecture and techniques to ensure the integrity and reliability of data. Processing a credit card transaction from a wireless point-of-sale terminal requires a high-availability, fault-tolerant architecture that can reliably and securely deliver data while ensuring its integrity. Transactions are performed by "ladder stepping," confirming and re-confirming each piece before moving on to the execution of the next piece. A wireless application that allows field workers to update a home office database will call for extra layers of database and record management routines. Conversely, a browser-based application that queries order status needs a much simpler architecture.

- *Data Format*   The client-side application will define acceptable data formats for presentation or display. Some data will fit these formats "as is" requiring no processing or manipulation. Other data will have to be re-formatted or re-purposed to match the device requirements. Browser-based client applications support different types of markup languages and server-side middleware (transcoders, re-purposers) must adjust and filter existing content to get it into a format acceptable to the device. These tools will extract unnecessary information, and consolidate and re-work large volumes of data to make it more compact and amenable to display on the small screens of client devices. Moreover, most of these tools are device and network agnostic, allowing them to work with whatever devices and networks your solution requires.

- *Data Latency*   All data has a useful "shelf life" after which time it is no longer useful or else is unreliable. Some data, such as stock prices or flight arrival and departure times, has a very short shelf life. Determine the latency of the data that will be used by your wireless application, and the ramifications of using out-of-date information, to figure out how often the data must be created or refreshed. For example, order status information accessed by salespeople in the field is useful if refreshed once or twice a day. Conversely, inventory data must be kept up-to-the-minute to allow a salesperson to promise shipment dates to customers and reserve quantities without conflicting with other salespeople. Application servers and back-end legacy systems are responsible for creating this data, however, your solution may require staging, consolidating, and integrating data from a variety of sources on a single server so that the data is available on demand to client devices. The latency of the data will determine the frequency with which the data stores on the server will need to be refreshed or repopulated.

- *Data Immediacy*   Closely related to latency is the concept of immediacy. How often and how quickly does data need to be exchanged? At a high level, immediacy affects the choice of network. To push updates to the field, and to transmit highly actionable information to the home office, it is best to use an "always on" network connection. But characterizing data by its immediacy also lets you limit the amount of data that will be wirelessly exchanged, avoiding network bandwidth limitations and, sometimes, minimizing airtime charges. Data that must be acted on or that is needed immediately is transmitted back and forth as soon as possible using a wireless connection. The rest of the data—routine or static information—will be stored and transmitted (either wirelessly or via a wired connection) at some later point in time. Client devices will need sufficient memory to store this data, and applications will have to call on the native file system or client databases to accomplish the task.

- *Data Synchronization*   If your wireless application will store any data locally on the client device, then synchronization issues arise. If the data will be uploaded to the server, how will it be integrated with existing data stores? Will it be added to them, replace them, or be merged with them? The same questions arise when downloading data from a server onto a client device. Depending on the complexity of your data model, the synchronization task can be complicated. In general, it is better to use one of the commercial synchronization products rather than to create your own routines. These products are quite powerful, and allow for individual customization.

12.3.5 Coding the Application

Behind the screens and navigational structure described above lies the actual code—the guts of the application. This code performs all of the operations and functions of the application, and it may be distributed between enterprise servers and client devices.

Wireless applications have two basic functions—displaying a screen and processing a "submitted" screen. Displaying a screen invokes presentation logic to determine how elements will be painted on a screen. A micro-browser or custom application residing on the device processes marked up or tagged data and displays it on the screen. But displaying a screen also includes a host of "behind the scenes" work to retrieve data, perform calculations, and integrate data sources. Before transmitting data to the device, it must be adjusted, filtered, and manipulated to put it into a format suitable for the device. This task is typically performed by transcoding or re-purposing middleware. To optimize the display, device profiles and user preferences may be consulted.

Most of the work to "process" a screen takes place on the server side. Processing a screen may be as simple as extracting and forwarding an e-mail message to performing a financial transaction or database update. As previously mentioned, a variety of systems and services including middleware, back-end applications, and messaging systems may be involved in servicing a request or transaction submitted from the field.

To perform these display and processing functions, wireless applications rely on standard programming techniques and principles, and some wireless-specific conventions and APIs. Given the constrained device and network environments, wireless applications must be resource-efficient. On the device side, applications must work within memory and processing power limitations, and conserve battery power. Programs must be compact and execute speedily. On the network side, applications must work within fairly tight bandwidths and spotty coverage. To varying degrees, wireless applications must deal with the following issues.

- *Ensuring Message Integrity* Many networks and devices limit the length of messages. If a message will exceed the prescribed length, an application can allow the message to be truncated or can shorten it in some manner. A message can be effectively shortened by breaking it into smaller chunks at the transmitting end and re-assembling it at the receiving end. Whenever a message is broken apart, integrity issues arise. Some networks have built-in mechanisms to check the integrity of messages, but developers may have to perform this task at the application-level. For example, the Motient network is responsible only for delivering a single packet of data safely to the device's radio. To ensure the integrity of an entire message, it is the responsibility of the application to check to see that the packets are intact and in the correct order.

- *Ensuring Transaction Integrity*   Even with the most reliable network, coverage lapses will occur. Well-designed applications will have the ability to store messages when the connection fails and re-try the network until coverage resumes. How severely coverage gaps will affect your application depends on what it is doing. If the application is performing browser-based lookups, then a disconnection is probably more of an annoyance than a problem. If, however, the application is in the middle of performing a database update, a disconnection could have serious ramifications unless precautions are taken. To deal with this threat to integrity, applications employ techniques like "laddering" where each "rung" or step of the transaction is confirmed and re-confirmed before proceeding to the next rung. A failure to confirm or re-confirm leads to a rollback of the transaction. Saving state information, or logging and auditing activities, are essential to resume and/or roll-back transactions. Database record locks must time out if sessions are not re-established. Incomplete transactions must be rolled back to prior states. In other cases, if coverage is not restored in a specified period of time, then the application should restart the user at some known place so that work can be resumed.

- *Ensuring Security*   Security is performed at the network, device, and application levels. Application-level security is usually employed where data is highly confidential, sensitive or valuable, and generally involves encrypting/decrypting data. There are many commercial grade cryptographic algorithms that wireless applications can use. Luckily, encrypting data does not appreciably increase the size of a transmission. If a device-resident application will either store or rely on data stored locally on the device, security risks are heightened. Mobile devices are easily lost or stolen, exposing this data to prying eyes. Storing data on the device is a tradeoff between user convenience, application performance, and security risks.

- *Coping with Bandwidth Limitations*   If the wireless application will exchange voluminous information between client devices in the field and the host server, it will likely run up against network bandwidth limitations. Although the bandwidth of data networks is improving, with current 2.5G networks operating at about the speed of a dial-up modem, it is still paltry compared to the transmission rates achieved over a wired or Ethernet connection. Until broadband wireless networks become reality, developers must use techniques that foster efficient data exchange. Limit the amount of data that will be transmitted by focusing application functionality on what users most need to know. As discussed in Chapter 3, Atlantic Envelope Company used the top ten queries that a customer was likely to ask to limit the amount of data exchanged. Use compression techniques where possible to further reduce message volume. Consider caching more data on the device as a way to avoid transmissions altogether, but be cognizant of latency requirements. Perform all high-volume data transfers using wired connections.

12.4 Application Approaches

Depending on the business objectives of the wireless project, and the variety of applications and content in an organization's existing portfolio, there are several different options for delivering functionality to mobile users. ArcStream Solutions, a wireless systems integrator based in Watertown, MA has created a helpful taxonomy of these options whose concepts are incorporated in the discussion below.[2]

First, if your company has existing back-end functionality, such as a CRM, SFA, or ERP system, that would serve mobile users well, consider leveraging that functionality and extending it to the field. Second, if your company has a web site or web content that it would like to offer to the mobile community, it is possible to do so relatively quickly using re-purposing middleware. Third, if your company intends to create a new wireless application, it must decide whether to extend the same functionality to wired users, which will complicate the development effort, or to restrict the functionality to the wireless community.

These approaches are not mutually exclusive, but may be combined to varying degrees. For example, a company may give its sales force access to CRM and SFA data contained in back-end systems by creating a mobile sales application running on a mobile server. At the same time, the company may give its mobile employees access to the company intranet from a variety of handheld devices using re-purposing middleware. Both approaches are viable, although they will require separate solutions.

12.4.1 Leverage Existing Back-End Functionality

Organizations often have existing application functionality—CRM, SFA, ERP, inventory, warehousing, etc.—that they would like to make available to workers using mobile devices. In this scenario, a new wireless application is not required per se. Rather, the organization can use the back-end application functionality already residing on its host servers and simply develop a separate path for delivering that functionality to mobile devices. This new path supplements, but does not replace, the existing path that delivers the functionality to users sitting at wired, desktop PCs.

---

2. "Wireless Application Architectures for Business: Key Implementation Considerations," Andrew Robertson, *Cutter IT Journal*, March 2001, Vol. 14, No. 3.

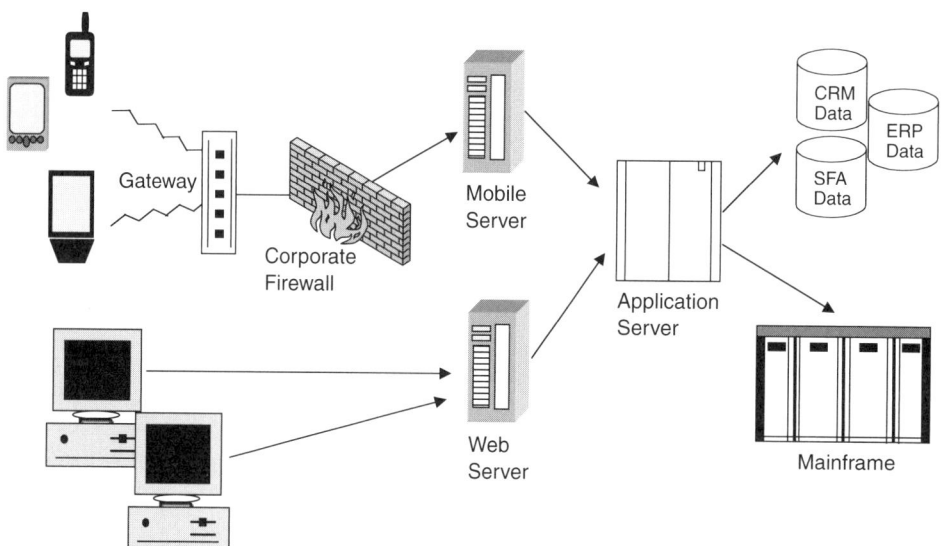

FIGURE 12.3
Leveraging Back-End Functionality

To extend functionality to mobile devices, an organization generally inserts a mobile application server, sitting between client devices and the servers containing the back-end application or business logic, as shown in Figure 12.3. This mobile server handles all interactions between client devices and back-end systems, including requests to access data, perform transactions, and format content.

If back-end systems are sufficient to meet mobile workers' needs, then extending that functionality to the field is a speedy implementation approach. It avoids the overhead of designing a new application from scratch, but still requires the design and development of code to sit between application servers and mobile users, and to present, format, and deliver content. The downside of this approach is that it relies on back-end functionality "as is." If mobile workers could benefit from a slight variation in that functionality, then this approach is not suitable, since it does not contemplate changes to existing systems.

12.4.2 Re-Purpose, Transcode, or Clip Web Content

Another quick way to move data and content out to client devices is to use existing web sites and applications housed on corporate servers, as shown in Figure 12.4. This approach has many names including web re-purposing, transcoding, or web clipping. The goal is to filter, adapt, and reformat web site content to make it suitable

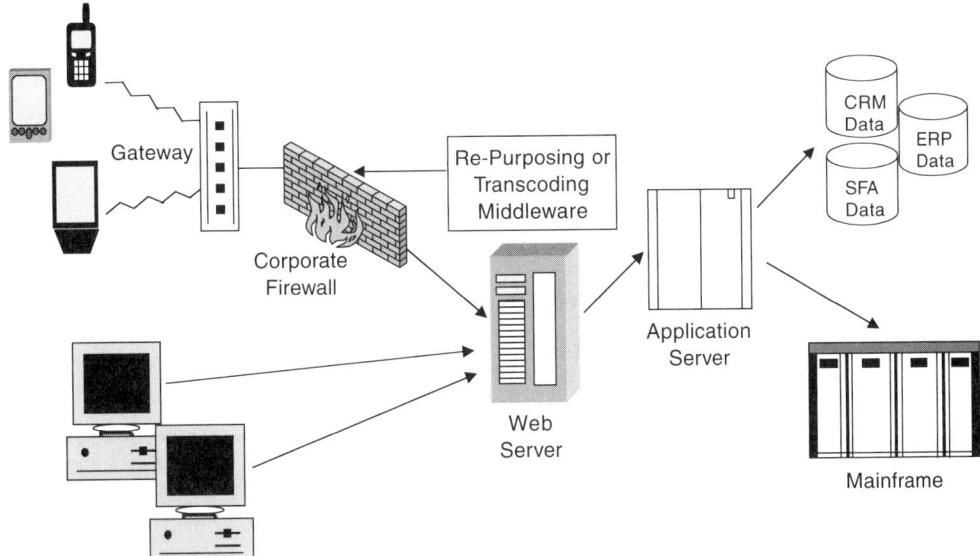

FIGURE 12.4
Web Re-Purposing

for presentation on a particular device type, over a particular network. To accomplish this goal, the HTML content comprising the existing web site is broken apart into segments and then converted into an alternative markup language, such as WML, HDML, Voice XML, XML, or iMode, that is supported by the device's resident browser or software. Elements that would not display well on the device, such as graphics and large-sized tables are stripped out. Images are either eliminated altogether, converted from one format to another (JPEG to GIF, for example), or re-scaled and adjusted to fit smaller screen sizes. Middleware products typically provide this functionality.

The advantage of web re-purposing is that it allows an organization to re-use existing work, deliver functionality to client devices quickly, and add support for new categories of devices relatively painlessly. It does not, however, offer the provision of entirely new wireless application functionality. Moreover, since it relies on an existing web structure and content, this approach cannot tailor or reconstruct web site flows or information to make them more responsive to unique mobile user needs. And, although supporting a new device type is purportedly as easy as plugging in a new device module, companies must still depend on their middleware providers to create these modules in a timely manner.

Companies can create static and dynamic re-purposed web sites. In addition to its regular HTML-based web site or intranet, a company can create alternate versions in markup languages supported by a range of handheld devices from cell phones to PDAs. Palm, for example, has its own web clipping service that allows Palm users to view web site content formatted in Palm OS HTML. Creating multiple, static versions of a web site is labor-intensive and adds to maintenance burdens, making it unsuitable as a long-term approach.

Middleware products such as IBM's WebSphere Transcoding product perform dynamic web re-purposing or transcoding. They convert HTML content on the fly into a format usable by the requesting device, eliminating the need to create a duplicate web site. IBM's WebSphere product contains "transcoders" corresponding to the different device types supported. These transcoders apply user, device, and network profiles to intelligently transform, adapt, and reformat content expressly for the requesting device. IBM's product also comes with tools that allow users to create their own stylesheets to specify how content, such as XML documents, will be presented on the client device. The product has an External Annotation Editor that enables organizations to specify which portions of a web page to re-format and send to a device and which portions to ignore.

12.4.3 New Application Functionality

Sometimes, neither existing back-end systems nor web sites will provide the functionality needed by mobile users, requiring companies to create or purchase entirely new application functionality. Although wireless access is the impetus for the new functionality, companies can choose to make the functionality available to wired users as well. If both user communities are supported, then a dual access approach is needed. If only mobile users will be accommodated, then a dedicated approach is feasible.

Dual Access Approach   When designing a new application from scratch, an organization has the luxury of building in support for both wired and wireless users and devices *in the same application*. Separate deployment paths, as described in Section 12.4.1, are not necessary. The application will contain the business and presentation logic required for both wired and wireless access and will not require a separate mobile server to interact solely with wireless devices, as depicted in Figure 12.5. As an example, Apriva, a wireless middleware service provider, offers a solution called Apriva Talk for the point-of-sale industry. Merchants can perform point-of-sales transactions using wireless devices in the field or through a wired, web-based interface that emulates the functions of a traditional point-of-sales device.

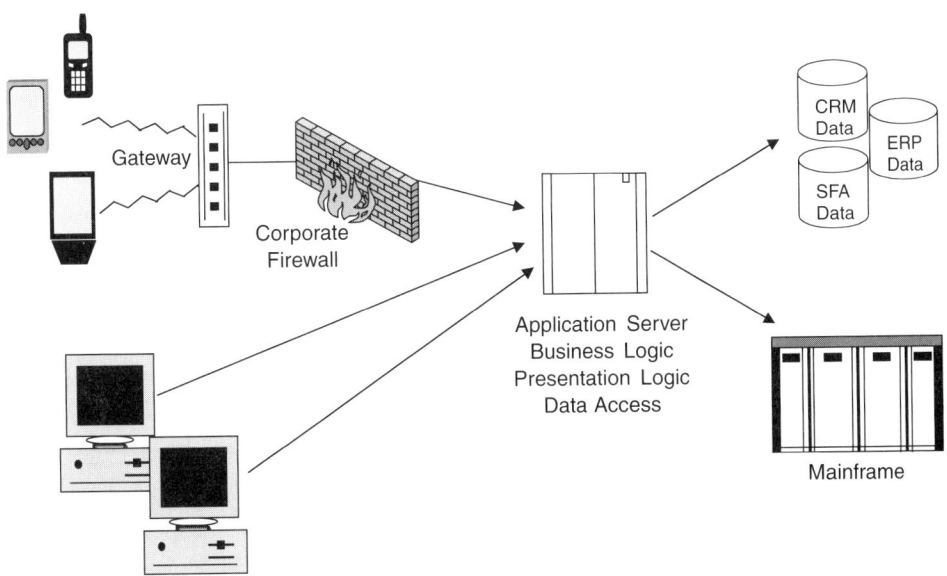

FIGURE 12.5
Dual Access Approach

A dual access approach enables maximum code re-use and avoids duplication, yet still permits customization of presentation and logic flow to suit the two classes of access. Centralizing functionality eases support burdens and prevents the divergence that inevitably occurs when multiple sets of code are maintained.

Dedicated Access Approach   A dedicated access approach is feasible if the goal is to support only mobile workers. These applications are developed expressly to exploit the native capabilities of the devices and networks used. On the host side, new applications and application servers are created to deliver the necessary functionality and integrate with legacy or back-end systems, as illustrated in Figure 12.6. Separate code is created for each device type supported. Both thick and thin clients are possible depending on how functionality is partitioned between client and server.

The advantage of native applications is that they deliver highly targeted solutions. But this advantage comes at a price. Since they are specific to device or network type, these applications are not amenable to re-use. They tend to be hard-coded, proprietary, and inflexible. Interfaces are more complicated and scalability is

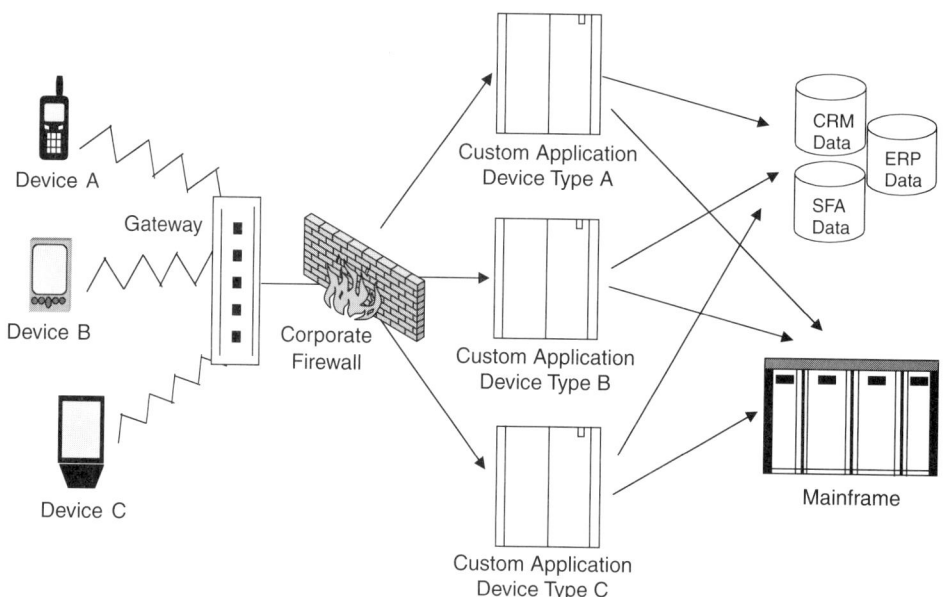

FIGURE 12.6
Dedicated Access Approach

limited. As new devices appear or as network technologies change, existing code must be modified and new code written. Multiple applications add support overhead, and allow divergent logic to develop over time.

Support

Take a peak inside a traditional wired computing environment—offices and data centers—and you'll find that supporting and managing users, software, and hardware is a well understood function, even a science. IT organizations have plenty of ammunition to help them in the task, from tools to procedures to training programs. Many of the support and management issues that arise are predictable, not show-stoppers. And with years of accumulated experience, IT organizations are more than prepared to deal with any unexpected issues that do crop up.

In mobile and wireless environments, the opposite is true. Surprises and new challenges abound. While it is true that support and management issues may be negligible for straightforward wireless solutions, such as giving executives e-mail access via RIM BlackBerry devices, they can also be quite acute for larger, more complicated wireless endeavors. If your company is implementing a groundbreaking solution, your best course of action is to start at ground zero and figure out how to deal with the technical, business, and human challenges that are apt to occur. Although many companies have managed to meet these challenges, as demonstrated in the examples cited in Chapter 3, they undoubtedly had to come up with creative solutions along the way.

What makes the wireless environment so unique? From a technical perspective, the technologies are relatively immature. They don't come with an assortment of independent, built-in development and support tools, battle-tested management and administrative software, detailed documentation, or even a base of wide industry experience. IT staffers may be relatively unfamiliar with wireless technologies and may never have used a handheld device before. Outsiders may have been engaged to develop and deploy wireless applications, with IT staff only peripherally involved.

Supporting and managing widely dispersed, mobile users, perhaps in multiple time zones and equipped with different devices, becomes complicated rather quickly. The range of problems that may arise is large and new. What do you tell a user whose handheld batteries have died in the middle of a client presentation? How do you restore data on a remote user's PDA after he has accidentally erased his files? The devices themselves may be quite diverse. Despite intentions to the contrary, devices will inevitably proliferate, presenting their own quirks and idiosyncrasies that must be mastered. Troubleshooting problems is tricky, given the number of components involved, some of which, like wide area networks, are outside the control of the IT organization.

Alongside these technical challenges are business and human challenges too. On the business side, what types of procedural and organizational changes are implicated by rolling out a wireless solution, and how do you accommodate those changes? Mobilizing a new set of users, altering the way that users perform their tasks, providing workers with access to new sources of information, or even asking them to perform entirely new functions like on-site billing and payment have significant repercussions. Companies must determine what policies, guidelines, and procedures are needed to address these job and workflow changes. If third parties are performing any related services, organizations must deal with contract negotiation and management issues, and perhaps proactively monitor the service provider's performance to service levels.

On the human front, a host of obvious and subtle issues will arise. Who will support mobile users and their wireless devices and applications? While the IT organization is the usual suspect, what happens when third parties are involved in creating the solution, managing devices, or supplying ongoing services? What is their involvement in supporting users once the solution is in production? Training users may be simple or complex depending on their locations, relative sophistication with using wireless technology, and openness to adapting their behaviors. Finally, the ways in which many wireless solutions are designed and deployed requires a greater level of user participation than is typically called for in a wired environment. Users may have to carry around accessories and spare batteries, have a heightened awareness of things like power consumption and battery life, perform security, synchronization or back-up tasks, and learn new methods of typing or data input. The amount of handholding, support, and management that users will require is often an unknown factor, yet is central to the success of the wireless solution.

This chapter highlights some of the support and management issues that your company can expect to encounter after it rolls out its wireless solution. Looking at things from a technical, business, and human perspective, the chapter will alert you to some of the more common challenges that will arise and offer guidance for addressing them. As mentioned previously, a limited but growing set of tools and software can help support and manage mobile and wireless solutions. In the meantime, companies may have to use a little ingenuity and homegrown tools to resolve problems that appear.

13.1 Technical Support Considerations _____

Wireless solutions present technical support and management issues, just like any other wired implementation. They also pose a separate set of headaches. Imagine hundreds of users working all over the place, in varied environmental conditions, with different levels of technical expertise, operating an unfamiliar device and application. Imagine that you have limited, if any, tools and resources to support and manage these users and that you are still relatively inexperienced dealing with the underlying technologies and infrastructure. Get the picture?

Distressing as this picture may appear, your company probably already has the technical support infrastructure in place that will handle your wireless solution. Coming up to speed may simply be a matter of training your resources in wireless issues and technologies. It could also mean contracting with a third party to help you provide the level of support that you need.

Although each organization may have its own way of classifying support and management functions, the following sections consider technical support issues from four angles as depicted in Figure 13.1: network management, device management, systems management, and change management.

13.1.1 Network Management

For purposes of this section, network management has two aspects: managing and supporting the network itself, and managing and supporting network access. Although these two areas are closely related, they concentrate on different things.

- *Managing and Supporting the Network* This function is concerned with the performance of the network, and characteristics like latency, coverage, bandwidth, and reliability. It examines how well the network performs under a range of

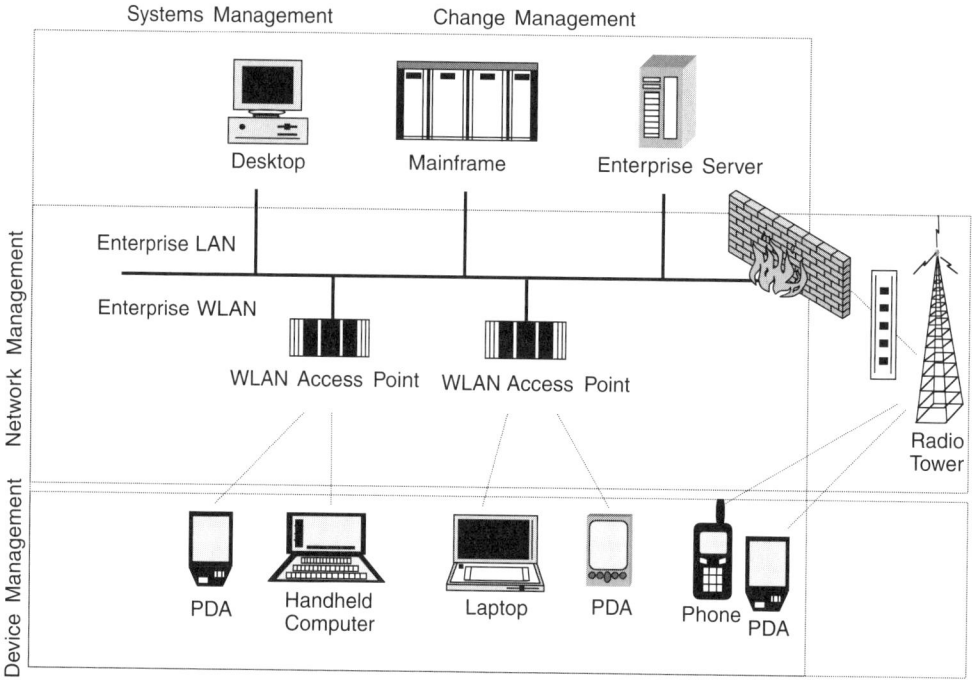

FIGURE 13.1
Technical Support

circumstances and makes adjustments to improve performance where necessary. It also deals with issues of capacity planning, load balancing, and throughput.

Depending on the type of wireless network that you are using, the network performance function can differ quite a bit. If your company purchases WAN services, the burden of supporting and managing the wireless network resides with your provider. Your role is to monitor network performance, either to a set of specific contractual service levels, or to the standard terms applicable to your deal. You will want to ensure that your users are obtaining acceptable or promised levels of coverage and bandwidth. You will also want to monitor usage levels (number of users, amount of data transferred, etc.) if the pricing of your deal is dependent on these factors. Your service provider will often have measurement programs in place to collect these performance metrics and reporting tools to analyze the data. If so, make sure that the tools and methodologies used are acceptable to your company. In some cases, depending on the importance of the wireless solution and the amount of money that you are willing to spend, you may want to purchase your own tools to verify performance data yourself.

If your company is using a Do It Yourself wireless network, the network management function will be much broader and encompass all support and maintenance tasks (unless these have been outsourced to a third party). In that case, you will be responsible for:

- Determining the network architecture
- Installing and maintaining the physical components of the wireless network (transceivers, access points, etc.)
- Monitoring and tuning the network to provide acceptable levels of service to end users (coverage and bandwidth)
- Performing capacity planning
- Physically securing the network (through optimal placement of components)
- Avoiding signal interference or network conflicts
- Verifying vendor claims (distances supported, throughput rates, etc.)

  For example, in a company using a WLAN, the network management group would be responsible for: mapping the placement of access points throughout the physical space; installing the equipment; monitoring the performance of the network and adding or moving access points as needed; adjusting the direction of antennas to prevent signal leakage outside the physical premises; perhaps installing network interface cards in computing equipment; and dealing with signal clashes between any Bluetooth-equipped devices and the WLAN.

- *Managing and Supporting Network Access*   Securing access to corporate networks is a number one priority for almost every company. Adding a wireless network—one that connects and offers an entree to your wired network—complicates that task. An intruder could theoretically gain access to your wireless network, and thereby get unfettered access to your wired network, if appropriate precautions are not put in place. Managing and supporting network access means constantly asking:

  - Who is getting on the network?
  - What are they trying to do?
  - Are they authorized to do so?

    To deal with these issues, the network management and support function performs tasks such as: setting up and administering network users; assigning IDs and passwords; registering and tracking end user devices; detecting and monitoring wireless end user devices connecting to the network; monitoring what network clients are attempting to do; and implementing security measures from installing security packages to activating built-in network security mechanisms.

Each type of wireless network has its own security strengths and weaknesses. Part of the job of the network management function is to understand what these issues are, and take proactive steps to minimize their risk. This task may include activating built-in security measures, purchasing and installing commercial security products, applying fixes, and establishing procedures to close security loopholes. Understanding the unique security issues and risks posed by all of the wireless devices used in your organization is also important for the network management group. For example, Bluetooth devices have reasonable built-in security features and pose fewer security risks, whereas 802.11x WLANs have had well-publicized security weaknesses in the past requiring organizations to be vigilant about applying fixes (128-bit versus default 40-bit encryption) and following the latest guidance proffered by the IEEE.

13.1.2 Device Management

Device management picks up where device deployment leaves off (see Chapter 7). Once the roll out of the devices and wireless applications has occurred, the device management function deals with monitoring, fixing, and securing devices, among other things. Device management also deals with the issue of device evolution.

- *Monitoring Devices*   Ideally, a company would like to be able to monitor its wireless devices in much the same way that it monitors its fixed, wired computing assets. Monitoring allows an organization to detect things about the hardware and software "states" of a device, check up on device health, and see what's happening with the operating environment. If something appears out of the norm, the organization can take responsive or corrective action. Monitoring a device assumes that there is a way for the host to "contact" the device (or vice versa) to share data.

  A company may want to monitor several aspects of a device, and the list below provides a few examples. For instance, a device that keeps having "out of memory" problems may need to be replaced with a more powerful unit, or may indicate that the user has installed unsanctioned applications on the device. Note that many of the vendors offering support and management tools claim to address some, or all, of these areas.

  - Versions of software (operating system and applications) running on the device
  - Programs and software loaded on the device, including any unauthorized software
  - Performance issues (out of memory, power problems, etc.)

- Sign-on passwords enabled
- Virus scanning enabled
- Record of log-ins

- *Fixing Devices*   Devices may fail or malfunction in the field due to breakage or other physical problems, or because of software problems. Your support group may be able to fix a software glitch over the phone or send updates to the device as described below in Section 13.1.4. But if a device's software becomes massively messed up, or if the device is broken, the user may need to send it in for repair or replacement.

  How will your company go about repairing and replacing devices? Is it necessary to send a repairperson on site to apply a fix, such as when a sensor breaks in a piece of HVAC equipment? Or can the user bring or send the device in for repair? If your central support organization is performing the device management function, will it serve as a repair depot to receive broken devices and distribute replacements? Having end users send devices directly to outside repair centers could pose problems if sensitive data is contained on the device. How big a pool of "spare" devices do you need on hand to cover anticipated replacement needs? The size of the pool may be a straight percentage of units in the field (e.g., 5% of units deployed), or may be nil if you will use a just-in-time replacement philosophy. Replacing a device may also require restoring data and programs, which assumes that a reasonably current back-up is available.

- *Securing Devices*   Securing devices is a combination of provisioning them with the proper software and hardware in the first instance, and ensuring that those items are activated and used regularly. The monitoring activities described above aim to ascertain how well, and whether, security components are functioning at the device level. As part of device management, your support organization may also maintain lists of user IDs and passwords, and ensure that users comply with procedures to change passwords. If the security of a device has been compromised—through theft or misplacement—the device management function will take steps to delete user IDs and revoke authorizations, and if possible, "push" commands to disable or lock the device and destroy all of its contents.

- *Dealing with Device Evolution*   Although your organization may start supporting and managing one device type, you'll inevitably end up supporting either multiple models of the same device (because of obsolescence or user preference) or multiple types of devices, from cell phones to pagers to e-mail appliances to PDAs. For each device in the stable, your support organization will need to educate itself about the features and characteristics of the device, understand its unique support requirements, and know what the best upgrade path is if a device becomes outdated or needs to be replaced.

13.1.3 Systems Management

The systems management function is concerned with the "soft" aspects of the wireless solution including systems, applications, and middleware. Normally, a wireless solution is composed of device-resident systems and host or server-resident systems. On the host side, the systems management function generally operates the same whether the systems are "wireless" or not. Things like server back-ups and software updates are performed in the same way. When it comes to the wireless device, however, the system management function diverges. In this section, we'll focus on two aspects of wireless systems management: back-ups and security.

- *Back-ups*   The systems management function is responsible for performing regular back-ups of computing resources like servers and PCs. Extending this responsibility to cover wireless devices is essential. Assuming that the software and data contained on a wireless device have some value, then they will need to be backed up. When a device is lost or stolen, or data accidentally deleted, the only way to avoid a crisis is to have a recent back-up available.

  Relying on users to perform their own back-ups is a risky proposition. They will either fail to do it or do it so infrequently that the back-up will be outdated when needed. Users can still back-up select data individually using Compact-Flash cards, but for system back-ups, the task is best handled automatically and overseen by a central systems management group.

  The ideal way to back-up the contents of a device is to synchronize the device automatically with a server over some type of wireless network connection. A client agent periodically uploads data to the server, and a server program knows what to do with the data and where to store it. These back-ups are incremental to minimize the time and bandwidth needed to accomplish the task. Once on the server, the data is available for restores, and will be further backed up along with the contents of the server. If a wireless connection is not available or inconvenient, then users should synchronize their wireless devices frequently with their desktop PCs, which in turn will be backed up according to company policies. The downside of this approach is that it is voluntary, and therefore apt to be done only sporadically. Figure 13.2 illustrates three different back-up methods.

  The flip side of backing up is the ability to do a restore upon demand. Be sure to test your restore ability before a real need arises. In the wireless environment, you may need the ability to restore data to a different device type, in the event a device is lost and an exact replacement is not available. Ideally, you should be able to perform a remote restore, i.e., "push" the data, to reload the contents of a device in the field.

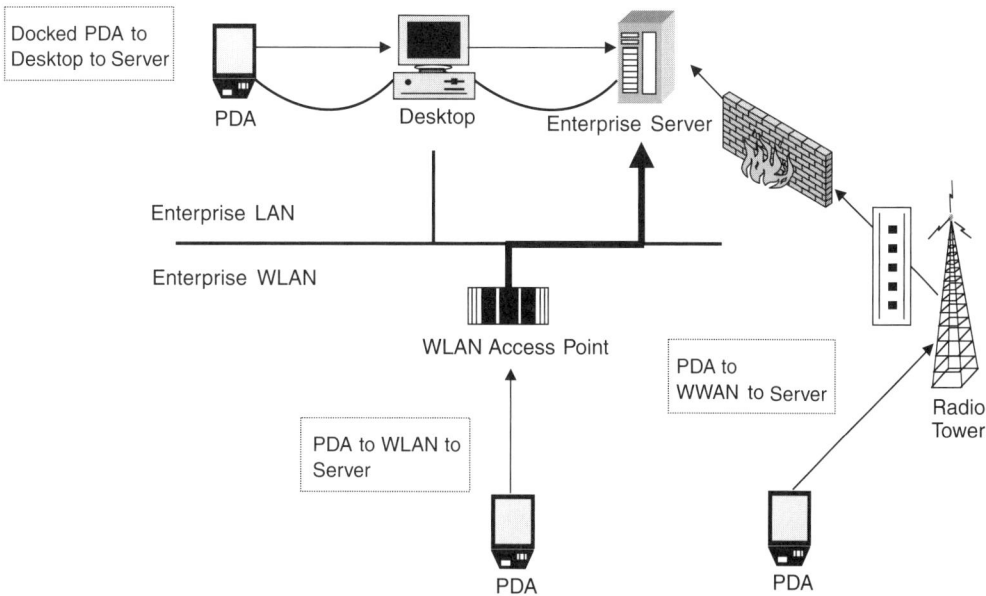

FIGURE 13.2
Backing Up Wireless Device Data

- *Security*   In a wireless solution, security is implemented at the network, device, application, and system levels. At the system level, security is enabled by installing and activating all of the appropriate security software on the wireless devices and back-end servers. Many wireless devices come with built-in security programs, which may or may not be stringent enough for your needs. Your organization may decide to supplement or completely replace these items with other, stronger security software commensurate with the sensitivity of the data involved in your solution. The package may include cryptographic software on devices and servers, setting up a virtual private network or simply applying patches regularly to fix known security holes in systems.

Besides securing systems from intrusion, your company will also want to prevent damage or corruption inflicted by viruses. You will need to install anti-virus software on your wireless devices or the desktop computers with which they synchronize, and keep it current and activated. Distributing virus scan updates to devices in the field is part of the responsibility of the change management function.

13.1.4 Change Management

The change management function is responsible for upgrading and updating data, software, programs, and content on devices and servers. As with systems management, the server-side aspects of change management do not materially differ for "wired" versus "wireless" components. For wireless devices, however, the change management function has some unique aspects.

Before selecting and deploying devices to the field, your company should consider how it will update, or keep current, the contents resident on the device. Once the devices have been distributed to users, it can be a hassle to implement new methods of change management. Let's consider a few different change management approaches.

- *Physical Retrieval*   In this approach, your central support organization takes physical possession of the devices to update their contents, apply fixes, etc. If your users are local and few in number, this approach is an option. The more widely dispersed your users or the more users you have, the more infeasible the approach becomes. The process of retrieving the devices, updating them, and returning them to users is unwieldy. The loss of productivity during the downtime can also become significant. If users will periodically congregate, however, at events like sales meetings, trade shows, or training sessions, then physically updating the devices may be possible.

- *Physical Synchronization*   Your users may be able to synchronize their wireless devices with their desktop PCs via a docking cradle. Assuming that the PC contains the current content for download to the device (or can get it from a server), the wireless device can be updated in this manner. Just as with backups, however, the success of this approach is entirely left to the user, who may synchronize less frequently than desired.

- *Automatic Provisioning of Changes*   The ideal way to accomplish change management is through an automatic approach that does not depend unilaterally on the user. For an automatic approach to work, a wireless network connection between host and client must exist and the host must be able to "push" updates out to the device as shown in Figure 13.3. In some cases, a device-resident agent will be needed to accomplish the task, software that should ideally be loaded onto the device before it is deployed in the field. Many support management tools have the capability of pushing updates to and configuring remote devices via a wireless connection, while others require a physical connection through a synching cradle. For example, as you may recall from Chapter 3, Southwest Gas updates the maps loaded in its repair truck terminals automatically overnight by pushing them over a WLAN that covers the parking lot. The ability to push information to devices is also the

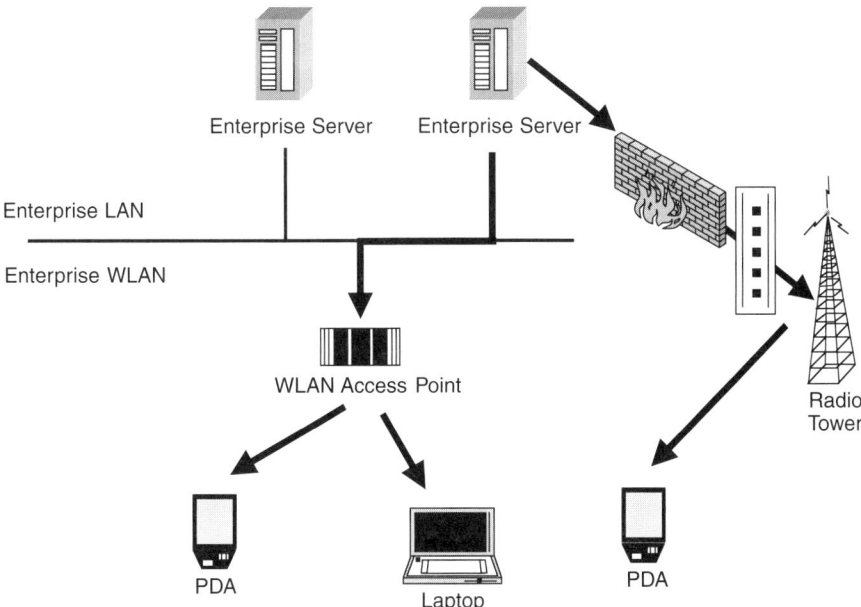

FIGURE 13.3
Pushing Changes

means by which companies can send commands to remove unauthorized software from a device, or lock the device and destroy contents in the event of loss or theft.

Don't forget that your ability to pipe updates to devices in the field will be affected by network bandwidth limitations and even time constraints. You don't want to tie up a user device with a contact list update during the middle of a client presentation, or try to send a 2 MB file update to a device over a WAN.

Having a viable approach to distributing updates is one thing; having software and programs that are amenable to updates is another. Clearly, items like contact lists, address files, and data files can be simply updated or replaced with a current version. Your ability to update software, however, depends on the vendor's approach to distributing fixes (e.g., is a patch available?) and the method of applying the fix. Most patches are installed in device RAM, and remain there until the device's batteries are removed or a hard reset occurs at which point they disappear. Keeping track of patches and re-applying them if they are lost is a tedious process. The ability to update operating system software depends on whether the device is equipped with flash memory (ROM). Some device manufacturers offer tools that allow operating systems to be upgraded by overwriting flash ROM. If a device does not have flash ROM, or if a company is not inclined to perform the necessary upgrades itself, then the only way to obtain a new operating system is to purchase a new device. Make sure you understand what your options are for updating software installed on your devices.

13.1.5 Support and Management Tools

A year ago, companies rolling out wireless implementations had virtually no tools to support and manage the solution. That situation has changed, and several vendors now offer tools, consoles, reporting software, and more to manage and support the proliferation of wireless solutions and devices. These offerings are still evolving, however, and companies must investigate claimed capabilities to ensure that they offer the value promised.

Support and management tools are available from:

- Traditional synchronization vendors seeking to expand their market
- Mobile management vendors with offerings specifically aimed at mobile assets like PDAs, cell phones, laptops, etc.
- Desktop management vendors with roots in managing and supporting desktop computing assets
- Enterprise network and systems management vendors that want to add mobile and wireless support to their enterprise-class products

Specific references to solution providers appear in Appendix B.

Generally speaking, these products attempt to "discover" or manage the stable of wireless devices, provide systems management functions like back-ups, and provide change management functions like automatic distribution of software and updates, and remote configuration management. Most solutions are comprised of a device-resident component—a client agent—and server-resident software to push updates, retrieve and analyze information sent by the device, and submit commands to the device. Some products offer a console to monitor device conditions and create reports. Others offer varying levels of network management capabilities.

13.2 Business Support Considerations _____

If your company has ever rolled out a major new system—an ERP system, an e-commerce system, or a supply-chain management system, for example—then you are well aware of the business upheavals that accompany these technology leaps. Some systems will change the way people perform their jobs, alter established procedures, spawn new processes, or impact organizational structures. Many wireless solutions fall into this category. Mobilizing workers, giving them new tools and methods of performing their tasks, providing them with access to fuller information, or asking them to undertake entirely new activities has ramifications throughout your business. Giving workers wireless e-mail access may not be a business-altering proposition, but giving emergency room doctors a wireless patient whiteboard is.

To get a sense of the types of business issues that may be in store for your company, consider the following categories.

- *Process Changes*   When new classes of workers are mobilized and when existing mobile workers are empowered through better access to tools or information, business processes will inevitably change. Wireless applications can automate data collection tasks in the field, eliminating paperwork and the need for manual data entry back at headquarters. Field-collected data may also be more accurate and timely, improving and speeding decision-making based on the data. Wireless solutions may also shift process steps from the office out to the field. Asking field service workers to generate invoices and accept payment on the spot moves a formerly internal function—billing—into the field and eliminates process steps in the accounting department. Wireless solutions may also collapse process steps and streamline the delivery of service. Sending insurance claims adjusters into the field to process accidents and issue checks on the spot is a profound change in the way the insurance company does business.

- *Organizational Change*   Where process change is likely to occur, so too is organizational change. Organizational change may be structural or a shift in culture or attitude. Shifting more or new workers into the field is an obvious change to organizational structures. Outfitting your ice-making equipment with wireless monitoring applications means having to setup an internal organization to support the application—to receive, review, and take action on incoming data, to warn customers of potential problems.

  Wireless solutions may also impact the way your company, as an organization, delivers its services, responds to its customers, or treats its employees. Allowing customers to track the performance of your trucking fleet in real time through wireless technologies, as Penske Logistics does, makes your company more service-minded. It affects all aspects of your service delivery, makes your company more accountable to its customers than ever before, and signals a willingness to accept a higher level of risk for the perceived higher level of rewards. Relying on your field workers to generate invoices and accept payment requires a higher level of trust and confidence on behalf of your company, and a better understanding of your company's billing practices on behalf of your employees.

- *Policies and Guidelines*   New types of mobile working conditions, new gadgets to operate, new processes, and significant organizational changes cry out for corporate guidance. As discussed in more detail in Chapter 8, wireless solutions may require changes to existing policies and guidelines, or the creation of new ones, to give employees advice, handle new situations, and even avoid legal liabilities. For example, if repairworkers have access to dashboard-mounted devices, your company may need to establish policies for safe device usage (hands-free, voice-activated, etc.) while operating the vehicles. Using wireless technology to change the tasks performed by unionized field workers may impact union regulations. Disabled employees may require adaptive wireless devices or special software to operate wireless equipment.

- *Contract Negotiation and Management*   Most wireless solutions will require some level of outside assistance, whether it's purchasing airtime from WAN providers, contracting for custom system integration work, or outsourcing device support to a third party. Depending on the needs of your wireless solution, its value to your company, and the value of the deal to your provider, the associated agreements may be boilerplate or customized. Your company will need to have the internal expertise (or engage a third party) to negotiate these contracts and to manage them over their lifetimes. Managing a contract may entail monitoring a vendor's performance to service levels, overseeing milestones and deliverables, and enforcing penalty and reward clauses. Make sure you have the organizational structure in place, staffed with the right resources, to perform these tasks.

13.3 Human Support Considerations _____

Last, but not least, comes the human side of the wireless equation. This aspect primarily covers the users of your wireless application, but also includes the employees that will provide them with back-office support. Not only must your users be trained and ready to work with the wireless application, so too must your staff of support workers. You will need to have the right resources, with the right expertise, as part of your support crew to make the wireless application a success in the field. This section considers three facets of human support and management: training users, supporting users, and user responsibilities.

- *Training Users*   A basic premise is that every new system roll out will require some level of user training. Learning to use a new application and equipment, and performing new activities and process steps, takes some practice. Wireless solutions are no different in this respect. If your user community is composed of business professionals, they are likely to be more technically savvy than their predecessors, but the wireless environment may still be completely foreign to them. Worrying about wireless network coverage, synchronizing handheld devices with PCs, and learning about power consumption are things that are unlikely to be on the radar screen of your average worker.

What method will your company use to train its users in using the wireless solution? As Chapter 3 detailed, Atlantic Envelope Company took advantage of a sales meeting to roll out its wireless application and devices to all affected salespeople and conduct training at the same time. Network and devices partners were on hand to answer questions, and a trainer and facilitators performed a walk-through of the application. Depending on where your users are located, central hands-on training sessions may be the ideal way to demystify the wireless concept, shorten the learning curve, and make your users maximally productive from the outset.

Factors such as the level of sophistication of your users and their willingness to learn new technologies will also affect the choice of training method. Very sophisticated users, or users already conversant with wireless technologies may be able to rely on FAQs or user guides to familiarize themselves with the wireless application. Web seminars and device-resident demos are also ways to disseminate information to dispersed users.

Where you have a highly heterogeneous user community (consumers, for example), your training options may be limited. Web tutorials and application demos can help you disseminate usage tips and techniques, but you may also have to rely on very intuitive application interfaces and straightforward application designs to lessen the learning curve and drive up acceptance rates.

Once you have trained your users and they have some experience using the wireless application and device, consider collecting tips, techniques, and best practices from your user community, particularly from your power users, and making these generally available over the web or some type of knowledge management tool. For example, a user may have figured out a way of operating the device to preserve battery power, a tip that your other users might appreciate knowing. Similarly, best practices for keeping data secure and for performing back-ups would be helpful for all of your users.

- *Supporting Users*   To help your users make optimal use of their wireless devices and applications, setup, staff, and properly equip an organization to perform the support function. The size and responsibilities of this organization will depend on the characteristics of your wireless solution, but the following considerations will apply:
  - *Who Will Support the Wireless Application, Devices, and Users?*   Support and management of production systems is typically vested in IT or a related organization. Will you support your wireless solution in the same

manner, and should you? Can your existing support organization handle the volume and types of calls expected from users? Will your support organization serve as a central repair depot where employees can send broken or malfunctioning devices? What kind of inventory of replacement devices is needed? Depending on whether, and how much, third parties were involved in creating and deploying your solution, does it make sense to rely on them to support the solution in production? What kinds of agreements will you need? Will you eventually want to transfer support back to your IT organization, and how will you do so? If your IT organization will support the implementation, do you have the right resources and expertise available? What kind of training, preparations, organizational structure, and procedures do you need?

- *What Will They Support?*   Once you know who will perform the support and management function, you must determine where their area of responsibility starts and ends. What is the scope of the support function? Will you want to outsource any facet of support (e.g., user devices) to a third party? Will you support all types of user devices, including those purchased by users themselves, or just officially sanctioned ones? In which situations is it appropriate for users to contact vendors or service providers directly with support issues? Or do you prefer that the central support staff run interference on every issue? What happens if a business department funds the development of a wireless application using the services of a third party? Will the central support group cover that application?

- *What Support and Management Services are Required?*   Support and management services run the gamut from network management to systems management to change management. Within these categories, what types and level of service are required? If your company is using one of the Do It Yourself network solutions, then it will be responsible for end-to-end support and management of the network, assuming it hasn't engaged a third party to provide those services. On the other hand, if a WAN is used, then the carrier will be responsible for much of the support, management, and maintenance issues, and your company will have to focus on monitoring service levels and/or evaluating performance. Do you need an automated way to push updates to devices in the field, or is it easy to retrieve and re-load them? Knowing the types of support and management services needed allows you to create a shopping list for tools and software that meet your requirements.

- *User Responsibilities*   At a high level, your wireless implementation may require your users to change the way they perform their jobs, learn new process steps, or shift their mindset, as discussed in the preceding section. At the operational level, what types of responsibilities are you going to offload to your users? Will you require users to proactively back-up their own device data, a practice that is inherently risky? Will you rely on your users to voluntarily enable security features on their devices, download software updates, or even send broken devices to an outside center for repair? Will you ask your users to learn new handwriting recognition schemes? Purchase their own batteries? Carry around accessories?

Care must be taken in delegating responsibilities to your end users. The more responsibilities delegated, the more training, guidance, and support you can expect to provide. In addition, some responsibilities are inappropriate for users. Relying on users to voluntarily back-up data, secure their devices, or perform software updates is risky because they will either forget to do it or do it only intermittently. Before offloading a responsibility to a user, ask what the downside is if the task is not performed at all. If the risk is too high, then find some other way to perform the task.

Before burdening your users with unnecessary responsibilities, determine whether a simple change in application design, component selection, or internal support practices could avoid the issue entirely. An interface that uses forms or templates simplifies data input requirements, and frees users from having to learn a handwriting recognition scheme (and the attendant support headaches). Choosing a network with a high level of coverage in the user's geographic location translates into fewer support calls for "out of coverage" situations.

13.4 Support and Management Resources _____

For information on support and management tools, please visit the solution providers' web sites referenced in Appendix B. In addition, please visit *www.justenough-wireless.com* for periodic updates on support and management issues.

Appendix A
Wireless Business Requirements Questionnaire

This questionnaire is designed to support the Five W's approach described in Chapter 4. This approach was developed to addresses the nuances of capturing business requirements for a wireless solution. It organizes questions into five categories: Why, Who, What, When, and Where, to consider the unique characteristics that shape device, network, information architecture, and application decisions. Three additional categories appear at the end of the questionnaire to handle issues that cross the five W's. These categories are Security Considerations, Implementation Considerations, and Cost Justification Considerations.

Why? _____

The first step towards defining a successful wireless solution is understanding why you want that solution in the first place. While the other 4 W's capture solution requirements, "Why" provides the mission statement for the overall effort. It defines the ultimate objective of the solution and becomes the screen for all decision-making. Keep going back to this objective when defining a solution. Whenever a question of direction arises, pick the path that best supports the solution's goal. For instance, if the main goal is to increase the sales force's responsiveness to customers, don't compromise that goal to improve performance tracking.

What are the goals of the wireless solution?

Select all that apply

- Generate revenues
- Increase efficiency
- Improve customer service
- Protect assets
- Curb losses

- Improve safety
- Expand market share/reach new markets
- Improve accuracy
- Other (list)

What is the potential ROI of the solution?

- How important is the wireless application?
- Can the potential value of the wireless application be quantified?
- Does this value justify the cost of the solution components under consideration?

What types of benefits are expected?

Estimate all that apply

- Tangible Benefits (Hard dollar benefits)
    - Annual increase in revenue
    - Annual savings
        - Labor costs
        - Error/loss reduction
        - Travel
        - Other (list)
    - Cost avoidance—(Savings over an alternate option)
- Semi-Tangible Benefits (Have financial benefits, but don't have current measures)
    - Improved safety
    - Fewer errors
    - Better use of people or equipment
    - More efficient operations
    - Faster access to information
    - Other (list)
- Intangible Benefits (Not quantifiable)
    - Better employee morale
    - Improved customer satisfaction
    - Gaining technology experience
    - Improved responsiveness
    - Enhanced corporate image

Who is going to fund the implementation?

- Will costs be shared among users?
- Will a single entity absorb the costs?
- What is the price sensitivity of the payer(s)?
- Is use of the solution optional or required?

What are the plans for future growth?

- Will this solution be a platform for additional functions and applications?
- How important is scalability?
- What kind of growth is anticipated?

Who? _____

This set of questions gathers information about the identities of users, their expected usage patterns including their relative levels of comfort and sophistication in using wireless devices, and their preferences and biases. These questions help to develop a profile of users, so that a device can be selected that meets the group's needs. For example, executives may not have the patience or inclination to work with a multi-featured device. Salespeople may avoid stylus-oriented devices for fear of losing the stylus while traveling. Price sensitivity may be an issue for users, like students, that must purchase their own devices.

Who will use the solution?

- Who will use the wireless solution?
  - A single individual or a group of users?
  - Executives? Salespeople? Field workers?
- Do these users have anything in common?
- Are there any social or cultural challenges that must be overcome to make the solution successful? *Examples: impact of device on their professional image, fear of "big brother" tracking movements, union work rules, etc.*

How do users currently do their work?

- What does the user do on a daily basis to perform his or her job? *Examples: Drive a car or truck? Visit clients? Sort packages on a ramp? Affix bar codes to boxes? Diagnose patients?*

- What actions do users perform to accomplish their jobs? *Examples: enter data, look up information, print invoices, etc.*
- How flexible are the process steps performed by the user?
    - Can these steps be conformed to specific application requirements?
    - Can the user adapt his or her behavior?
- What types of tools, equipment, and accessories do users currently use to perform these tasks?
- Does the user have the carrying capacity to take along accessories? Is carrying and attaching accessories more or less convenient than using a single, integrated device?

What is the users' level of technology experience?

- Are the users familiar/comfortable with technology in general?
- Are the users familiar/comfortable with mobile devices or wireless technologies?
- What are their biggest complaints about the technologies that they use?
- Are they willing to experiment with and learn new techniques and technologies, such as handwriting recognition schemes?
- Do users already own any devices? If so, which ones?
    - Do they use mobile phones?
    - Do they have PDAs?
    - Do they use laptops?
    - Other (list)
- Who owns, pays for, and supports these devices? Who will own the new devices?
- Do users have any preferences or biases for a particular device type?

Will the user face any physical constraints in using the application?

- How will users be physically situated while using the application? *Examples: Sitting down, standing, moving?*
- Are there any impediments to using the application on a mobile device?
    - Is hands-free or single-handed usage needed?
    - Do users already know how to type?
    - Are they already versed in handwriting recognition schemes?

- How do users anticipate using the application?
    - Will they be mobile?
    - What range of motion must be supported?
    - Are there any restrictions or limitations on mobility?
    - Will the user interact with other equipment while using the application?

Note: Additional environmental questions appear in the Where section.

What are the personal traits and preferences of the users?

- Given the user's anticipated range of activities, how does the user prefer to carry the device? Examples: On their person? In a hand, on a wrist or on a hip? In a pocket or briefcase? Stowed in a vehicle?
- Does the solution have to support multiple languages?
- Do users have a preferred mode of entering information? Examples: voice, keyboard, stylus.
- Do users have a preferred mode of receiving information? Examples: voice, display, printed.
- Is there value in having user customizable device display and navigation interfaces?

What will it take to support and manage users?

- Where are users typically located?
    - Are they dispersed?
    - Do they meet regularly in a single location?
- Is it feasible to retrieve devices from users on a regular basis?
- How experienced are the users with the planned devices, software, and their usage? What are their anticipated training and support needs?
- How will users be trained when the application and devices are rolled out?

How will the user deal with a loss of power?

Unlike wired desktop computers, battery life is an issue for every type of mobile device. Many factors affect battery life. The needs of the application (multimedia and wireless communications require a great deal of power), the expected amount of use (constant or intermittent), the power needs of accessories, availability of a power source to recharge, the downtime needed to recharge, the convenience of recharging and the availability of spare batteries. In most cases, it is difficult to gauge power needs beforehand, and users will have to learn by trial and error. However, if loss of battery power could cause significant harm or problems, then carefully consider battery options and risk mitigation strategies beforehand.

- Given anticipated usage patterns, how long does a device have to operate between charging periods?
- What are the ramifications of a loss of battery power? Will the application be used in front of a client? To track assets?
- What amount of recharging downtime is tolerable for the user? Is the user willing or able to carry spare, swappable batteries?
- Are sources of power convenient to the user's location?
- Is the user likely to store valuable data on the device? For how long? Will it be backed up?
- How high are the applications' anticipated power requirements? *Simple message-based applications use little power, display and calculation intensive applications use more, solutions that have to power long-range transmissions and other devices use the most.*

What? _____

What does the user of the wireless solution need to accomplish? The purpose of this query is to gain a user-centric understanding of the types of functions and features that the solution must provide to meet its objectives. This understanding feeds directly into many of the solution's design requirements. The questions in this subsection consider the services the wireless application will have to support, the characteristics of the information it will exchange, how that information must be processed, and the ways in which the user will interact with the application. The answers drive the solution's application design, information architecture, device selection, network selection, and its presentation/user experience design.

To determine what your wireless solution needs to accomplish, first learn how the desired activities are currently performed. Note the specific activities performed by the user and ask what they need to know to perform their actions. Pay close attention to the nuances of how they work and the constraints that they face. Actual observation is the best way to avoid mistaken perceptions.

What is the purpose of the wireless application?

- Describe the primary purpose of the application.
- What types of functions does it have to accomplish?
    - Communicate
    - Access information
    - Collect data
    - Update information
    - Transact *What type of work does the user have to perform on the spot? Will the solution have to produce invoices, create proposals, calculate payments, submit orders, make reservations, issue tickets or perform other types of specialized processing?*
        - Is the integrity and accuracy of transactions paramount? Is it essential to ensure that data is not corrupted or lost during transmission?
        - Can a transaction be restarted, rolled back, or resumed if a disconnection occurs?
    - Monitor
    - Locate/track

What kinds of inputs and outputs does the application require?

- What volume of data is the application expected to produce and/or require?
    - Limited data capture of bar code information?
    - Transmission of stock quotes?
    - Complex database updates?
    - Simple form processing?
    - Large file uploads/downloads?
- How important are data input capabilities? Will users have to enter a large volume of data?
- Is the application interactive? What kinds of user inputs are required?
- Will the application depend on automatic data capture through sensors or tags?
- Will location-based contextual information be needed?

- Will the application predominantly be used for voice communications or data processing?
- What kinds of data will be presented to the user?
    - What kind of screen resolution and size is required?
    - Does the application require a color display?
    - Does the solution have to be visually appealing? *For example: will it be used in a board room or in front of customers?*

What other things might the user want to do?

- What types of additional functions or capabilities might the user want?
    - The ability to make voice recordings?
    - Scan bar code information?
    - Send and receive e-mails?
    - Perform navigation and location functions?
- Is Internet access necessary or desirable?

What kinds of additional applications are desirable?

- What kinds of bundled software, if any, are needed to support the application?
- Does the application have particular operating system requirements or need special utilities or middleware?
- What kinds of third-party software, if any, are desired? Are they limited to any particular platform(s)?

What kinds of processing requirements does the application have?

- How much processing power is the application likely to need?
- What other applications will be running simultaneously on the device?
- How much memory will the application need to run?
- What other applications will make demands on the device's memory?
- How much information is the user likely to store on the device?

When?

Providing 24x7, universal access to up-to-the-minute information is tough and sometimes impossible. Precisely because wireless technologies are constrained, you need to consider **when** users will need to have access to the wireless solution and to the underlying information. To understand these timing requirements fully, a two-part inquiry is needed.

First, knowing when users will typically use the wireless solution helps to determine the timing of various other events and aids in capacity planning. The hours of operation of back-end or supporting systems, the timing of back-ups and other maintenance and support tasks are affected by when the wireless application will be used. Application usage, especially concurrent usage, affects system load and capacity planning.

Second, people often tend to think that data must be one hundred percent current to be useful and reliable. In reality, up-to-the-second, accurate information is needed in only certain cases, usually where an irrevocable, life-altering or financial decision is made based upon the data. Understanding issues of immediacy, latency, synchronization, and integrity can help avoid costly over-engineering, while ensuring that essential needs are covered.

What hours of the day will users use the application?

- Will the application be used primarily during business hours, daytime hours, or at any time of the day or night?
- Does the application have to be available on weekends or holidays?

In what time zones are users located?

- Will users be located in the same or multiple time zones?
- What type of collaboration must be supported?

Will the application be used concurrently or intermittently?

- How many users will use the application at the same time?
- Will they be doing the same or different things?
- Will the application be used randomly, by different users? Or at predictable times by the same users?
- Will the application be used only periodically or at special events such as trade shows, registration periods, or high-volume shopping times?

How quickly must information be delivered? (Immediacy)

- When information is needed, how quickly must it be delivered?
- Does information need to be "pushed" to a user as soon as it is available, or can it wait until the user asks for it?
- Can information be stored and sent at a later, more convenient time?

How long does data remain valid? (Latency)

- How fresh and current does data have to be?
- Is data perishable?
- What is an acceptable "lag" time between data creation and dissemination?
- Is real-time access to information needed?
- How frequently must data be refreshed or repopulated for it to be reliable?

How many versions of the same data exist? (Synchronization)

- Where is information created?
- Where does it reside?
- How closely do sets of data residing in different places have to match?
- What effort must be expended to synchronize data?
- Can data simply be overwritten or does it have to be compared, matched, and merged?

How important is data quality? (Integrity)

- Is a transaction, and the data involved, so sensitive, valuable, or fleeting that it must be safeguarded against corruption, interruption, or loss?
- Can data be re-created, and at what effort or expense?
- Is it necessary to recover and resume interrupted transactions or can users start again from scratch without any adverse effects?

Where? _____

Knowing **where** the user expects to use the application has implications for network coverage and device characteristics. The network must have sufficient coverage to allow the user to operate the application in all anticipated locales. **Where** questions also gather useful information about the environment, which will influence the working as opposed to theoretical range of the wireless network and uncover possible reception and interference problems. For example, the range of an infrared network goes to zero if there are obstructions between communicating devices. Understanding the physical environment also helps delineate required device attributes such as durability, displays, data input mechanisms, and battery life. The physical surroundings determine how rugged the device needs to be, and whether it will be subject to adverse temperatures or conditions. It also relates to dexterity in using the device. A person wearing gloves may find it difficult to use a tiny embedded keyboard or a physical location may rule in or rule out voice or audio capabilities. Conditions like dampness, humidity, and dust call for a device with good sealing. The potential for drops requires a hardy device made from rugged materials. Extreme lighting conditions, or even dim conditions, will require special kinds of displays to enable the user to view the screen. The proximity of moving equipment, conflicting radio signals, obstructions, and metal walls or floors will also influence device characteristics such as cabling and network options.

Where do users intend to use the application?

- Consider:
    - Where is the activity performed now?
    - Where would you like the activity to be performed once the solution is available?
    - Where might the activity be performed in the future?
- Are users local and confined to the premises of a corporate office building, facility, school classroom?
- Are users local and confined to the immediate areas surrounding the building, facility or classroom, such as a parking lot, campus, dealership lot, etc.?
- Are users free to roam far from any building or facility?
    - How far? Locally; in a state; in a region; nationwide?
    - In what types of areas? Metropolitan? Rural? Remote?
    - Will users be stationary? In motion? Or in a moving vehicle?
    - Outside? Or inside any number and type of buildings?

If use is primarily local and within a company building, describe the premises.

- Are users located in individual offices?
- Will they congregate in certain places, like a conference room or library, when they use the application?

- Are anticipated places of use crowded with furniture, people or equipment?
- Is the space open, like a warehouse or factory floor?
- Are users near moving equipment?
- Do users need to move about while they use the application?
- What materials are used in the ceilings, walls, floors?
- Are there other devices or equipment operating within the space that use radio frequencies to communicate?

If users will roam, describe the expected locales.

- Will the user be in open space or surrounded by buildings?
- Will the application be used outside?
- Will the application be used inside buildings?
- In which specific areas will users roam?

What are the conditions like in the user's environment?

- What are the lighting conditions like? Do those conditions vary?
- What is the temperature?
- What is the noise level?
- Will the user be wearing any type of protective gear?
- What are the ground and air conditions like? Hot or cold? Hard surface (concrete or tile) or soft surface (carpet)?
- Will the device be subject to vibrations or shocks?
- Will the device be mounted in a vehicle?
- Is a convenient source of power available?
- Are there any distractions?
- Will the device be exposed to hazardous conditions such as toxic chemicals, contact with germs, spills, or other damaging materials?
- Will the user be near any moving equipment?

Security Considerations _____

Depending on the sensitivity of the data stored on a device or exchanged between a device and back-end system, security is an issue. Many devices come with rudimentary security, such as power-on user IDs and passwords. If data is confidential, sensitive, or highly valuable, then it is wise to supplement this base level of security with measures taken at the device, application, and network level. These measures may include cryptography, authorization, virtual private networks, biometrics, and more. Ensuring security takes vigilance and monitoring, and intelligent use of state-of-the-art tools, software, and techniques. Companies are also advised to adopt policies regarding secure usage of mobile devices, applications, and networks to limit their exposure.

How valuable, sensitive, or confidential is the data?

- Is the data exchanged between device and back-end systems confidential, sensitive, or highly valuable?
- Will any data be stored on the device?
- What is the potential downside if the data were lost, stolen, or intercepted?
- If security cannot be assured, is a wireless solution still an option?

What type of security already exists?

- What security measures come bundled with the device, application, or network?
  - Are these security mechanisms sufficient?
  - If not, can they be supplemented with commercial products or with tougher measures in the other layers?
- Are existing security measures sufficient to protect the wired network from potential intrusions through a wireless network?

Are commercial security products available?

- What methods and products are commercially available to protect sensitive data on a device or in transit, or to prevent unauthorized access to back-end resources?

Implementation Considerations _____

The following questions help to scope the implementation to determine resource needs, software and hardware platforms, and ongoing support and management requirements.

Is in-house expertise available to deploy, maintain, and support the solution?

- Who will perform solution development and deployment?
- Does the company have sufficient internal resources to handle the task?

- Does internal staff have sufficient training and skills?
- Are external resources needed to deploy and support the solution going forward?

Are standard platforms or preferred vendors already in place?

- Are there corporate policies regarding standards or future directions?
- Does the organization already have preferred device vendors or preferred carriers for wireless services?
- Are there preferred platforms or vendors in place?
    - What kinds of platforms or vendors are currently used by the organization?
    - Do they have associated mobile offerings?
- Do users already have mobile devices?

How will devices be supported and managed?

- Is commercial software to support and manage the devices available? How good is it?
- What is involved in upgrading an application? A device?
    - Can the operating system, software, accessories, etc. be upgraded independently, or do new devices have to be purchased?
    - Can upgrades be pushed to devices in the field?
- How difficult is it to develop custom applications for the device?
- Are development tools and/or a development environment available?
- Is software expertise generally available for the particular device platform?

Cost Justification Considerations _____

This set of questions is designed to gauge cost concerns and price sensitivity. A variety of suitable device and network options, with a range of features, may be available at different cost points. Knowing the correct ones to choose may hinge on the amount of money budgeted for the solution.

What is the overall budget for the project?

- How much money is likely to be available for:
    - Capital costs?
    - Software development/acquisition?

- Consulting fees?
- Internal labor costs?
- Ongoing fees and maintenance?

What are the estimates for initial and ongoing device costs?

- What is the per unit cost of candidate devices?
- How many users will need devices?
- What types of accessories will need to be purchased?
- What will it cost to support and manage these users and devices?
- What types of development tools, synching software, and middleware will need to be acquired?
- What are the options for acquiring these items (WASP, system integrator, etc.)?
- What will it cost to maintain the devices and related components? Can devices or components be individually upgraded?

What are the estimates for initial and ongoing network costs?

- What kind of network equipment (access points, routers, etc.) will need to be purchased, if any?
- How many devices will need upgrades to communicate with the network?
- What service and/or usage fees are associated with the network, if any?
- What will it cost to support and manage the network?
- What types of administrative, security, performance, and management tools will need to be acquired?

Information Infrastructure Cheat Sheet

Data Source

    Single Source _____

    Multiple Sources - Merge _____

    Multiple Source - Process _____

Type of Data Access *(select for each data source)*

    View Only - Host Data

    Update - Host Data

    View Only - Device Data

    Update - Device Data

Volume *(select for each data source)*

    Very Low *(Examples: Short Messages, Alerts)*

    Low *(Examples: text e-mails, Internet clippings, simple forms)*

    Moderate *(Examples: Transactions, complicated forms, e-mail attachments)*

    High *(Examples: Large files, program uploads and downloads)*

    Very High *(Examples: Video files, Videoconferencing)*

Format *(select for each data source)*

    As is   Style Sheet   Extract   Needs Processing

Latency

    Shelf Life of Data *(Estimate time)*_____

Immediacy

    Device to host

        Data must be sent immediately

        Data can be sent periodically *(approximate timing)*_____

    Host to device

        Data must be sent immediately

        Data can wait for user request

Data Integrity

    Not Important   Important   Critical

Synching Requirements

    One way - Device to Host   One way - Host to Device   Two way

Security

    Importance *(check highest applicable need)*

        Not Important

        Low - *data has little value to outsiders*

        Moderate - *sensitive data*

        High - *highly confidential or valuable data*

    Exposure *(check all appropriate areas of risk)*

        Authorization

        Transmission

        Caching/Storage on Device

Network Selection Cheat Sheet

Type *(check all applicable)*
>   Voice
>   Data

Coverage/Range Requirements *(check all applicable)*
>   Close Proximity (under 3 feet)
>   Within an Office
>   Within a Building
>   Campus or Office Complex
>   Regional
>   National - Urban
>   National - Full Coverage
>   International

Bandwidth Requirements *(check highest applicable need)*
>   Very Low *(Examples: Short Messages, Alerts)*
>   Low *(Examples: text e-mails, Internet clippings, simple forms)*
>   Moderate *(Examples: Transactions, complicated forms, e-mail attachments)*
>   High *(Examples: Large files, program uploads and downloads)*
>   Very High *(Examples: Video files, Videoconferencing)*

Latency *(check applicable in each category)*

User to Host
>   Access when possible
>   Access when convenient
>   Access on demand
>   Always connected
>   No fail, always connected

Host to User
>   Reach when possible
>   Reach on demand
>   Always connected
>   No fail, always connected

Reliability *(check highest applicable need)*
>   Minor importance - Can start over
>   Moderate - Must recover from point of interruption
>   Cannot fail

Security *(check highest applicable need)*
>   Not important
>   Low - data has little value to outsiders
>   Moderate - sensitive data
>   High - highly confidential or valuable data

Interoperability *(check all applicable)*
>   Not important - single platform; highly homogenous
>   Important - several platforms; some divergence
>   Highly Important - highly dispersed users; multiple platforms; highly heterogeneous

Cost *(Note approximate limits)*
>   Cost of installation tolerance _____
>   Cost of usage tolerance _____

Wireless Application Cheat Sheet

Functionality *(check all that apply)*

Communicate   Access Information
Transact   Collect Data
Monitor   Update
Locate/Track

Approach

Build   Extend

Source

Established (List)_____
Buy Package
Need to Research
Custom Build

Device/Network Support Strategy

Agnostic   Native

User Experience

Navigation *(circle appropriate level)*

[1] *Simple*   [2]   [3]   [4]   [5] *Power user*

Level of Interaction *(circle appropriate level)*

[1] *"Hold my hand"*   [2]   [3] *"I'll tell you"* [4]   [5] *Automatic*

Importance of Visual Presentation *(circle appropriate level)*

[1] *Not important*   [2]   [3]   [4]   [5] *Critical*

Interface

Languages Supported? *(List)*_____
Voice Recognition?
Support User Preferences

Not necessary   Important   Critical

Host Platform *(list characteristics)*

Hardware platforms_____
Operating systems_____
System Software *(security, network, etc.)*_____
Databases_____

Security

Authentication

Not necessary   Important   Critical

Authorization

Not necessary   Important   Critical

Device Selection Cheat Sheet

Voice

| None | Communications (telephone) | Voice processing (recording, speech recognition, etc.) |

Size/Weight/Portability

| In a hand | In a pocket or on a belt | In a briefcase | In a vehicle | Not an issue |

Display

Display Size

[1] *Single line of text*   [2]   [3]   [4]   [5] *Full screen display*

Resolution/Color

[1] *Text*   [2]   [3]   [4]   [5] *Color multi-media*

Performance in variable lighting

[1] *Office lighting*   [2]   [3]   [4]   [5] *Variable outdoor lighting*

Data Input

Manual
    Large volume - keyboard
    Small volume *(circle options)* Keyboard Keypad Touchscreen Handwriting recognition
Automated data capture

Processing Power

| Low | Medium | High |

Memory

| Low | Medium | High |

Durability

[1] *Office use only*   [2]   [3]   [4]   [5] *Rough outdoors use*

Battery Life - *Importance of long life*

[1] *Not important*   [2]   [3]   [4]   [5] *Essential*

Built-in Capabilities

Wireless communications (check all that apply)
    WLAN
    Bluetooth
    Infrared
    Wide-Area Networks
Special features
    Bar code scanning
    Credit card processing
    Voice recording
    Other *(list)*_____

Extensibility

[1] *Not important*   [2]   [3]   [4]   [5] *Essential*

Device Selection Cheat Sheet *(Continued)*

Bundled Software		
Not important	Nice-to-have	Specific Needs *(list)*

Availability of Third-Party Software		
Not important	Important as proof point	Specific Needs *(list)*

PIM		
Not important	Nice-to-have	Essential

Internet Access		
Not important	Limited access useful	Essential

E-mail Access		
Not important	Limited access useful	Essential

Cost *(Note approximate limits)*		
Not an issue	Important	Major concern

Approximate number of units required _____

Maximum acceptable cost per unit _____

Cost parameters (list) _____

Appendix B
Solution Providers Offering Wireless Functionality

The solution provider list presented in this appendix is offered as a service to readers. It includes examples of solution providers who offer the types of software, hardware, and services mentioned in the book. It by no means represents all of the wireless solution providers on the market. Inclusion of a provider does not imply an endorsement by the author, nor does exclusion imply any negative opinion.

This list is meant as a starting point for research. The author recommends that readers supplement this list with their own market research. The wireless technology market is highly volatile, and since starting this book, the author has seen many providers enter and exit the market, get acquired, change their position and offerings, and even rename themselves. Furthermore, changes in a provider's technology, management, or strategy can dramatically affect the attractiveness and viability of its solutions against its competition.

As long as warranted by reader interest, the book's associated web site, *www.justenoughwireless.com*, will maintain and periodically update an on-line version of this list. Solution providers who wish to be added, removed, or update their reference in this list should contact the author through the web site.

Device and Accessory Manufacturers

Casio

www.casio.com
Casio makes Pocket PC devices including the Cassiopeia line, and a variety of peripherals and accessories.

Compaq (merged with Hewlett-Packard)

www.compaq.com
Compaq makes Pocket PC devices, including its iPAQ line.

Ericsson

www.ericsson.com

Through an agreement with Sony and Flextronics, Ericsson markets mobile phone handsets and smart phones.

Fujitsu

www.ftxs.fujitsu.com

Fujitsu sells various user devices (especially pen tablets), including the iPAD and TeamPad models, and accessories.

Garmin

www.garmin.com

Garmin produces GPS navigation and communications products for the aviation and consumer markets.

Handspring

www.handspring.com

Handspring produces Palm OS-based handheld PDAs including the Visor line and the Treo Communicator.

Hewlett-Packard

www.hp.com

Hewlett-Packard makes Pocket PCs, including the Jornada line, and notebook computers including the Omnibook line.

IBM

www.ibm.com

IBM offers notebooks (ThinkPad) and handheld devices running both Palm (WorkPad) and Microsoft operating systems.

Intermec

www.intermec.com

Intermec sells ruggedized, intelligent handheld devices for automated data collection with RF capabilities aimed at vertical industries including manufacturing, warehousing, transportation, and health care.

Itronix

www.itronix.com

Itronix makes handheld PCs (Fex21), PDAs, and rugged notebook computers including the GoBook MAX line, all with integrated wireless communications capabilities. Aimed primarily at field workers.

Kyocera

www.kyocera.com

Kyocera produces mobile phone handsets and smart phones.

Lowrance

www.lowrance.com

Lowrance sells GPS devices and accessories.

Magellan

www.magellangps.com

Magellan makes a variety of handheld GPS devices, vehicle navigation systems, and GPS accessories.

Motorola

www.motorola.com

Motorola offers pagers, two-way radios, accessories, phone handsets, GPS products, embedded computer products, mobile data terminals (both handheld and in-vehicle), and a variety of public safety applications.

NEC

www.nec.com

NEC makes notebook computers and handheld devices.

Nokia

www.nokia.com

Nokia manufactures mobile phone handsets including the Communicator series of smart phones.

Palm

www.palm.com

Palm manufactures a variety of handheld PDA devices including its new i705 model, and a variety of accessories. A separate company develops the Palm operating system.

Panasonic

www.panasonic.com

Panasonic sells ruggedized and regular notebook computers, including the Toughbook line.

Psion

www.psion.com

Psion offers a variety of specialized handheld devices and PDAs; popular in Europe.

QUALCOMM

www.qualcomm.com

QUALCOMM offers closed-loop fleet management solutions that include satellite or digital communications capabilities, application software, handheld computers (OmniTRACS, OmniExpress), and in-vehicle units (MVPc).

Research in Motion (RIM)

www.rim.com

RIM makes wireless handheld devices, including the BlackBerry wireless e-mail solution. RIM sells its devices as part of a package that includes e-mail software, synchronization software and real time, and nationwide wireless network access.

Samsung

www.samsung.com

Samsung makes Windows CE compatible handhelds and organizers and mobile phones.

Sharp Electronics

www.sharp-usa.com
Sharp manufactures notebooks, handheld PCs, PDAs, and the Zaurus organizer.

Sierra Wireless

www.sierrawireless.com
Sierra Wireless makes the AirCard product line for wireless connectivity to wide area wireless networks.

Socket Communications

www.socketcom.com
Socket Communications provides a variety of cards for handheld devices and notebook computers, including WLAN Bluetooth, Ethernet, modem, and serial cards.

Sony

www.sony.com
Sony offers handheld devices, including the Clié and notebooks including the Vaio line.

Symbol Technologies

www.symbol.com
Symbol Technologies makes a variety of handheld terminals, pen computers, and wearable units with built in wireless communications capabilities. Strength is bar code scanning technology and automated data collection devices.

Think Outside

www.thinkoutside.com
Think Outside makes an attachable folding keyboard, the Stowaway, for PDAs.

Toshiba

www.toshiba.com
Toshiba is a leading maker of notebook computers, and also offers a handheld Pocket PC device.

Wide Area Network Carriers and Operators (Voice, Data, Paging, Satellite)

Aeris

www.aeris.net
Aeris provides wireless connectivity and control of remote devices (telemetry).

Arch Wireless

www.archwireless.com
Arch Wireless provides wireless two-way data services including e-mail and instant text messaging. The company was in the midst of a Chapter 11 reorganization at the time of writing.

AT&T Wireless

www.attwireless.com
Covering 98% of the U.S. population, AT&T Wireless offers wireless data, and voice services using GSM/GPRS and TDMA technology.

Cingular Wireless

www.cingular.com
A joint venture between SBC and BellSouth, Cingular Wireless provides wireless data services using TDMA and GSM technology to over 93% of the urban business population in the U.S.

Globalstar

www.globalstar.com
Globalstar offers satellite data services to track, monitor, and manage remote assets. The company is working through a Chapter 11 bankruptcy at time of writing.

Hughes Network Systems

www.hns.com
Hughes Network Systems provides broadband satellite network solutions, including Internet access services, for businesses and consumers.

Inmarsat

www.inmarsat.com

Inmarsat is a mobile satellite communications operator. Service is offered through resellers.

Iridium

www.iridium.com

Using satellites operated by Boeing, Iridium provides mobile satellite data and voice solutions.

Metrocall

www.metrocall.com

Metrocall offers two-way messaging services including SMS, e-mail, and paging.

Motient

www.motient.com

Motient provides wireless data services across America using its terrestrial and satellite network.

Nextel

www.nextel.com

Using iDEN technology developed by Motorola, Nextel provides wireless digital services to 182 of the top 200 markets in the U.S.

Orbcomm

www.orbcomm.com

Orbcomm offers satellite data services to track, monitor, and manage remote assets. Service is offered through resellers.

Skytel

www.skytel.com

A subsidiary of WorldCom, Skytel provides traditional paging, text messaging, interactive two-way messaging, and wireless telemetry services.

Sprint PCS

www.sprintpcs.com
Covering over 85% of the U.S., Sprint PCS offers wireless digital PCS services using CDMA technology.

Teledesic

www.teledesic.com
Teledesic is building a satellite-based network to offer broadband Internet access and expects to offer service through resellers starting in 2005.

Verizon Wireless

www.verizonwireless.com
Covering nearly 90% of the U.S. population, Verizon Wireless offers wireless digital cellular services using CDMA technology.

VoiceStream

www.voicestream.com
A subsidiary of Deutsche Telekom AG, VoiceStream offers wireless digital cellular services using GSM technology throughout most of the U.S.

Wide Area Network Infrastructure and Tool Providers

Ericsson

www.ericsson.com
Ericsson supplies infrastructure, terminals, applications, and consulting for 2G and 3G network technologies.

Lucent

www.lucent.com
Lucent's INS group supplies metro and long haul optical networks, core, edge and access networks, circuit-to-packet convergence solutions, and network management software for network service providers.

Motorola

www.motorola.com
Motorola offers software-enhanced wireless telephone and messaging, two-way radio products and systems, and networking and Internet-access products for network operators, consumers, and private and public sector companies.

Nokia

www.nokia.com
Nokia provides mobile, broadband, and IP networks and related services for network operators and Internet Service Providers.

Nortel Networks

www.nortel.com
Nortel Networks supplies network technologies for 2G and 3G networks.

QUALCOMM

www.qualcomm.com
QUALCOMM supplies CDMA chipsets and software.

Radioscape

www.radioscape.com
Radioscape provides semiconductor intellectual property for the wireless and digital communications market. Its RadioLab 3G 2.0 toolset, designed for 3G UMTS/W-CDMA wireless solutions, enables software developers to validate ideas and designs prior to coding.

WLAN Providers

3Com

www.3com.com
3Com offers WLAN (AirConnect) and Bluetooth products as well as accessories.

Agere

www.agere.com

Agere makes the ORiNOCO family of WLAN products, and a variety of components for all types of networks—WAN, Bluetooth, and WLAN.

Alvarion

www.alvarion.com

Alvarion offers the BreezeNET family of WLAN products.

Apple

www.apple.com

Apple makes the Airport 802.11b WLAN product for Apple computer users.

Cisco

www.cisco.com

Cisco offers the Aironet series of WLAN solutions including access points, adapters, and bridges for enterprise, medium, and small sized businesses.

Enterasys

www.enterasys.com

Enterasys offers a variety of networking equipment including its RoamAbout WLAN components.

Intel

www.intel.com

Intel offers the PRO/Wireless LAN product line, in addition to chipsets used in a wide variety of wireless devices and equipment.

Intermec

www.intermec.com

Intermec offers the MobileLAN family of WLAN products.

Linksys

www.linksys.com

Linksys provides home, small business and office networking kits.

Proxim

www.proxim.com
Proxim sells a variety of WLAN solutions including RangeLAN2 and Harmony (802.11a and 802.11b) product lines.

RadioLAN

www.radiolan.com
RadioLAN provides non-802.11b indoor WLAN solutions and long-range bridges between separate LANs. Primarily for campus use.

Raylink

www.raylink.com
Working in conjunction with Raytheon RF Networking, a subsidiary of Raytheon Company, Raylink provides WLAN products.

Symbol Technologies

www.symbol.com
Symbol Technologies offers the Mobius (802.11a) and Spectrum (802.11b) families of WLAN products.

Middleware Providers (Enabling Services, Platforms, and Tools)

@Hand

www.hand.com
@Hand offers its M-Tier Platform consisting of the M-Tier Studio, a suite of tools to develop and deploy mobile applications, and the M-Tier Server to manage communications and transaction functions.

Adobe Systems

www.adobe.com
Adobe provides software, including its well-known Acrobat Reader, to download and view PDF files on Palm OS-based devices.

Air2Web

www.air2web.com

Air2Web offers a wireless application development and delivery environment called the Mobile Internet Platform that provides access to back-end systems and data.

Aligo

www.aligo.com

Aligo offers a Java-based M-1 Server software platform to develop and deliver mobile applications to a host of client devices.

AlterEgo

www.aego.com

AlterEgo offers a mobile server to customize content for mobile devices, and also provides system integration for financial services, health care, and telecommunications.

AvantGo

www.avantgo.com

AvantGo offers the M-Business Server to develop and deploy mobile and wireless applications. The company also offers SFA and Pharma solutions. Companies with existing web sites can also use AvantGo software to create web content accessible by a variety of mobile devices.

Brience

www.brience.com

Brience provides a Mobile Processing Server that provides content processing, application processing, and multichannel presentation, as well as tools and integration adapters to permit development and deployment of wireless applications.

Broadbeam

www.broadbeam.com

Broadbeam offers the Axio platform that provides wireless application development and deployment with messaging, content, development, and back-end connection capabilities.

Extended Systems

www.extendedsystems.com
Extended Systems offers software for file and database synchronization, software distribution, and asset management. Also offers Bluetooth and IR SDKs.

Firepad

www.firepad.com
Firepad offers a mobile application platform to develop and view wireless applications and images on Palm OS devices.

IBM

www.ibm.com
IBM offers various software products to assist in creating a mobile application environment. Its WebSphere Transcoding Publisher product adapts and delivers web content to mobile devices.

iConverse

www.iconverse.com
iConverse provides a mobile platform to develop and deploy wireless applications. Its Mobile Studio features a drag-and-drop visual development tool and voice recognition technology.

Jacada

www.jacada.com
Jacada's Integrator and Interface Server products web-enable legacy applications and provide transaction access (via XML, COM, and Java) as a first step to deployment on mobile devices.

JP Mobile

www.jpmobile.com
JP Mobile offers the SureWave Mobile Server to synchronize and connect mobile devices to enterprise legacy platforms and applications.

Mapinfo

www.mapinfo.com
Mapinfo supplies location-based software and technology for mobile applications.

Microsoft

www.microsoft.com

Microsoft supplies the operating system software for Pocket PC devices. It also offers a Mobile Information Server that serves applications and content to mobile devices.

MobileQ

www.mobileq.com

Acquired by ViaFone, MobileQ offers XMLEdge, a development and a hosting platform based on XML that mobilizes secure and customized data applications, information services, and transactions.

Pumatech

www.pumatech.com

Pumatech offers a variety of synchronization tools to deliver content to mobile devices.

River Run Software Group

www.riverrun.com

River Run offers MobileSphere, a web browser-based set of software products to develop and deploy mobile applications for a variety of device and network types. Aimed primarily at field workers.

Sun Microsystems

www.sun.com

Sun offers Java 2 Micro Edition (J2ME), a programming language for mobile devices.

Synchrologic

www.synchrologic.com

Synchrologic offers the iMobile Suite to manage mobile devices and supply them with enterprise applications, e-mail, PIM data, web pages, and intranet content through synchronization.

Wavelink

www.wavelink.com
Wavelink offers a platform to develop and manage wireless applications. It also offers applications to monitor and manage WLANs, and to push updates to wireless devices.

Xcellenet

www.xcellenet.com
Xcellenet offers its Afaria software to deploy and manage mobile devices, applications, and contents from a central location.

Wireless Application Providers _____

All of the vendors listed below offer some type of wireless application functionality. They may license the software to enterprises, host it themselves, or offer a combination of the two.

724 Solutions

www.724solutions.com
724 Solutions provides software to enable secure mobile financial transactions over mobile devices.

Aether Systems

www.aethersystems.com
Aether provides wireless data software infrastructure, applications, and services across industries including financial services, transportation, and government. Aether's Fusion platform gives enterprises wireless delivery and application management capabilities.

Antenna Software

www.antennasoftware.com
Antenna offers and develops software to integrate web, wireless, and supply-chain technologies. Its ServiceTools product is a field service management solution that connects service managers, technicians, inventory managers, and customers in real time.

Apriva

www.apriva.com

Apriva offers its Apriva Talk environment to enable companies to design, develop, and deploy wireless solutions on a variety of network and devices. In addition, Apriva offers a solution for the point of sale industry that provides wireless connectivity to the mobile merchant plus a host of value-added services to merchants, card acquirers, and transaction processors.

Datria Systems

www.datria.com

Datria Systems offers voice enabled mobile solutions to collect data and information in the field. It also offers an application studio to develop and deploy voice-based applications.

Everypath

www.everypath.com

Everypath provides software that delivers SFA, field force, and other business applications and data to mobile devices. It also offers an application development platform.

FieldCentrix

www.fieldcentrix.com

FieldCentrix offers field force automation software that assists technicians in diagnosing and solving problems.

iMedeon

www.imedeon.com

iMedeon offers mobile workforce management solutions. Its iM:Work product enables work-order scheduling, routing, dispatching, field performance, communications, and data collection.

Infowave

www.infowave.com

Infowave provides the Wireless Business Engine to connect the mobile workforce to corporate application such as e-mail, web-based applications, the Internet, corporate intranets, and client/server applications.

Mobilesys

www.mobilesys.com

MobileSys provides wireless infrastructure services to deliver data worldwide from business applications to any wireless device. The MobileSys Network is an outsourced solution for deploying, managing, and tracking wireless messages. MobileSys MX is a two-way messaging engine with custom or plug and play integrators.

Oracle

www.oraclemobile.com

Oracle's products include a business-oriented mobile portal, a wireless application server, and an eBusiness suite. The products provide remote wireless access to e-mail, databases, and Oracle's ERP and CRM software.

Syclo

www.syclo.com

Syclo offers software to extend enterprise systems including e-mail, contracts, and calendars, out to mobile devices. Its Smart software gives technicians access to work orders, job plans, and inventories.

Thinque

www.thinque.com

Thinque supplies mobile workforce automation software. Thinque MSP is a field force application for Pocket PC devices and Thinque SP automates the entry of customer information.

Vaultus

www.vaultus.com

Vaultus offers wireless CRM, ERP, SFA, and field force applications as well as a Java-based mobile platform to develop and deploy mobile applications.

ViaFone

www.viafone.com

ViaFone offers OneBridge, software that provides mobile access to CRM, SFA, and supply-chain applications.

Viryanet

www.viryanet.com

Viryanet offers mobile workforce software aimed at the field service business—to schedule and dispatch field workers.

Wireless Internet Service Providers (WISPs) _____

Boingo

www.boingo.com

Boingo offers its subscribers WLAN-to-Internet access in public "hot spots" such as airports, hotels, and coffee shops.

GoAmerica

www.goamerica.net

GoAmerica offers wireless Internet, e-mail, and corporate intranet access.

OmniSky

www.omnisky.com

Omnisky provides wireless e-mail and Internet access services to subscribers.

Wireless System Integrators _____

Many system integrators provide some form of wireless consulting services. The companies below have specialty wireless integration practices, and often use proprietary components in their consulting assignments to jumpstart development efforts.

ArcStream Solutions

www.arcstream.com

ArcStream develops applications for mobile workers in the health care and pharmaceutical industries, and has a mobile sales workbench aimed at the mobile sales force. Its offerings include strategy consulting, systems development, and integration.

Ascend Mobility

www.ascendmobility.com
Ascend offers the gamut of design, implementation, and integration services, as well as business consulting and project management services.

Cap Gemini Ernst & Young

www.usa.capgemini.com
Cap Gemini offers mobile business model consulting, research and analysis, wireless channel integration, mobile payment and trading systems integration, and security and authentication.

Deloitte & Touche

www.dttus.com
Deloitte provides systems integration and strategy consulting as part of its mobile and wireless practice.

IBM Global Services

www.ibm.com
IBM provides mobile and wireless consulting and integration services.

Mobilocity

www.mobilocity.com
Mobilocity is a wireless systems integrator and application developer focusing on the financial services, health care, service, retail/manufacturing, and mobile workforce industries and sectors.

Questra

www.questra.com
Using its own development platform, Questra creates custom mobile applications for its clients that focus on monitoring, diagnosing, replenishing, managing, and servicing remote, intelligent devices.

Stellcom

www.stellcom.com
Stellcom creates mobile workforce applications focusing on SFA, field force automation, CRM, and business process applications.

Appendix C
Glossary of Terms

1G   First generation, analog cellular networks.

2G   Second generation, digital cellular networks relying on circuit-switched technologies.

2.5G   Interim digital cellular networks, between 2G and 3G, notable for increased bandwidth and data packet-switched technologies. 2.5G network technologies include GSM/GPRS and EDGE.

3G   Third generation digital cellular networks offering high data transmission rates up to 2 mbps, relying on data packet-switched technologies. 3G network technologies include W-CDMA and CDMA2000.

802.11x   A series of standards for WLANs endorsed by the IEEE.

Access point   Hardware that serves as a hub on a WLAN, allowing wireless clients to communicate with the wired network.

API   Application Programming Interface. Routines that allow an application to control the functions of a device or interact with the operating system.

Application   The software that operates on a wireless device, a host server, or a combination of the two to provide specific functionality to the device's user. Applications automate information capture, extract, and format data for display, perform calculations, and handle error checking and navigation between functions.

Bandwidth   Describes the transmission capacity of a network. The higher the bandwidth, the more data that can be sent over a given period time.

Bluetooth   A short-range network technology relying on radio frequencies. Ideal for creating a personal area network or permitting mobile devices to share information without physical cables.

Broadband	Describes a network or other communications medium capable of transmitting large volumes of data in a given time.
Browser	Software that permits the viewing of HTML documents. On a mobile device, browsers are commonly termed, micro-browsers.
Carrier	A company such as AT&T Wireless, Cingular Wireless, or Verizon Wireless that provides telecommunications services such as voice, data, or SMS.
CDMA	Code Division Multiple Access. Refers to a digital network communications technology used by certain carriers on their 2G networks.
CDPD	Cellular Digital Packet Data. Refers to a network technology that allows data to be transmitted over voice-based, analog cellular networks. Carriers must upgrade their networks to provide these data transmission services.
Circuit switched	Voice-based network communications technology that relies on the establishment of a dedicated circuit between the parties prior to communication.
CompactFlash	A physically small type of card used to store data or software, used in the expansion slot on handheld devices.
CRM	Customer Relationship Management. Refers to systems that help companies manage and improve relationships with their customers. From a functional perspective, may include customer self-service, support, sales, and other applications.
Device	Equipment that can send (transmitters), receive (receivers), or send and receive (transceivers) wireless information. Devices include mobile phones, pagers, PDAs, personal computers, laptops, tablets, remote sensors, handheld units, or other types of equipment that capture, display, store, manipulate, print or relay voice, or electronic data.
Encryption	Part of a cryptographic algorithm, encryption refers to the process of disguising a message, using some type of key, so that only an authorized party holding the matching key can decode the message.
ERP	Enterprise Resource Planning. Refers to an integrated set of systems covering all or some aspects of business operations including human resources, financial, manufacturing, inventory, etc.
Firewall	Describes a layer of security and protection that companies establish between their corporate networks and the outside world.

Gateway	A gateway performs some type of "translation" function, perhaps translating data from one protocol or standard to another, enabling a sender and receiver relying on different technologies to communicate.
GPRS	General Packet Radio Service. A 2.5G network technology superseding GSM to enable high-speed data transmission (up to 115 kbps) via an "always on" connection.
GPS	Global Positioning System. A network of satellites that provide exact geographic coordinates to land-based receivers. Used as an enabling technology for location-based tracking and monitoring applications.
GSM	Global System for Mobile Communications. Popular in Europe and Asia, GSM is similar to TDMA technology. A digital cellular technology not widely used in 2G networks in the U.S., but offering a migration path to GPRS.
HTML	HyperText Markup Language. A type of markup language used to identify the format of text (underline, bold, italic, etc.). Browsers interpret HTML to display web pages, for example.
iDEN	A network technology used by Nextel in the U.S. Similar to GSM.
IEEE	Institute of Electrical and Electronics Engineers, a standards-setting body.
IR	Infrared, a short-range network technology that uses light waves to communicate.
IrDA	Infrared Data Association. A group of wireless infrared product vendors that set standards and promote interoperability between products.
kbps	Kilobytes Per Second. Refers to the number of bytes, in thousands, of data transmitted per second over a network.
Latency	The "shelf life" of information, measured in time. Also refers to the amount of elapsed time that it takes a network to transmit data from sender to receiver.
mbps	Megabytes Per Second. Refers to the number of bytes, in millions, of data transmitted per second over a network.
Mobile	Refers to a person, device or thing that is not stationary but that is in motion at some point in time. A "mobile application" or a "mobile device" are used by people that are commonly in motion rather than in a fixed location such as an office.

Network	The software and hardware used to transmit information between devices. This communication can occur at short-range using infrared technology, at a wider range using a high-speed wireless LAN within a building, at regional or national levels using wide area cellular technology, or at extra-terrestrial distances using satellites.
OS	Operating System. Refers to the low-level routines and systems that control the operation of a device or piece of computing equipment.
Packet switched	A network technology ideal for transmitting data. Senders and receivers are always connected to the network and do not need to establish a circuit to communicate. Data is sent in packets to enable efficient "timesharing" of the network.
PAN	Personal Area Network. Used to describe the type of spontaneous network that a user can establish with other closely located devices, such as a printer, desktop computer, and handheld device.
PCMCIA	A type of memory card used in laptops or notebook computers, larger than a CompactFlash card.
PDA	Personal Digital Assistant. Refers to a class of handheld devices such as Palm OS-based units and Pocket PCs.
Piconet	An ad hoc network created by at least two and no more than eight Bluetooth-enabled devices.
PIM	Personal Information Management. Refers to personal productivity tools and functions such as managing calendars, contact lists, address books, etc.
Pocket PC	A handheld device running a Microsoft operating system such as Windows CE, Pocket PC 2002, etc.
RIM	Research in Motion, a manufacturer of handheld devices such as the BlackBerry line of e-mail devices.
SDK	Software Developer Kit. A set of tools typically provided by device manufacturers or carriers to enable developers to write application code specific to the product platform. May include a developer's guide, APIs, development utilities, sample source code, and an emulator.
SFA	Sales Force Automation. Refers to applications and systems to help better manage the sales process, with tools for sales managers and salespeople alike.
Smart phone	A phone handset that incorporates more than voice capabilities, approaching the functionality of a PDA.

SMS   Short Message Service. Refers to brief text messages that users can send and receive using their mobile phones over digital cellular networks.

TDMA   Time Division Multiple Access. Refers to a digital communication technology used by certain carriers on their 2G networks.

Telematics   Wireless applications that operate on in-vehicle computers, used in conjunction with location and tracking technologies to provide roadside assistance, maps, and navigation.

VPN   Virtual Private Network. A "tunneling" technology that uses encryption to create a private, secure connection through a public network like the Internet.

WAN   Wide Area Network. Refers to wireless networks relying on radio frequency technologies and having a national footprint such as 2G and 2.5G digital cellular networks.

WAP   Wireless Application Protocol. A standard for delivering Internet content to the micro-browsers of constrained mobile phones and other devices, relying on more efficient markup languages such as WML.

WECA   Wireless Ethernet Compatibility Alliance. A body that certifies compliance of products with the 802.11x WLAN standards issued by the IEEE.

Wi-Fi   An acronym used by WECA to signify compliance with 802.11x WLAN standards. Vendors seek the Wi-Fi stamp of approval to demonstrate that their devices, networks, and other products inter-operate with other 802.11x devices, networks, and products.

Windows CE   An operating system built by Microsoft for the handheld device platform.

Wireless solution   Refers to a combination of device(s), application(s), network(s), and data source(s) that enable the mobile execution of one or more business functions.

Wireless technology The hardware and software that allows information to be transmitted between devices without the use of physical (wired) connections.

WLAN   Wireless Local Area Network. A medium-range network technology that relies on radio frequencies. Ideal for office or campus use to connect users to wired networks without physical cables.

WML   Wireless Markup Language. A variant of HTML designed to provide more efficient delivery of content to constrained WAP-compatible mobile devices.

XML   Extensible Markup Language. A markup language used to identify data content rather than data format.

Index

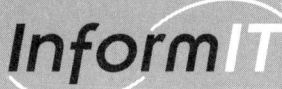

Solutions from experts you know and trust.

| Articles | Free Library | eBooks | Expert Q & A | Training | Career Center | Downloads | MyInformIT |

Login   Register   About InformIT

Topics

Operating Systems
Web Development
Programming
Networking
Certification
and more...

Expert
Access

Free
Content

www.informit.com

✓ Free, in-depth articles and supplements

✓ Master the skills you need, when you need them

✓ Choose from industry leading books, ebooks, and training products

✓ Get answers when you need them - from live experts or InformIT's comprehensive library

✓ Achieve industry certification and advance your career

Visit *InformIT* today
and get great content
from PH
PTR

Prentice Hall and InformIT are trademarks of Pearson plc /
Copyright © 2000 Pearson

Prentice Hall: Professional Technical Reference

`http://www.phptr.com/`

P R E N T I C E   H A L L

Professional Technical Reference
Tomorrow's Solutions for Today's Professionals.

Keep Up-to-Date with
PH PTR Online!

We strive to stay on the cutting edge of what's happening in professional computer science and engineering. Here's a bit of what you'll find when you stop by **www.phptr.com**:

@ Special interest areas offering our latest books, book series, software, features of the month, related links and other useful information to help you get the job done.

Deals, deals, deals! Come to our promotions section for the latest bargains offered to you exclusively from our retailers.

$ Need to find a bookstore? Chances are, there's a bookseller near you that carries a broad selection of PTR titles. Locate a Magnet bookstore near you at www.phptr.com.

! What's new at PH PTR? We don't just publish books for the professional community, we're a part of it. Check out our convention schedule, join an author chat, get the latest reviews and press releases on topics of interest to you.

Subscribe today! Join PH PTR's monthly email newsletter!

Want to be kept up-to-date on your area of interest? Choose a targeted category on our website, and we'll keep you informed of the latest PH PTR products, author events, reviews and conferences in your interest area.

Visit our mailroom to subscribe today! **http://www.phptr.com/mail_lists**

www.phptr.com